SPEECH TECHNOLOGY FOR TELECOMMUNICATIONS

BT Telecommunications Series

The BT Telecommunications Series covers the broad spectrum of telecommunications technology. Volumes are the result of research and development carried out, or funded by, BT, and represent the latest advances in the field.

The series includes volumes on underlying technologies as well as telecommunications. These books will be essential reading for those in research and development in telecommunications, in electronics and in computer science.

SPEECH TECHNOLOGY FOR TELECOMMUNICATIONS

Edited by

F.A. Westall
Formerly of British Telecommunications Research Laboratories
UK

R.D. Johnston
British Telecommunications Research Laboratories
UK

and

A.V. Lewis
British Telecommunications Research Laboratories
UK

CHAPMAN & HALL

London · Weinheim · New York · Tokyo · Melbourne · Madras

Published by Chapman & Hall, 2–6 Boundary Row, London SE1 8HN, UK

Chapman & Hall, 2–6 Boundary Row, London SE1 8HN, UK

Chapman & Hall GmbH, Pappelallee 3, 69469 Weinheim, Germany

Chapman & Hall USA, 115 Fifth Avenue, New York, NY 10003, USA

Chapman & Hall Japan, ITP-Japan, Kyowa Building, 3F, 2-2-1 Hirakawacho, Chiyoda-ku, Tokyo 102, Japan

Chapman & Hall Australia, 102 Dodds Street, South Melbourne, Victoria 3205, Australia

Chapman & Hall India, R. Seshadri, 32 Second Main Road, CIT East, Madras 600 035, India

First edition 1998

© 1998 British Telecommunications plc

Printed in Great Britain by T.J. International Ltd, Padstow, Cornwall

ISBN 0 412 79080 7

A catalogue record for this book is available from the British Library

∞ Printed on permanent acid-free text paper, manufactured in accordance with ANSI/NISO Z39.48-1992 and ANSI/NISO Z39.48-1984 (Permanence of Paper).

Contents

Contributors

J A S Angus — Department of Electronics, University of York

S Appleby — Systems Research, BT Laboratories

D J Attwater — Voice Services, BT Laboratories

P A Barrett — Speech Coding and Processing, BT Laboratories

A E Bass — Recognition Software Development, BT Laboratories

A P Breen — Spoken Language Systems, BT Laboratories

J Bridges — Formerly Speech Recognition Algorithms, BT Laboratories

B M G Cheetham — Department of Electrical Engineering and Electronics, Liverpool University

G Cosier — Advanced Perception, BT Laboratories

S N Downey — Recognition Software Development, BT Laboratories

M Edgington — Speech Synthesis and Analysis, BT Laboratories

D J Franklin — Software Consultant

M P Hollier — Multi-modal Perception, BT Laboratories

S A Hovell — Speech Recognition, BT Laboratories

D M Howard — Department of Electronics, University of York

P Jackson — Speech Synthesis and Analysis, BT Laboratories

S H Johnson — Recognition Technology, BT Laboratories

R D Johnston — Speech Technology, BT Laboratories

E Kaneen — Natural Language Research, BT Laboratories

J A Lear — Formerly Recognition Technology, BT Laboratories

P R Lee	Software Consultant
A V Lewis	Speech Coding, BT Laboratories
A Lowry	Formerly Speech Synthesis Algorithms, BT Laboratories
R M Mack	Speech Coding Algorithms, BT Laboratories
B P Milner	Speech Recognition, BT Laboratories
S Minnis	Natural Language Research, BT Laboratories
D G Ollason	Formerly Acoustic Modelling, BT Laboratories
J H Page	Speech Synthesis and Analysis, BT Laboratories
M Pawlewski	Speech and Speaker Recognition, BT Laboratories
K R Preston	Natural Language Research, BT Laboratories
K J Power	Formerly Speech and Speaker Recognition, BT Laboratories
S P A Ringland	Speech and Speaker Recognition, BT Laboratories
F Scahill	Speech Recognition Algorithms BT Laboratories
A D Simons	Formerly Spoken Language Systems, BT Laboratories
X Q Sun	Department of Electrical Engineering and Electronics, Liverpool University
J E Talintyre	Formerly Speech Recognition, BT Laboratories
S V Vaseghi	Queens University, Belfast
R M Voelcker	Human Perception, BT Laboratories
F A Westall	Formerly Speech Technology, BT Laboratories
S J Whittaker	Voice Services, BT Laboratories
S H Williams	Formerly Natural Language Research, BT Laboratories
W T K Wong	Formerly Speech Coding, BT Laboratories
P J Wyard	Natural Language Research, BT Laboratories

Preface

In the beginning, the means of communicating was the grunt, the growl, the gesture and possibly the grin, all of which contained some semantic information — whether to indicate possession, anger, acceptance or to indicate contentment, however brief.

We have moved on from the grunters, growlers and gesticulators to the age of the mass mediators driven by the convergence of communication, computing and content. This revolution is set to globally transform the fundamentals not only of commerce and culture but of power and politics too, redefining the way we all live, work, learn and play.

Speech and language — the very essence of social development — have become the key components lying at the humanising heart of this information revolution. Ironically, the principal attributes of 21st century telecommunications will be just those qualities the telephone possessed in abundance at its birth, but subsequently lost. In the 1880s, making a telephone call was a very personal and intuitive thing - customers asked for the person they wanted to contact. The latest advances in speech technology will not only restore this natural and intuitive style of communication between people at a distance, but for the first time extend it to interactions between people and machines and thereby provide assistance to some people with disabilities.

BT have made notable contributions to the field, ranging from the early development and deployment of PCM, through Topaz, the first hands-free speech recogniser for use in mobiles, to the pioneering trial with the Royal Bank of Scotland, and Skyphone, the world's first aeronautical telephony system. More recent developments include Laureate, the world-class text-to-speech system, voice activity detection for GSM, large-vocabulary telephony speech recognition, the speech application platform, and Callminder, the first national network-based interactive answering service. The theme of this book focuses on the developments of the last five years, bringing together the work of many speech technologists based at BT Laboratories. It continues the work described in two earlier volumes (Wheddon C and Linggard R (Eds): 'Speech and Language Processing', Chapman & Hall, 1990 and Westall F A and Ip S F A (Eds): 'Digital Signal Processing in Telecommunications', Chapman & Hall, 1993).

It has been my sincere pleasure to be associated with many of the authors in this book and to have personally contributed to this exciting subject. I confidently expect that the reader will find the book to be authorative, informative and stimulating.

Chris Wheddon *OBE*
Director, Systems Engineering
BT Networks and Systems

Introduction

This book is dedicated to speech technology for telecommunications — a commercially important subject of rapidly growing significance throughout the world. Tutorial material is included throughout, together with an applications perspective from the 'leading edge' of each technology area.

For readers who are new to the subject, the first chapter provides a broad overview, explaining terminology in preparation for the more detailed treatment of subsequent chapters. As a fundamental introduction, the second chapter reviews the basic mechanisms of human speech and explains the key techniques of speech analysis that are central to all speech technology.

A group of twelve chapters then focuses on the core enabling technologies, designed to give a broad and balanced perspective of each topic. Developments in low bit rate speech coding are described in two chapters, highlighting increasingly important applications in both fixed and mobile networks. Four chapters describe state-of-the-art developments in speech synthesis systems and their performance, with particular reference to the BT Laureate text-to-speech system. The next four chapters offer a tutorial review of speech recognition techniques, a description of enhancements to a large-vocabulary telephony speech and speaker recognition system, a discussion about implementing speech recognition in practice, and a presentation of the fundamental difficulties in properly characterizing speech recognition performance. Techniques for signal enhancement and important issues in the human perception of both speech and mixed-mode signals are described in two further chapters.

The subsequent three chapters form a second group, intended to provide a unique insight into the challenges of integrating a range of speech technologies into effective commercial solutions that meet customer needs. These chapters debate key issues in interactive speech platform and application design and in spoken language processing. The final chapter boldly takes a leap forward to speculate on the future of telephony.

Speech technology is a dynamic, exciting and commercially relevant field, encompassing a multi-disciplinary blend of skills from the hard and soft sciences — ranging from mathematics, engineering and computer science to human perception, linguistics and phonetics. The editors hope this book will demystify a complex subject while conveying the excitement they feel in being part of a revolution in human communications. Speech technology is poised to dramatically change the way we all interact in the 21st century.

Finally, a great deal of effort goes into the production of a book like this, and the editors would like to thank all the authors and reviewers, too numerous to mention by name, for their tremendous dedication, skill and sustained effort.

Fred Westall
Denis Johnston
Alwyn Lewis

1

SPEECH TECHNOLOGY FOR TELECOMMUNICATIONS

F A Westall, R D Johnston and A V Lewis

1.1 INTRODUCTION

Humans have been talking to each other using some form of speech for at least 50 000 years so it is not surprising that most of us take it for granted. Yet speech is an extremely complex process, and the simplest sentence contains a world of information besides its literal content. Spoken language is communicated through the air via a longitudinal pressure wave. It is an extraordinarily effective method of transferring information between humans and has the capacity to reliably communicate complex ideas and emotions under constantly varying and hostile environmental conditions. A statement can convey such nuances as the mood, approximate age, education, background and gender of the speaker. Not surprisingly, getting a machine to generate such subtle speech is somewhat of a challenge; making machines which can understand it is even more daunting. It is this challenge — to understand the mechanisms of speech production and perception — that forms the focus of this book.

This chapter introduces the key technologies that underpin the field, and sets the scene. It provides a snap-shot of current interactive speech applications, systems and issues, and concludes with some personal predictions on future trends and challenges. The second chapter provides an overview of speech science, and the basic processes of speech production and perception.

The subsequent chapters form a number of logical groupings — coding (Chapters 3 and 4), synthesis (Chapters 5, 6 and 7), recognition (Chapters 9 and 10), performance assessment (Chapters 8 and 12), and enhancement (Chapter 13). The chapters on recognizer implementation (Chapter 11), spoken language processing (Chapter 17) and on large-vocabulary interactive speech systems (Chapters 15 and 16) provide an insight into the challenges of integrating speech technologies to provide network solutions that meet customer requirements [1]. Chapter 14 addresses important aspects of human perception of speech and other signals, and Chapter 18 boldly takes a leap forward to speculate on the future of telephony.

In each group, tutorial material is given to help the reader, a perspective on applications is provided, and a view from the 'leading-edge' of each technology area is presented.

1.2   EVOLUTION

Although digital techniques have been pre-eminent in recent years, speech processing itself is not a new subject. Early voice coders, such as Dudley's VODER, date back to the late 1930s, and Von Kempelen made a voice synthesizer as early as 1791. The mathematics on which many current speech processing systems are based was established over 20 years ago, such as with the work of Atal, Makhoul and Itakura-Saito on linear prediction in the 1960s and early 1970s.

However, it was the advent of the digital signal processing microcomputer in the late 1970s that helped to convert research into practical, cost-effective systems, and heralded an upsurge of interest in new telecommunications applications [2, 3].

Figure 1.1 illustrates the evolution of digital signal processing (DSP) devices, measured in millions of operations per second (note the logarithmic scale). Also shown are the processing requirements for some 'popular' speech algorithms. Apart from the large-vocabulary speech recognition systems, most speech (and an increasing number of video) algorithms could be implemented on a single device by 1995. It is interesting to note that, if aeronautical engineering had progressed at the same pace as device technology, it would now take under a minute to fly across the Atlantic!

Telephony applications are today in the vanguard of this minor revolution, just as, to a lesser extent, modems were a few years ago. The latest digital mobile, aeronautical and multimedia terminals all depend on speech coding. The BT intelligent network answering service, CallMinder™, uses speech recognition to allow customers without a touch-tone phone to access the service, and speech coding to store efficiently the vast amounts of data generated. Speech echo cancellers are regularly used on international calls to facilitate two-way, simultaneous, speech conversations. Synthesized speech is used for services as diverse as the speaking clock, operator services, and reading e-mail over the phone.

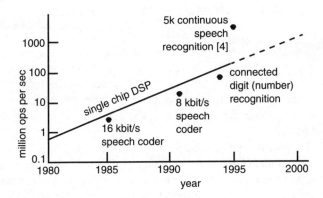

Fig. 1.1   Evolution of DSP technology.

Speech technology is an amalgam of many disciplines, as shown in Fig. 1.2, and real-world speech applications increasingly require a broad-based, holistic approach to realize systems that are acceptable to the public at large. Engineering and computing need to be complemented by expertise in man-machine interaction, human perception, psychology, acoustics, linguistics, natural language processing, and many other disciplines.

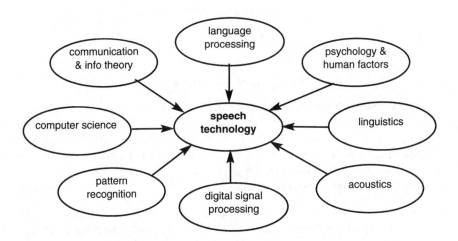

Fig. 1.2   Speech technology disciplines.

1.3   THE MUTABILITY OF SPEECH

For many years speech was considered to be composed of linear sequences of elemental speech particles, or phonemes, put together rather like a string of beads. This misleading interpretation of the speech signal led engineers to imagine that machines could recognize speech simply by decoding the individual phonemes as they appear through time, in the same way that a modulated data signal could be detected.

In a similar vein, it appeared that synthetic speech could be produced by simply storing an example of each of the phonemes of a language and then sticking them together to form the new word. Unfortunately this proved not to be the case. While it is convenient to visualize speech as a set of discrete symbols, in reality linguistic information is not encoded in discrete packets of time as imagined by this model, and traditional methods of dealing with coded signals will not work with speech. Our familiarity with text also encourages us to think of speech as 'atomic'; but, convenient as this picture is, speech is not like text — see Fig. 1.3.

Therearenogapsbetweenwords.

Sometim esthereareg aps inthe

mid dleofwords. **Acc**ents,

dialects and Stress *eggs*ist

Fig. 1.3   Example of speech as text.

Compared with many other areas of digital signal processing, the challenges presented by speech processing are immense. Although the public imagination has been fired with images like 'HAL' in Stanley Kubrick's '2001 — a Space Odyssey', the goal of producing completely natural language interfaces between humans and machines is still a long way off. The difficulty arises due to the variability in factors that dictate performance:

- the great variety in the signal characteristics, when the word or phrase is uttered by different speakers and even when repeated by the same speaker;

- the wide variety of channel characteristics through which the speech signals are sent;

- the nature of the accompanying background noise.

Above the level of the acoustic signal, ambiguities can occur at a higher level of linguistic abstraction. The phrase 'It's easy to recognize speech' could be interpreted as 'It's easy to wreck a nice beach' at a conference of surfers! To make matters worse, people do not always say what they mean or mean what they say. This mutable behaviour is unavoidable and must be tolerated by a speech processing system which aims to provide a seamless and natural interface between people and machines.

1.4 APPLICATIONS AND OPPORTUNITIES

Speech technology provides significant revenue opportunities by introducing new speech processing services which can be accessed easily over the telephone, twenty four hours per day, seven days per week [4, 5]. Speech is BT's core business, accounting for over 90% of revenue and providing access to 26 million customers in the UK and around 700 million telephone users world-wide. Speech technologies can add value by:

- stimulating additional call revenues (e.g. the BT network-based call answering service, CallMinder™);

- creating new speech services (e.g. aeronautical telephony, talking e-mails);

- providing differentiation in existing services (e.g. speech enhancement, noise control);

- reducing operating costs (e.g. by partial or full automation of operator services and call centres);

- allowing control of systems by voice to make them easier to use and hence improving customer satisfaction;

- extending system usage to areas where other digit entry systems fail, where hands-free input is necessary, or where users require easily remembered access codes (e.g. names);

- providing a simple means of getting large amounts of variable text data to users over the telephone (e.g. text-to-speech).

Current activities are primarily directed towards applying the technology to the telephone network for interactive speech services such as:

- Voice store and forward
  — CallMinder;

- Finance
  — banking,
  — stocks and shares,
  — insurance quotations,
  — credit card transactions;

- Entertainment
  — betting,
  — horoscopes,
  — games;

- Information services
  — timetables,
  — Yellow Pages,
  — News;

- Telemarketing
  — promotions;

- Teleshopping/reservations
  — theatres,
  — airlines,
  — catalogue shopping;

- Field operations
  — data operation and retrieval,
  — field personnel job despatch,
  — voice access to electronic mail;

- Automatic operator
  — network services,
  — call centres.

1.5   INTERACTIVE VOICE RESPONSE SYSTEMS

As indicated above, recent developments in speech technology have enabled a new generation of interactive voice response (IVR) services operating over the telephone network. These range from telco-type services such as automation of directory enquiries, to customer-handling and information-retrieval applications which can offer commercial opportunities for many companies.

A major activity is the integration of speech technology with existing databases, IT processes and call-centre capabilities. These developments range in scale from small bespoke systems involving single-line PC solutions to embedded network systems capable of supporting many hundreds or thousands of simultaneous calls. The larger scale applications can significantly re-engineer the way in which customers interact with providers of information, goods and services.

Until recently, the performance of commercial speech recognizers restricted their use to relatively small vocabularies unsuited to applications where information such as names and addresses is required. However, the recent emergence of accurate large vocabulary recognition at acceptable cost has increased the range of practical applications.

A typical interactive speech system is shown in Fig. 1.4.

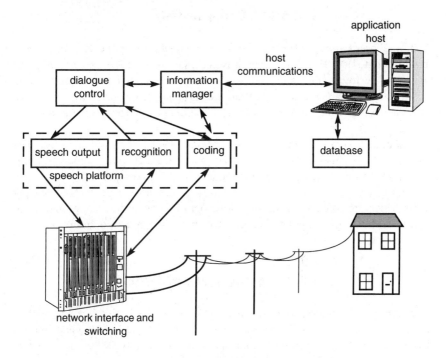

Fig. 1.4   Interactive speech system.

Speech input is achieved via the recognizer (which may be supported by DTMF or dial-pulse input), and speech output is by means of the synthesizer (which can be based on stored concatenated speech or text-to-speech). For a network-based application serving thousands of customers, a speech coder may be necessary in order to make efficient use of expensive storage media. The user aims to extract information from the system database (or voice store) by means of spoken commands.

The information manager interfaces with the host application or database and acts as a mediator for high-level dialogue requests, for example, building recognition vocabularies based upon the database content. The role of the information manager is discussed in Chapter 16.

The dialogue controller ensures that the required information is obtained from the user in a controlled and structured way. It acts as a kind of gearbox matching the recognizer performance (engine) to the application (the wheels). Like any gearbox it can, to some extent, overcome deficiencies in the engine, but only if the engine is basically sound. Dialogues are not a substitute for inadequate or inappropriate speech technology.

1.5.1   Dialogue design

From the user's point of view the system should be friendly, reliable and comfortable to use. To achieve this, it is necessary to ensure each component of the system performs adequately, and aligns with the customer's expectations. It is a considerable challenge to orchestrate all the constituent technologies to best meet the expectations of the whole user population under all operating conditions.

Dialogue design is often the key to successful IVR speech system design (see Chapter 16). It defines the system interface which the caller will encounter, guiding and interpreting each user interaction and selecting the appropriate response. The dialogue controller also ensures that the system recovers gracefully from any errors which are detected in the speech recognition, either by asking the user to repeat, or by modifying the dialogue to obtain the required information by rephrasing the question.

1.5.2   Example of a dialogue

Within the UK, Directory Assistance (DA) is a chargeable service with a database of over 20 million residential, business and government entries involving nearly half a million distinct surnames, and nearly 30 thousand distinct localities. An example of the dialogue for such an experimental application is given in Chapter 16. Partial spelling of the first N letters helps the recognizer distinguish between confusable names and homonyms (words which sound the same but which are spelt differently) — typically 30% of names may be homonyms or synonyms of others.

In current IVR systems, the flow of the dialogue is controlled primarily by the system, not the user. Information is garnered one item at a time and a rigid dialogue structure enforces certain strategies for confirmation, rejection, error detection, correction and reprompting.

It can be frustrating for users to find themselves 'steered' through a complex dialogue with a machine only to find either that they are lost or that they have ended up with the wrong piece of information. This calls for a different approach to dialogue design, and systems are beginning to emerge where it is possible for the caller to respond much more spontaneously, flexibly and take control of the interaction. For example, the next generation of interactive voice response system might have a 'conversation' with you, as shown in Fig. 1.5.

How many messages do I have from Denis?
There are two messages from Denis Smith and one from Denis Crowe
Can I hear the ones from Smith?
The first message from Denis Smith is about 'Budgets', it's 20 seconds long and a summary is '...
(Interruption) ....'
Give me the next one....
The next message is about 'The future of speech recognition', it's 10 seconds long and a summary is ..
OK, — uhm — read the previous one back to me in full.

Fig. 1.5   Example of possible future interactive voice system 'conversation'.

This example illustrates a number of features which are desirable in future interactive voice applications. The system does not need to constantly prompt the caller for explicit information. It uses discourse and domain knowledge to resolve ambiguous expressions such as 'give me the next one' and is able to understand references to earlier information. The caller can interrupt (barge in) at any stage, and the interrupting phrase is recognized. The caller controls whether the full message is heard (having been told beforehand how long it is), or if a summary is given. The system discards paraverbals such as 'uhms' but is able to identify words of significance wherever they occur (wordspotting).

Such a dialogue allows the initiative in the conversation to migrate from the computer to the user. More fluent interactions of this type will inevitably place greater demands on the underlying speech technology and require more language knowledge than would be the case with more structured dialogues. However, spoken language will be an essential vehicle for searching and retrieving information, and may help to make it accessible to all users, of all ages, and not just the computer literate.

For example, cellphones are already too small to support a keyboard. For such applications, interactive dialogue-based systems will require both high-performance speech recognition and language understanding in order to properly identify and respond to user's requests. These, and other aspects of spoken language processing are discussed in Chapter 17.

1.5.3   The speech applications platform

To address network-based applications for interactive voice systems, BT has developed the speech applications platform (SAP). This is a modular system based on industry standards for hardware, software, interfaces and protocols. The

SAP can be scaled from single line systems to large embedded network and intelligent network installations and supports a range of exchange signalling and network integration options. Figure 1.6 shows the speech applications platform connected as a network element.

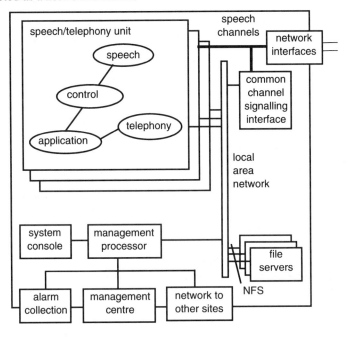

Fig. 1.6   SAP as a network element.

The SAP provides a wide variety of speech and signal processing facilities such as large-vocabulary speech recognition, text-to-speech, speech coding, tone and loop disconnect recognition. The system is closely integrated with BT's billing, maintenance, customer and network management systems.

Dynamic allocation of individual speech functions on call establishment ensures that efficient use is made of available DSP resources, and that these are made available to other channels when no longer in use. Applications may be written for the SAP using an object oriented programming interface or through an integrated graphical dialogue creation tool [1, 6].

1.6   SPEECH TECHNOLOGY

The key underlying speech technologies for IVR systems are now reviewed, starting with the most mature of the technologies — speech coding.

1.6.1   Speech coding

This is the process of converting speech into digital bit streams for efficient storage and/or transmission over bandlimited channels. It can exploit redundancy in speech signals to reduce the transmitted bit rate while at the same time exploiting the known properties of human speech perception to reduce the coding distortion to acceptable levels. Speech coders for particular applications are selected according to a trade-off between coding complexity, bit rate and signal quality. The design trade-offs are further complicated by additional factors such as:

- codec delay (which makes echoes more audible);

- transparency (which effects the ability of the codec to pass non-speech signals);

- coder robustness to transmission errors and background noise;

- tandeming ability (or the ability to mix different coders in a transmission path without the accumulation of distortion to unacceptable levels).

Speech coding has been subject to considerable international standardization activities in recent years. Standards now exist for 32 kbit/s (G.721), wideband (7 kHz) speech coding (G.722), 16 kbit/s (G.728), 8 kbit/s (G.729), variable-rate speech coding for circuit-multiplication equipment (G.726), and broadband packet networks (G.727).

Speech coders are in demand where there is a need to conserve radio spectrum. Examples include aeronautical telephony via satellite (Skyphone — 9.6-kbit/s coder), digital cellular (GSM — 13-kbit/s coder) and digital cordless systems (CT2/DECT — 32-kbit/s coder). Chapter 4 provides a comprehensive review of speech transmission requirements and techniques for digital mobile radio applications (using GSM as a case study).

Coders have also been applied in customers' private network applications, e.g. for speech-and-data multiplexers, in PSTN and ISDN videophone applications for the audio channel (e.g. VC8000 [7]), and in interactive network-based voice-messaging applications (e.g. CallMinder) where cost of storage is still a critical factor.

A comprehensive review of speech coding algorithms and applications is provided in Chapter 3. Coders are usually classified into the following three types:

- waveform coding;

- vocoding;

- hybrid coding.

The aim of waveform coders, as the name implies, is to reproduce the original waveform as accurately as possible. As these coders are not speech specific, they can deal with non-speech signals, such as background noise, music and multiple speakers, without difficulty. However, the cost of this fidelity is a relatively high bit rate. Examples include PCM, adaptive differential PCM (ADPCM) and sub-band coding [8] techniques.

In contrast, vocoders (voice+coders) make no attempt to reproduce the original waveform, but instead derive a set of parameters at the encoder, which are transmitted and used to control a speech production model at the receiver. Typically, linear prediction coding (LPC) is used to derive the parameters of a time-varying digital filter, which models the dominant resonances of the vocal tract. Speech quality, although intelligible, tends to be synthetic and variable between speakers. Hence vocoding is not used for telephone network applications.

Hybrid coders combine features from both waveform coders and vocoders to provide good-quality, efficient speech coding. At rates between about 16 kbit/s and 4 kbit/s, good-quality coding is achieved using 'analysis-by-synthesis' techniques, as shown in Fig. 1.7.

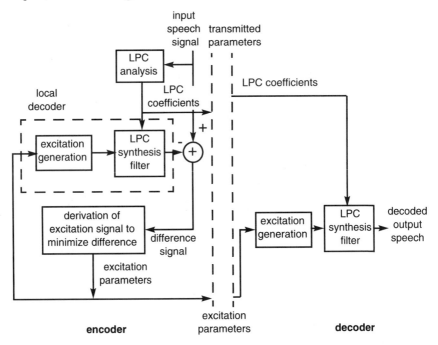

Fig. 1.7   Analysis-by-synthesis coding scheme.

The objective is to derive an excitation signal, like that produced by the glottis, such that the difference between the input and synthesized speech signals is minimized according to some suitable perceptual criterion. A good description of the operation of this type of technique can be found in Boyd [9]. This technique is the basis of the coding schemes in Skyphone (9.6 kbit/s), the ITU low-delay coding standard (16 kbit/s), and the recent ITU standard for toll-quality coding at 8 kbit/s.

As shown in Fig. 1.8, 'toll' or telephony-quality coders currently operate down to bit rates of around 8 kbit/s, with the prospect in sight within a few years of rates around 4 kbit/s. A new class of coders, based on sinusoidal transform coding and waveform interpolation techniques appears to be especially promising at these lower rates. Chapter 3 reviews these techniques.

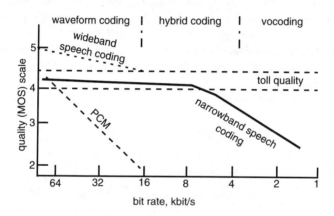

Fig. 1.8   State-of-the-art in speech coding.

In the longer term, more significant gains are theoretically possible if one considers that the maximum rate of articulatory movement is limited by inertia of the physiological structures of tongue and jaw to about ten discrete sounds per second. Given the 44 phonemes of English (see Chapter 2), this gives a theoretical minimum of around 100 bit/s (even allowing for non-verbal cues such as emotion and phrase emphasis to be included).

There is also considerable current interest in coding enhanced quality wideband speech (typically between 7 kHz and 20 kHz bandwidth) at rates of around 1 bit/sample for speech and 2 bit/sample for audio (including music), for ISDN applications such as teleconferencing, for CD audio compression (MPEG-audio), and for commentary channel transmission.

1.6.2   Speech recognition history

Today, interaction with computers normally involves the use of more artificial methods, such as keyboards, touchtone, touch-screens or mouse-driven menu systems. Speech recognition can improve the user interface by allowing spoken commands for accessing network services or for call routeing, e.g. voice dialling over mobile phones.

The earliest attempt at voice recognition was the voice-operated writing machine described by Flowers in 1916 [10]. The first paper on electronic speech recognition was published by Bell Labs in 1953 and described a system which could recognize single digit utterances spoken in isolation by a single speaker. Figure 1.9 shows this system, which used zero crossings from two frequency bands as features, and a form of vector quantization of the track traced by the oscilloscope beam for the recognition decision. Later in the 1950s the technology was extended to recognition of isolated syllables and phonemes in isolated words at RCA, in the USA, and University College, London. This pioneering work provided statistical information concerning possible phoneme strings in English and was the forerunner of stochastic language modelling used in modern large vocabulary systems. An early trial of speaker-independent recognition was held at MIT Lincoln Labs in 1959.

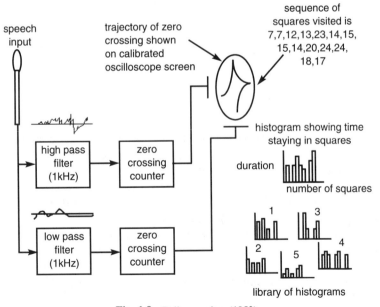

Fig. 1.9   Bell recognizer (1953).

During the 1970s, dynamic programming (also known as dynamic time warping (DTW)) emerged to cope with the time expansion and contraction found in normal speech. Connected-word recognition was also developed, including the two-stage dynamic programming procedure, continuous speech recognition (CMU and IBM) and speaker-independent recognition (Bell Labs).

During the 1980s there was a shift from template (reference-pattern) matching to statistical modelling methods, such as hidden Markov modelling. Stochastic methods started to be used for language modelling in place of rule-based methods, and a recognition system based on word concatenation probabilities was investigated at IBM.

Also in the 1980s the Defense Advanced Research Programme Agency (DARPA) drove forward continuous speech recognition with large vocabularies (1000+ words) at establishments such as SRI, MIT, Carnegie Mellon University, Bolt Beranek and Newman, and Lincoln Labs.

BT began work on speech recognition in the early 1980s resulting in Topaz — the first repertory dialler for hands-free voice dialling in cars. Early speaker verification and small vocabulary speaker-independent recognizers were also deployed in the pioneering trial of automated banking with the Royal Bank of Scotland in 1988 [10]. BT first demonstrated rapid vocabulary generation together with spelt connected recognition in 1993. More recently, large-vocabulary recognition has been implemented on the network-based speech applications platform (see section 1.5.3) [6].

Currently telecommunications provides one of the largest market sectors for speech recognition, with applications as diverse as voice dialling, credit card validation, and access by voice to the information services indicated in section 1.4.

1.6.3   Recognition technology

A speech recognizer is designed to recognize one of a set of words or phrases specified in a vocabulary. This is a difficult task, and not even a person is 100% accurate.

Figure 1.10 shows the components of a speech recognition system. The front-end stage performs feature extraction, taking segments of the speech at regular intervals and transforming them to facilitate pattern recognition. Part of the process involves nonlinear filtering operations, which are assumed to occur in the cochlea of the inner ear. The resulting frames of features are then processed to identify either words or parts of words (phonemes).

In a more complex scenario, knowledge of the context can be used to aid sentence recognition by exploiting semantic (meaning), syntactic (grammar) and dialogue (discourse) constraints.

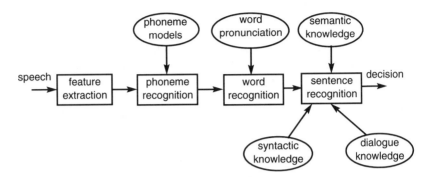

Fig. 1.10   Modern speech recognition system.

Many current speech recognizers use the technique of hidden Markov model-ling (HMM) for pattern matching. An HMM is a type of model based on a statistical representation, and this helps the recognizer to cope with most of the variability in the way people speak. The recognizer aims to identify, from observing a sequence of speech features, which of the stored Markov models is most likely to have produced these observations. A good 'layman' description of speech recognition techniques, including HMMs, is provided in Chapter 9. For those who wish to delve deeper into HMMs, a good theoretical treatment of the subject can be found in Cox [11].

A connected-word recognizer has to match utterances to strings rather than to single words. This is usually achieved with a finite-state network of word models representing the whole vocabulary. The shortest path through this network which matches the utterance is selected as the recognized string.

Recognition performance depends on the vocabulary to be recognized and the number of users. A business dictation system has a large vocabulary (typically 20 000 to 30 000 words), but can be trained for a particular speaker and exploit tight constraints on language and word statistics.

In contrast, speech recognition for telephony is likely to use only a small vocabulary, since the recognition task is far more challenging — with much greater signal, network and speaker variability. Also context and grammar constraints tend to be weaker, as with surname and address recognition. As an example, BT's latest speaker-independent recognizer is capable of identifying several thousand words and any connected grammar over the telephone network. Chapter 11 addresses the implementation aspects of telephony-based speech recognition.

In the past, the production of new vocabularies took several months, requiring many examples of the words from a hundred or more speakers. Also the whole process had to be repeated if a change or addition to the vocabulary was subsequently needed. Recently, a method has been developed to allow new vocabularies to be generated very quickly, simply by typing the words to be recognized. Word models may be built from phonetic transcriptions using either a dictionary or a set of letter-to-sound rules. Because of the various approximations, recognition accuracy for text-generated vocabularies are generally inferior to those that have been trained on actual speech utterances. In applications where the vocabulary is changing rapidly, such as directory or customer identification applications, the system must be able to dynamically update its active vocabularies. This aspect is discussed in detail in Chapter 15.

Speaker-dependent repertory dialling (e.g. 'Call Bob'). has been available for some time. Typically, this requires the user to repeat the word or phrase a small number of times, under the guidance of the dialogue controller. Recently, this concept has been extended to speaker-independent operation, as described in Chapter 9.

1.6.4   Speaker recognition

Speaker recognition is used to identify or validate a speaker from a sample of his/ her voice.

One of the first applications of speaker recognition was forensics. In 1660, speech was used as the key to detect a criminal for the first time in the case concerning the death of Charles I. However, at that time, the decision was made subjectively. In the 1940s, the sound spectrograph was invented by Potter at Bell Labs, and it became possible to draw voice prints (visible speech).

The first paper on speaker recognition was by Pruzansky of Bell Labs in the 1960s. From then until the early 1970s speaker recognition research was started at several laboratories, including IBM, TI and NTT, and both text-dependent and text-independent methods were investigated. In the 1980s, speaker recognition research also started to use stochastic approaches such as HMM.

The first network-based use of speaker recognition by BT was as part of a trial of automated banking with the Royal Bank of Scotland in 1988 (using dynamic time warping) [10].

Speaker verification technology can be used to validate the claimed identity of a person from his or her voice-print. It is often combined with a general-purpose connected speech recognition system to recognize a spoken PIN, and then to check that the PIN was spoken by the authorized person (see Chapter 10).

Speaker identification can be used to recognize a member of a closed user group (e.g. a family) directly from a known spoken word (text-dependent) or from any spoken utterance (text-independent).

Current applications include secure access control to information, banking, computer networks, PBXs and work areas. The technology can also be used to provide access to a range of network services according to customer profiles, such as name dialling and travel information.

1.6.5   Speech synthesis

The evolution of speech synthesis is shown in Fig. 1.11.

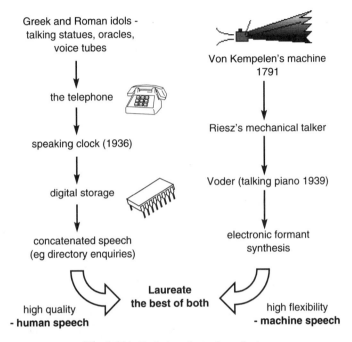

Fig. 1.11   Evolution of speech synthesis.

In 1791 Wolfgang Ritter Von Kempelen constructed a talking machine (Fig. 1.12); it consisted of a bellows, a mouth shape, nostrils and whistles. The machine included a compressible leather tube and an air chamber equipped with a reed leading to a soft leather resonator which could be manually shaped for the formation of the vowel sounds. Consonants were created by holes which the 'player' closed by movement of the fingers. The Von Kempelen machine could produce about twenty different sounds.

The earliest electronic synthesis of speech was achieved by Dudley in 1939 when he demonstrated a manually controlled speech synthesizer known as VODER (Voice Operated Demonstrator) at New York's World Fair. After the second world war, the development began in the British Post Office of techniques to analyse and synthesize natural speech. This led to a variety of synthesis systems based on the source-filter model of speech production of which the formant synthesizers developed by the Joint Speech Research Unit [12] in the UK and at MIT in the USA are perhaps the best examples.

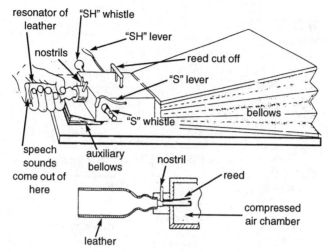

Fig. 1.12   Wheatstone's replica of Von Kempelen's machine.

Speech output from computer-based equipment has been commonly achieved by generating messages from stored speech fragments (as with the BT speaking clock). These messages require the recording by a speaker and although a natural sounding voice output is achieved, significant amounts of data storage are required. An additional constraint on system design and extension arises when the messages need to be changed or updated if the original speaker is not available. There are many applications where the versatility of a full text-to-speech system is the only practical solution.

BT's synthesis system, known as Laureate, is designed to convert unrestricted textual input into speech. Laureate differs from conventional speech synthesis systems in that it does not attempt to artificially generate sounds to mimic human utterances. Rather, it constructs the speech from elemental components of a person's recorded voice. This preserves the voice characteristics of the original speaker and hence the system can be easily modified to effect local accents, and other languages. Laureate has been engineered to commercial software standards, is portable between different platforms, and is designed for multichannel operation. Chapter 5 provides a review of the Laureate architecture and related applications.

In a text-to-speech system (Fig. 1.13), the speech output is derived directly from an unconstrained textual input. The first stage of processing involves expanding the text by a normalization stage which deals with abbreviations and acronyms, such as 'St. John St.'. This is followed by a stage of syntactic parsing, which aims to resolve ambiguities between words such as 'lives' or 'convict' which have both a verb and a noun form. A semantic analysis may also be carried out which provides a representation of the meaning of text as an aid to word pro-

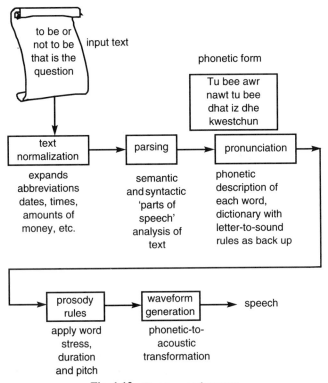

Fig. 1.13   Text-to-speech system.

nunciation. Next, a phonological analysis of the words is made to derive an appropriate word pronunciation, based on a large dictionary and a set of letter-to-sound rules, where the dictionary look-up fails (e.g. for proper nouns). Prosody rules are then applied to provide the required emphasis on duration, intonation and major word stress. Finally the speech is synthesized by looking up the appropriate speech segments, and concatenating them in as seamless a fashion as possible.

Chapter 6 provides a comprehensive review of text processing techniques (the first three blocks in Fig. 1.13), while Chapter 7 provides a review of processing techniques for prosody and waveform generation (the last two blocks in Fig. 1.13).

An example of an application for text-to-speech synthesis is a telephone-based catalogue ordering service, where the system can respond with a full description of the item as well as saying the name and address of where the item is to be sent. Here the quantity of information would be too large to directly record for the average catalogue. A good example of a system where the output information must change rapidly is access to news. Here large amounts of text can be updated regularly and made available over the telephone with no manual intervention. The ability to speak e-mail messages over the telephone is another important application, enabling people to check their messages when away from a computer terminal.

1.6.6   Speech enhancement

This is the process of enhancing the perceived quality of a speech signal received over a telephone link, as described in detail in Chapter 13. This area covers a wide range of applications such as noise and echo control, including the cancellation of echoes on long international circuits to enable simultaneous two-way speech communications. Echoes occur when signals undergo delayed reflections, and can affect both talkers and listeners on a call. The effects of telephony echo on customers are complex, depending on prevailing electrical and acoustic factors, as well as the motivation of both customers [13].

Echo suppressors have been used for many years and conceal echo by detecting when the distant customer is speaking and the near customer is silent.

The alternative of echo cancellation was first proposed by Sondhi of AT&T Bell Labs, and the first single-chip custom-VLSI implementation appeared in 1969.

As shown in Fig. 1.14, a replica of the echo signal is synthesized by an adaptive filter which models the echo path and is subtracted from the send signal. In this normal mode of use the echo canceller is only required to model the relatively short duration echoes returned by the national network. In general, it is desirable to cancel echoes as close to their source as possible. Cancellers offer somewhat better quality than suppressors on long-delay circuits, because they have greater signal transparency and less intrusive speech clipping effects.

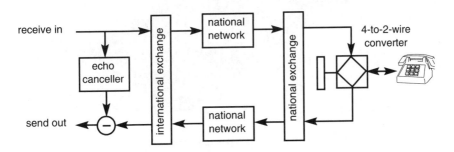

Fig. 1.14   Location of network-based echo cancellers.

Interest has also grown in applications for adaptive noise suppression where the speech and noise typically have overlapping spectra. In such a system adequate performance is dependent on obtaining a good estimate of the noise spectrum either from a separate reference microphone, or from noise-only periods in the incoming speech, obtained using a voice-activity detector (see Chapter 4).

Applications include very noisy environments where the signal-to-noise ratio is low, such as telephone kiosks in busy concourses, or in financial dealer rooms. Figure 1.15 shows an example of a noise-suppression system in action, in this case cancelling motorcycle and car noise from a telephone kiosk by the side of a busy road [14].

Signal processing is also employed in speeding up (or slowing down) the playback of recorded messages (without introducing noticeable distortion to the

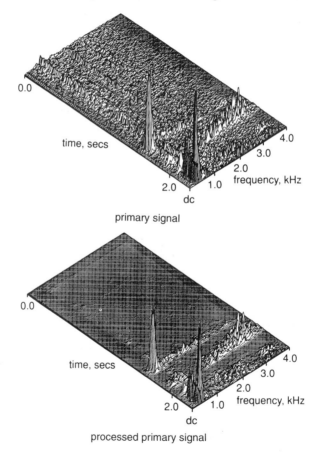

primary signal

processed primary signal

Fig. 1.15   Adaptive noise suppression.

speech), and to eliminate speech-clipping and other nonlinear distortions that are occasionally encountered in public switched telephone networks.

1.7   SPEECH TECHNOLOGY ASSESSMENT

Depending on the system being considered, subjective assessment is performed by either a conversation or a listening test [15]. In a listening test, people (who are called subjects) are placed alone in a controlled environment and asked to listen to speech transmitted through a series of randomly presented network connections (which are called conditions). At the end of each condition the subject is asked to vote on their opinion of the speech heard. A 5-point mean-opinion-score scale is used (5 = excellent, 4 = good, 3 = fair, 2 = poor, and 1 = bad). For example, a speech coder suitable for switched telephone network operation would have a mean-opinion score rating above 4.0.

During a conversation test, two subjects are placed in separate rooms and asked to converse over randomly presented conditions. As an aid to conversation they will be given a simple task to perform. At the end of each condition they each vote on the condition without discussing it with each other. Chapter 8 provides a comprehensive review of subjective testing methods, and describes a novel methodology for comparing text-to-speech systems.

In some cases where the requirements of the final system are not well understood or if a suitably accurate prototype of the recognition sub-system is not available, 'Wizard-of-Oz' experiments are sometimes conducted with a trained human operator in place of the speech recognizer. This allows the focus of the experiment to be maintained on the aspects of dialogue interaction of interest without the complication of errors introduced by the recognizer.

Comparative objective tests are conducted on in-house and commercial speech systems. Figure 1.16 shows the results of such an assessment on several commercial speech recognizers (averaged results). As can be seen the talker variability is greatest, with the telephone instrument contributing least to recognition errors. Chapter 12 discusses the testing of speech recognition systems and presents results to show steady progress in reducing error rates as the network has improved.

Recent studies on human perception of speech that has been subjected to nonlinear network distortions have been encouraging, with a high degree of correlation between the objective measures derived and people's subjective opinion of the impairment (see Chapter 14). There is research interest in extending this modelling approach to joint perception of mixed mode (video and speech) signals for emerging multimedia terminals.

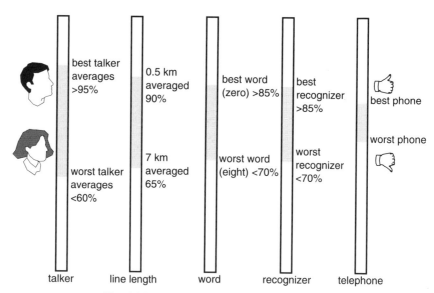

Fig. 1.16 Factors affecting recognizer performance.

1.8   FUTURE DIRECTIONS IN SPEECH TECHNOLOGY

1.8.1   Interactive voice systems

An essential ingredient in the success of any business is access to information sources, and modern telecommunications can give direct access anywhere in the world within seconds or minutes. With the spread of personal communications products (many of them too small to support a keyboard), speech and language processing, incorporating some degree of understanding, will play a necessary role in providing access to, transforming, and delivering information from these vast resources (see Chapter 17) [16].

    As interactive speech systems are deployed operationally they will generate new field experience and customers' service perception data. Such 'real-world' data will allow a new R&D focus on technology improvement. Advanced data visualization techniques will also help researchers to interpret this volume of data. By examining large numbers of interactions, statistical models of dialogues can be generated and used to optimize deployed dialogues that have been developed using traditional techniques. This offers the potential for a unified understanding from low-level signal processing to higher level semantics.

1.8.2   Speech coding

As optical-fibre systems provide virtually limitless bandwidth in terrestrial telecommunications networks, the need for speech coding may diminish on circuit-switched point-to-point connections. However, with the growth in corporate 'virtual' networking and wideband packet networks, the added flexibility afforded by variable-rate speech coding will be in demand for competitive, managed-bandwidth schemes (see Chapter 3).

Bandwidth will also remain under pressure in mobile and personal communications systems. In such areas, the demand for speech coding will continue to grow strongly over the next decade.

This demand will be driven by rising customer expectations on quality, international standardization, and by technical advances in the on-chip integration of digital and analogue functionality.

There is also considerable interest in using coding techniques to enhance the speech transmitted over networks and for CD-audio compression for future music and multimedia distribution systems.

1.8.3   Speech synthesis

The current drive to improve the naturalness of text-to-speech systems will continue with the emphasis in areas such as those described in Chapter 7:

- improving the 'natural rhythm' of synthesized speech (phonotactic variation);

- providing more choice in selecting the 'personality' of the speaker (current systems tend to adopt a single speaking style);

- providing more choice of language and accent with the associated tools to provide such flexibility in text-to-speech systems.

As full-motion video requires much bandwidth and/or suffers processing delays, there is growing interest in synthetic model-based image processing coupled with text-to-speech systems. In such a system, a wire-frame model of the person's head is created from photographs and a synthesized voice synchronized to the lips of the model (Fig. 1.17). By typing in text the synthetic persona can be made to 'speak'. Such technology shows promise for new services such as customized database access systems, 'continuous presence' conferencing, and to provide a research framework for studying fundamental mixed-mode signal interactions.

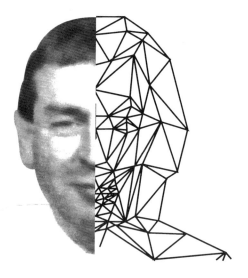

Fig. 1.17   Synthetic persona (talking head).

1.8.4   Speech recognition

Undoubtedly the current emphasis on accuracy and processing efficiency for large-vocabulary, speaker-independent recognition will continue. To make such systems scalable to national network applications, more work will be needed on improving noise robustness and out-of-vocabulary rejection. These aspects are comprehensively covered in Chapter 10. There is also interest in developing systems that can adapt to the characteristics of the user's voice and the transmission environment. Techniques such as 'barge-in' and 'wordspotting' will also grow in importance in the drive to make speech systems more natural, and hence more acceptable (see Chapters 9 and 10).

To enhance the performance of recognizers significantly, more basic knowledge will be needed in areas such as feature extraction, and building in of more knowledge, as it becomes available, of the physiology of the ear and brain as a coupled 'system'. Improvements to modelling of speech can be expected in a drive to reduce the limitations of the existing statistical model-based approaches. New paradigms based on mixed-mode (acoustic, speech, visual and gesture) cues may offer one way forward.

There will also be an increasing interest in multilingual recognition and the associated ability to quickly recognize the language of the speaker. Ultimately, the ability to identify the topic domain as a means of focusing the recognition resources will be developed.

1.8.5   Person-to-person — the future of telephony

Much of today's speech technology is concerned with man-to-machine interfaces using voice — but what about the future of person-to-person communications? Chapter 18 presents a personal and light-hearted view of the development of the telephone, and its future evolution.

Today, videoconferencing is limited to viewing the participants through a small rectangular window. Images are captured through a single camera and sound by means of a single microphone. In 'hands- and eyes-free' mode the speech often sounds clipped or as if it originated in a bathroom.

There is interest in acoustic echo cancellation, adaptive beam-steering (and talker location) and intelligent loudspeaker technology for 'last metre' processing to remove reverberation and noise in hands-free conferencing systems — to improve the naturalness of such interactions.

Early work in this area at BT Laboratories has shown the potential of this technology to provide improved naturalness and utility for loudspeaking telephones [17].

In the research environment teleconferencing can be made very realistic, as with electronic work spaces having eye-to-eye videoconferencing that maintains gaze awareness. When augmented by directional and speaker-tracking audio reproduction the illusion of 'being there' can be almost complete [18].

Speech and language processing has the potential to make such interfaces more natural and easy to use through the exploitation of visual, acoustic and gesture cues. Herein lies the true destiny of speech processing — not just one mode of communications, speech, but all the senses orchestrated to meet the real underlying communications needs of people.

1.9   CONCLUSIONS

Speech communication is, and will continue to be, key to the use of the network, currently accounting for well over 90% of revenue. Speech is the most natural way for users to communicate from person-to-person and, in the future, where appropriate, from person-to-system. Having the right applications and services is key to revenue generation. Understanding the technology behind them is vital to designing and operating the network efficiently, to the future of network intelligence and operational support systems, and to reducing costs, while presenting BT as a technologically advanced service provider.

Despite the enormous effort expended on speech technology research over the last few decades, still surprisingly little is known about speech signals — or rather how humans perceive and interact using them. For example, in areas like speech recognition, much more fundamental knowledge is needed to allow machines to be made that approach the capability of humans. Arguably, there has been too much recent emphasis in the digital signal processing (DSP) community on solving the problems of **digital processing** and not enough on understanding the subtlety of **signals** and human perception.

As speech technologies are commercially exploited, different skills and expertise will be required at different stages in their evolution, from research/prototyping in the early stages to downstreaming and support of real-world applications in the later stages. The effective coupling of research to delivery of applications is critical. In particular, the problem of scalability, from laboratory demonstrator to national network service, is fundamental, and one in which BT can be expected to play a key role [1].

Speech processing is a dynamic, exciting and commercially relevant field, overlapping the traditional subjects of mathematics, electronics and computer science. Speech technology is well poised to affect dramatically the way we all communicate and interact in the 21st century.

REFERENCES

1. Brooks R M (Ed): 'ISAP and its applications', BT Technol J, 14, No 2 , pp 9-83 (April 1996).

2. Lee E A: 'Programmable DSPs: A brief overview', IEEE Micro Magazine, pp 14-16 (October 1990).

3. Westall F A and Ip S F A (Eds): 'Digital signal processing in telecommunications', Chapman & Hall (1993).

4. Rabiner L R: 'Applications of voice processing to telecommunications', Proc IEEE, 82, No 2, pp 199-230 (1995).

5. Strathmeyer C R: 'Voice in computing: an overview of available technoloiges', IEEE Computer Magazine, 23, No 8, pp 10-15 (August 1990).

6. Rose K R and Hughes P M: 'Speech systems', British Telecommunications Eng J, 14, Part 1 (April 1995).

7.   Midwinter T: 'Desktop multimedia terminals', in Whyte W S (Ed): 'Multimedia Telecommunications', Chapman & Hall, pp 189-209 (1997).

8.   Jayant N S and Noll P: 'Digital coding of waveforms, principles and applications to speech and video', Prentice-Hall, Signal Processing Series, NJ (1984).

9.   Boyd I: 'Speech coding in telecommunications', in Westall F A and Ip S F A (Eds): 'Digital signal processing in telecommunications', Chapman & Hall, pp 300-325 (1993).

10. Wheddon C and Linggard R (Eds): 'Speech and language processing', Chapman & Hall (1990).

11. Cox S J: 'Hidden Markov models for automatic speech recognition: theory and application', BT Technol J, $\underline{6}$, No 2, pp 105-115 (1988).

12. Holmes J N: 'Formant synthesizers: cascade or parallel?', Speech Communication, $\underline{2}$, pp 251-275 (1976).

13. Lewis A V: 'Adaptive filters — applications in telephony', in Westall F A and Ip S F A (Eds): 'Digital signal processing in telecommunications', Chapman & Hall, pp 111-138 (1993).

14. Connell J M, Xydeas C S and Anthony D K: 'A hybrid bandsplitting acoustic noise canceller', in Westall F A and Ip S F A (Eds): 'Digital signal processing in telecommunications', Chapman & Hall, pp 373-399 (1993).

15. Guard D R and Goetz I: 'The DSP network emulator for subjective assessment', in Westall F A and Ip S F A (Eds): 'Digital signal processing in telecommunications', Chapman & Hall, pp 352-372 (1993).

16. Matsuoka T and Minami Y: 'Acoustic and language processing technology for speech recognition', NTT Review, $\underline{7}$, No 2 (March 1995).

17. Lewis A V and South C R: 'Extension facilities and performance of an LSI adaptive filter', Proc of ICASSP, San Diego (March 1984).

18. Travis D et al: 'Working together in the electronic Agora', British Telecommunications Eng J, $\underline{14}$, Part 2 (July 1995).

2

INTRODUCTION TO HUMAN SPEECH PRODUCTION, HUMAN HEARING AND SPEECH ANALYSIS

D M Howard and J A S Angus

2.1 INTRODUCTION

Human communication is a highly complex process, basic to life itself. Our main mode of communication is acoustic, through the organs of speech and hearing. From the initial formulation of the message to what is finally understood by the listener, this process involves many complex, variable and interacting factors. In the twentieth century, the widespread development and use of electronics has rapidly been exploited in speech research, producing tools for transmission, analysis, synthesis, recognition and speech training.

Advances in the study of how sound is perceived (psychoacoustics) have begun to provide new insights into the capabilities of the human hearing system, reshaping our approach to more efficient and effective design in speech processing systems. In particular, the recent development of inexpensive digital computers and advances in digital signal processing now make it possible to process signals in real time using highly complex algorithms. Speech technology is rapidly growing in commercial significance as a result. Yet many engineers treat speech as a simple signal, to be processed and transmitted with little appreciation of which aspects are important at the receiving end — the human ear.

This chapter describes and explains some important principles in relation to the production, perception and analysis of speech, and gives insight into the benefits of psychoacoustic research for speech technology developments of the future. Human speech production is first described from a basic acoustic viewpoint. The sounds of English are then described, in terms used by phoneticians, in order to explore their acoustic nature in a structured manner and to help appreciate those aspects of the speech signal that are of key importance in conveying the intended message. The human peripheral hearing system is then introduced, to aid understanding of modern analysis techniques that exploit advances in psychoacoustic knowledge. Lastly, some key techniques for the spectral analysis of speech signals are presented and their limitations explained.

2.2   HUMAN SPEECH PRODUCTION

In this section, human speech production is introduced in terms of the sound elements of English, analysed in terms used by phoneticians. These enable their acoustic properties to be categorized and explored. Fuller information on acoustic features can be found in Borden and Harris [1], Fry [2], Baken [3], Baken and Danilov [4], and Kent and Read [5].

2.2.1   Acoustic features of speech

Human speech production requires three elements — a power source, a sound source and sound modifiers. This is the basis of the source-filter theory of speech production [6]. The power source in normal speech results from the compressive action of the lung muscles. The sound source, during voiced and voiceless speech, results from the vibrations of the vocal folds and turbulent flow past a narrow constriction respectively. The sound modifiers are the articulators, which change the shape and therefore the frequency characteristics of the acoustic cavities through which the sound passes.

The main structures are shown diagrammatically in Fig. 2.1 and in an idealized, functional form in the right hand section of the figure. The arrows illustrate those elements which are adjustable and require appropriate control during the production of voiced sounds — lung action (power source), the position of the vocal folds (sound source), and the shape of the vocal tract (sound modifiers).

2.2.1.1   The power source

The power source in normal speech sounds is in the form of a flow of air from the lungs, known as an egressive airstream. In situations of particularly heightened

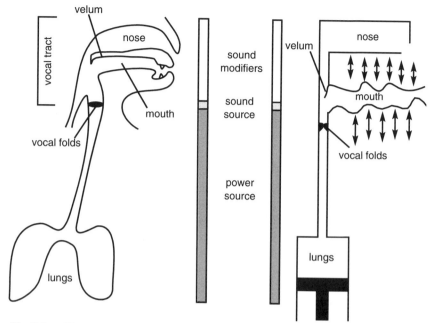

Fig. 2.1   The power source, sound source and sound modifiers during the production of voiced speech sounds.

stress, when it may be absolutely imperative to convey a message in a very short time, voice production can also occur while breathing in by using an ingressive airstream. Air is drawn in by enlarging the lung spaces through muscular action (equivalent to pulling the piston in Fig. 2.1 downwards) thereby creating a lung pressure that is lower than atmospheric so that air flows into the lungs.

2.2.1.2   The sound source

Sound is produced during speech either as a result of the vibration of the vocal folds (voiced excitation), or as a result of acoustic noise resulting from turbulent airflow (voiceless excitation), or a combination of these efforts (mixed excitation).

To initiate vibration of the vocal folds during voiced speech, they are brought closer together horizontally, or adducted, and air is expelled from the lungs, passing between the adducted folds via the glottis. The velocity of airflow must increase as it crosses the glottis, due to the narrowed airway at this point. This increase in air velocity reduces the pressure exerted on the surrounding structure (the Bernoulli effect), which for the glottis is the vocal folds themselves, tending to pull the folds towards each other. This action narrows the glottis still further and the air flows more rapidly, resulting in a further reduction in the pressure exerted on the vocal folds. The folds are thereby accelerated towards each other until they meet and collide at the mid-line of the larynx, snapping together rapidly as the glottis closes. At this instant, the airflow via the glottis is cut off and a rapid pressure drop results immediately above the larynx providing an essentially pulse-like acoustic excitation to the vocal tract.

Once the folds are in contact, two effects contribute to their parting — firstly the sub-glottal pressure is higher than super-glottal pressure, and secondly the elastic nature of the folds. These forces combine to part the folds, accelerating them through their equilibrium position until they decelerate to reach a maximally open position and return towards their equilibrium position to complete a cycle. This description is basic to the myolastic aerodynamic theory of vocal fold vibration [7], but it does not fully explain vocal fold excitation when the folds do not snap together, as during breathy voice production. Valuable insights into more recent work on vocal fold vibration can be gained from Titze [8].

An idealized illustration of a cycle of vocal fold vibration is shown in Fig. 2.2, with associated glottal airflow and glottal pressure waveforms. It can be seen that the reduction in airflow when the folds snap together (closing phase) is more rapid than the increase in airflow as the folds part (opening phase). The glottal pressure waveform is derived as the rate of change of this airflow which can be observed by comparing the waveforms in Fig. 2.2. This provides the acoustic excitation to the vocal tract. Notice the large negative-going pressure pulse resulting from each snapping together of the vocal folds, when the flow of air from the lungs is very rapidly cut off. Since the vocal folds are at least partially open throughout the closing and opening phases, they are referred to together as the open phase (see Fig. 2.2). Variations in closed phase have been associated with the efficiency of voice production both theoretically [8] and experimentally [9]. The average spectrum of the sound source during voiced speech is also shown in Fig. 2.2. It consists of harmonics whose amplitudes change with increasing frequency at a rate of -12 dB per octave.

In voiceless speech the vocal folds do not vibrate, but there is a constriction made somewhere in the vocal tract, through which air flows with sufficient velocity to create turbulence. This turbulent airflow produces acoustic noise that is essentially white over the frequency range of interest, and can therefore be considered as having a flat continuous spectrum. For some sounds, such as the non-vowel part of 'zoo', 'the', and 'vee', mixed excitation is used with turbulent airflow being produced at the point of constriction (the fricative part) while the vocal folds are vibrating (the voiced part). During these sounds the airstream is regularly interrupted by the closing and opening of the vocal folds and the acoustic noise produced at the point of constriction is pulsed on and off.

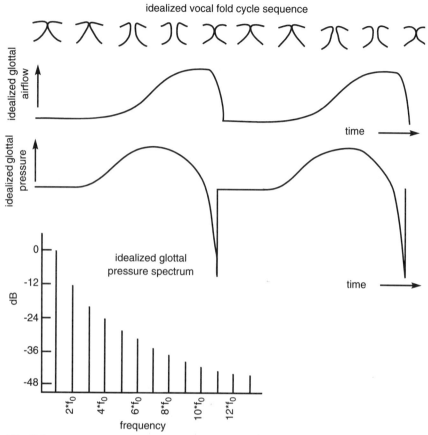

Fig. 2.2 Schematic sequence to illustrate vocal fold vibration sequence, idealized glottal airflow waveform and the derived glottal pressure waveform for two vocal fold vibration cycles. The open and closed phases are indicated.

2.2.1.3 The sound modifiers

The sound modifiers in voiced speech are primarily the cavities between the sound source and the outside world — the oral and nasal cavities which together form the vocal tract. The shape of the vocal tract (but not the nasal cavity) can be altered during speech production which changes its acoustic properties. The velum (see Fig. 2.1) can be raised or lowered to shut off or couple the nasal cavity. The acoustic frequency responses for the three vowels in 'bee', 'baa', and 'boo' are illustrated in Fig. 2.3 along with an illustration of the shape of the vocal tract in each case. During these vowels, the velum is raised to shut off the nasal cavity. Each vowel is characterized by a series of peaks in the vocal tract frequency response curve, known as formants. Three are shown in the figure and labelled first formant (F1), second formant (F2), and third formant (F3) in order of ascending frequency from the lowest frequency peak. The typical ranges of the first three formant frequencies for men, women and children are given in Table 2.1.

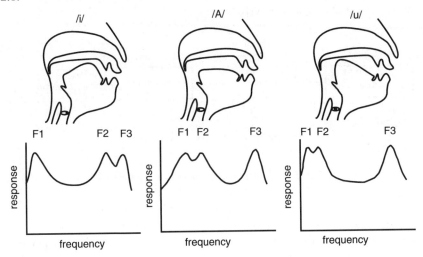

Fig. 2.3   Idealized vocal tract frequency responses for the vowels in bee (left), baa (centre), and boo (right).

Table   2.1   Typical first three formant frequency ranges of f_0 mean and ranges values of conversational speech of men, women and children (formant values from Kent and Read [5], f_0 values from Fry [2]).

Parameter	Men	Women	Children
F1 range	270 Hz-730 Hz	300 Hz-860 Hz	370 Hz-1030 Hz
F2 range	850 Hz-2300 Hz	900 Hz-2800 Hz	1050 Hz-3200 Hz
F3 range	1700 Hz-3000 Hz	1950 Hz-3300 Hz	2150 Hz-3700 Hz
f_0 mean	120 Hz	225 Hz	265 Hz

The average frequency domain effects in speech production are summarized in Fig. 2.4. Human speech exhibits an approximate 6 dB per octave roll-off with increasing frequency. The sound source and sound modifiers make the following contributions to this spectrum. For voiced speech, all harmonics are present with an average spectral roll-off of 12 dB per octave (see Fig. 2.2), and for voiceless speech there is a flat continuous spectrum. The sound modifiers encompass the time-varying formants but their average spectrum is flat. The radiation of the acoustic pressure waveform via the lips imparts a +6 dB per octave tilt with increasing frequency. This results in an average overall spectral variation with increasing frequency during voiced speech of −6 dB per octave, and of +6 db per octave during voiceless speech. Since the amplitude of voiced speech is usually very much greater than that of voiceless speech, the average spectral shape of speech tends to be close to −6 dB per octave.

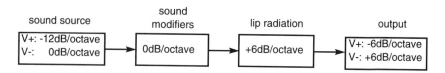

Fig. 2.4   Average spectral trends of the sound source, sound modifiers and lip radiation during voiced (V+) and voiceless (V-) speech.

2.2.2   Phonemic features of speech

There is a minimum set of sounds that is required to distinguish the words of any language, and these sounds are known as the phonemes of that language. Individual words of the language can be distinguished by changing one phoneme for another, and phonemes are therefore said to be contrastive with one another. For example, the English words *fit, sit, hit, pit, kit, wit, lit, nit*, indicate that the initial sounds are contrastive with each other since changing them produces different words of the language. Therefore these sounds *f, s, h, p, k, w, l*, and *n* are phonemes of English, and they are usually represented phonemically between '/' characters as /f/, /s/, /h/, /p/, /k/, /w/, /l/ and /n/ respectively. Similarly, the sounds that distinguish *bat, bit, but, bet*, as well as *boat, bait, bite, bought, beat, boot, Bert* are also phonemes of English. Notice that there is no one-to-one correspondence between phonemes and the spelling of a word.

Wells [10] developed a phonetic alphabet for European languages, known as SAMPA (speech assessment methodologies phonetic alphabet), using only ASCII characters as listed in Table 2.2. Every phoneme can be uniquely described in terms of a three-part label comprising its voice, place and manner. These are described in the following section, along with the general acoustic nature of any given phoneme in these terms. Further details can be found in Abercrombie [11], Ladefoged [12], Gimson [13], Wells and Colson [14], O'Connor [15] and Catford [16].

2.2.2.1   Voice

The voice label is a binary one and it refers to whether or not the vocal folds in the larynx are vibrating during its production, so that the sound is either voiced or voiceless. Several methods exist for confirming whether a sound is voiced, for example:

Table 2.2   SAMPA symbols [10], example word and its SAMPA transcription for the 44 phonemes of English.

Consonants			Vowels		
p	rope	/r@Up/	i	neap	/nip/
b	buoy	/bOI/	I	jib	/dZib/
t	tide	/taId/	E	red	/rEd/
d	deck	/dEk/	{	anchor	/{Nk@//
k	cabin	/k{bIn/	A	hard	/hAd/
g	galley	/g{li/	Q	locker	/lQk@/
T	thwart	/TwOt/	O	port	/pOt/
D	weather	/wED@/	U	foot	/fUt/
f	fog	/fQg/	u	food	/fud/
v	vang	/v{N/	V	rudder	/rVd@/
s	sea	/si/	3	stern	/sT3n/
z	zenith	/zEnIT/	@	tiller	/tIl@/
S	ship	/SIp/	eI	weigh	/weI/
Z	treasure	/trEZ@/	aI	light	/laIt/
h	heeling	/hilIN/	OI	oilskin	/OIlskIn/
tS	chain	/tSeIn/	@U	row	/r@U/
dZ	jibe	/dZaIb/	aU	bow	/baU/
m	mast	/mAst/	I@	pier	/pI@/
n	main	/meIn/	e@	fare	/fE@/
N	rigging	/rIgIN/	U@	fuel	/fU@l/
w	winch	/wIntS/			
r	rain	/reIn/			
l	lee	/li/			
j	yacht	/jQt/			

- trying to sing the sound;

- placing the thumb and index finger of one hand on either side of the throat at the level of the Adam's apple and feeling for vibration;

- putting hands over the ears and listening for a loud buzzing sound.

For example, the consonants in 'Sue' and 'zoo' are contrasted by the voice of the initial consonant. The /s/ in Sue is voiceless and the /z/ in zoo is voiced. A number of the phonemes of English are paired by their voice, for example /s/ and /z/, /f/ and /v/, /T/ and /D/, /p/ and /b/, /t/ and /d/, /k/ and /g/ (see Table 2.3).

2.2.2.2   Manner

The manner of articulation refers to how the sound is articulated. The sounds /p/, /b/, /t/, /d/, /k/ and /g/ are known as plosives because a complete closure is made in the vocal tract and the soft palate is raised. This completely prevents the escape of air during the hold stage of the plosive, resulting in an increased air pressure within the vocal tract. When the constriction is released an impulsive short and well-defined transient burst of acoustic energy results, termed plosion. The sounds /m/, /n/ and /N/ are known as nasals and while their place of articulation is similar to that of the plosives, the soft palate is lowered to allow air to escape via the nose. Acoustically, nasals are produced via a branched tube which has the effect of adding zeros to the spectrum of the sound modifiers. Sounds that do not involve a complete closure, but allow a narrow gap through which air passes to produce acoustic noise, such as /f/, /v/, /T/, /D/, /s/, /z/, /S/, /Z, are known as fricatives. The two consonants /tS/ and dZ/ have double symbols to indicate that they have a contrastive role as individual phonemes in their own right. Their manner is that of a plosive followed by a fricative, known as affricate. Acoustically, affricates have a sharper noise onset than their fricative counterparts. The four sounds /w/, /r/, /l/ and /j/ are known as semi-vowels because they are vowel-like sounds in themselves but they appear structurally in English as consonants (consider their positions in words such as: you, zoo, woo, right, light, bight, white). All vowels are articulated with a similar vowel manner of articulation.

2.2.2.3 Place

The place of articulation refers to where in the vocal tract either a complete or the greatest constriction is made. The main articulators and places of articulation used during the production of English consonants are shown in Fig. 2.5 and listed in Table 2.3. The sounds /p/, /b/ and /m/ have the same place of articulation, where the lips are brought into contact, known as bilabial articulation. The sounds /f/ and /v/ are produced by making contact between the lower lip and the upper teeth and their place of articulation is labio-dental. The sounds /T/ and /D/ are produced with contact between the tongue tip and the upper teeth, known as dental articulation. For /t/, /d/, /n/, /l/, /s/ and /z/ the tongue tip or blade makes contact with the alveolar ridge, which can be found by placing the tongue tip behind the upper teeth and moving it back in the mouth, and their place of articulation is known as alveolar. The /r/ sound is usually produced with tongue contact further back in the mouth than for the alveolar sounds, and its place is therefore known as post-alveolar. The sounds /S/, /Z/, /tS/, /dZ/ and /j/ are produced with close approximation between the front of the tongue and the area between the alveolar ridge and hard palate and are known as palato-alveolar. The back of the tongue makes contact with the soft palate or velum for the production of /k/, /g/ and /N/ and their place is known as velar. The /h/ sound is produced with close approximation between the vocal folds, and it is known as glottal.

Table 2.3   Voice, place and manner descriptors for the consonants of English.

SAMPA symbol	voice (V+/V-)	place	manner
p	V-	bilabial	plosive
b	V+	bilabial	plosive
t	V-	alveolar	plosive
d	V+	alveolar	plosive
k	V-	velar	plosive
g	V+	velar	plosive
T	V-	dental	fricative
D	V+	dental	fricative
f	V-	labio-dental	fricative
v	V+	labio-dental	fricative
s	V-	alveolar	fricative
z	V+	alveolar	fricative
S	V-	palato-alveolar	fricative
Z	V+	palato-alveolar	fricative
h	V-	glottal	fricative
tS	V-	palato-alveolar	affricate
dZ	V+	palato-alveolar	affricate
m	V+	bilabial	nasal
n	V+	alveolar	nasal
N	V+	velar	nasal
w	V+	bilabial	semi-vowel
r	V+	alveolar	semi-vowel
l	V+	alveolar	semi-vowel
j	V+	palatal	semi-vowel

A variation in the place of articulation results in a change to the shape of the acoustic cavity and hence the spectrum of the sound modifiers. The place of articulation of fricatives moves from the lips to glottis in /T/, /f/, /s/, /S/ and /h/, the front cavity becoming larger and its main resonances moving lower in frequency (e.g. Baken and Danilov [4]). Similarly for plosives, the centre frequency of the burst changes with place of articulation and is high for alveolar, low for bilabial and intermediate for velar articulation. As the articulators move, the formant frequencies vary (formant transitions), and the transition of F2 turns out to be an important acoustic cue for place of articulation.

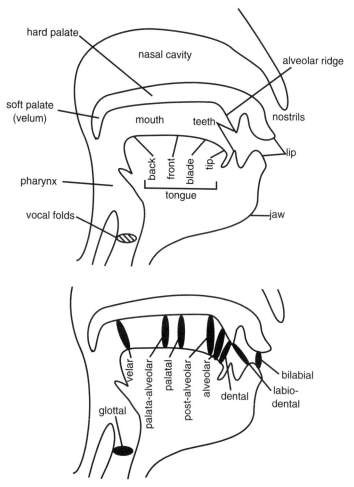

Fig. 2.5   Cross-section of the vocal tract illustrating the articulators (upper) and the main places of articulation used in English (lower).

2.2.2.4   Vowels

Steady English vowels, or monophthongs, are generally described in terms of: how narrow the constriction is between the tongue and the roof of the mouth, whether it is the front, centre or back of the tongue which makes the closest constriction, and whether the lips are rounded or not. The first two variables can be described as 'close to open', and 'front to back' respectively, and they are often depicted as a vowel quadrilateral (see Fig. 2.6). The position of each vowel on the quadrilateral is indicative of the highest point of the tongue, and the vowels of English are plotted on a quadrilateral in Fig. 2.6. The open/close and front/back axes are labelled. Lip rounding can be readily observed visually. Thus /i/ is a front, close and unrounded vowel, /{/ is front, open and unrounded, and /u/ is back, close and rounded. Diphthongs are vowels during which the sound quality changes, due to tongue and/or lip/jaw movement, and they are represented on the vowel quadrilateral by arrows indicating the starting and ending monophthongal qualities (see Fig. 2.6). They are described as either centring or closing depending on the direction of the arrow, as illustrated. Monophthongs have steady formants, diphthongs have slowly varying ($\approx$250 ms) formants, or formant transitions, and semi-vowels have more rapid ($\approx$100 ms) formant transitions.

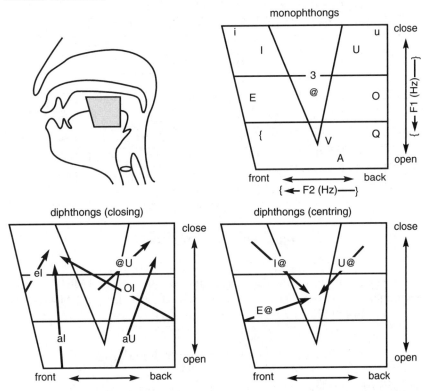

Fig. 2.6   The approximate relationship between the vowel quadrilateral and the vocal tract (upper left) and English monophthongs (upper right) and diphthongs (lower) plotted on vowel quadrilaterals (SAMPA symbols used throughout).

Average formant frequencies for a selection of vowels are given in Table 2.4 (from Peterson and Barney [17]). It can be shown that the vowel quadrilateral representation is directly related to the first and second formants, such that an F1 frequency change from low to high is equivalent to a close to open variation and an F2 frequency change from low to high is equivalent to a back to front variation. This can be inferred from Table 2.4, looking at the variation of F1 and F2 as the vowel changes around the quadrilateral in table order: /i/, /I/, /E/, /{/, /A/, /O/, /U/ and /u/. This effect is observed for men, women and children. Thus the axes of the quadrilateral could be replaced with F1 and F2 as indicated in Fig. 2.6.

Table 2.4   Average formant frequencies in Hz for men, women and children for a selection of vowels (from Peterson and Barney [17]).

	Men			*Women*			*Children*		
vowel	*F1*	*F2*	*F3*	*F1*	*F2*	*F3*	*F1*	*F2*	*F3*
/i/	270	2300	3000	300	2800	3300	370	3200	3700
/I/	400	2000	2550	430	2500	3100	530	2750	3600
/E/	530	1850	2500	600	2350	3000	700	2600	3550
/{/	600	1700	2400	860	2050	2850	1000	2300	3300
/A/	730	1100	2450	850	1200	2800	1030	1350	3200
/O/	570	850	2400	590	900	2700	680	1050	3200
/U/	440	1000	2250	470	1150	2700	560	1400	3300
/u/	300	850	2250	370	950	2650	430	1150	3250
/V/	640	1200	2400	760	1400	2800	850	1600	3350
/3/	490	1350	1700	500	1650	1950	560	1650	2150

2.3   HUMAN HEARING

This section introduces the key capabilites of the human hearing system. Descriptions of its basic anatomy can be found in Pickles [18], Moore [19], Rosen and Howell [20] and Howard and Angus [21].

2.3.1   The ear as a spectral analyser

The acoustic pressure waveform of incoming sounds is modified by the frequency response of the outer ear, or pinna, and of the auditory canal, leading to the eardrum, or tympanic membrane. The mechanical vibrations of the eardrum are transmitted via the three small bones, or ossicles, to the oval window of the cochlea. Running down the length of the cochlea is the basilar membrane, and the frequency analysis of incoming sound results principally from the mechanical properties of this membrane.

In shape, the basilar membrane is narrow and thin at one end becoming both wider and thicker towards the other end. It responds mechanically best to high frequency components towards the narrow and thin end, to low frequency components towards the wide and thick end, and to intermediate frequency components at different positions or places along its length. The overall frequency range of human hearing is generally taken as 20 Hz to 20 kHz, with the upper end reducing with the natural ageing process, an effect that is more rapid for men than women. The place of maximum response on the membrane varies as the logarithmic distance along the membrane with input frequency [22]. Thus the ear carries out a frequency analysis of the components in the incoming sound, and movements of the basilar membrane are converted into neural firings by hair cells in the organ of corti, for transmission to higher centres of the brain.

2.3.2   Dynamic range of the ear

The range of sound pressure levels which the average human ear can detect varies approximately between the threshold of hearing at 20 µPa and threshold of pain at 20 Pa. Sound pressure level (SPL) is usually represented in decibels relative to the threshold of hearing (20 µPa) by means of the following equation.

$$dB(SPL) = 20*\log\left\{\frac{P_1}{P_0}\right\} \qquad\qquad ... (2.1)$$

where P_1 = measured sound pressure level, and P_0 = 20 µPa

Thus the range 20 µPa to 20 Pa is equivalent to 0 dB(SPL) to 120 dB(SPL). These levels vary as a function of frequency and Fig. 2.7 illustrates the average nature of this variation for the thresholds of hearing and pain. The approximate average sound level (at 0.5 m) and frequency range for average normal conversational speech is also indicated on the figure. The ear is most sensitive in the 3-5kHz range and sounds at the lowest and highest frequencies have to be at considerably greater sound pressure levels to be heard.

2.3.3   Frequency selectivity of the ear

The frequency selectivity of the hearing system is usually described as a bank of a large number of band-pass filters, operating in parallel (e.g. Moore [19]). The responses of these filters are not symmetrical in the frequency domain, but their bandwidths can be defined in terms of the bandwidth of a hypothetical filter with a rectangular frequency response, which passes equal power and has the same response maximum as the biological filter in question. This is known as the equivalent rectangular bandwidth or ERB, summarized by the following equation [23], which enables an ERB value in Hz to be calculated for any given filter centre frequency (f_c) in Hz:

$$ ERB = \left\{ \left[6.23 * 10^{-6} * f_c^2 \right] + \left[93.39 * 10^{-3} * f_c \right] + 28.52 \right\} Hz \qquad \text{... (2.2)} $$

where $(100 \text{ Hz} < fc < 10 \text{ kHz})$.

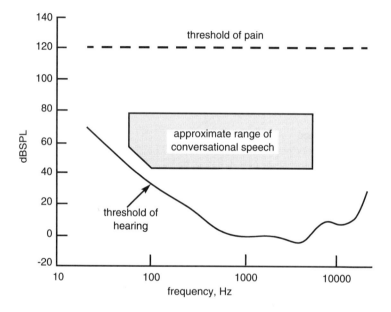

Fig. 2.7   Idealized average human hearing dynamic range between the threshold of hearing and threshold of pain as well as the area within which normal conversational speech takes place (adapted from Howard and Angus [21]).

The bandwidth of ERB-based analysis filters increases with frequency as plotted in Fig. 2.8. Therefore the hearing system exhibits fine frequency response at low frequencies and fine time response at high frequencies. Third-octave (four semitones) filters are often employed for sound processing because they approximate closely to the variation of the ERB, or critical bandwidth. This bandwidth is not the only factor relating to human discrimination of frequency, since the minimum frequency change, or just noticeable difference (JND), that humans can detect turns out to be approximately one thirtieth of the ERB [24] which is far smaller than one semitone.

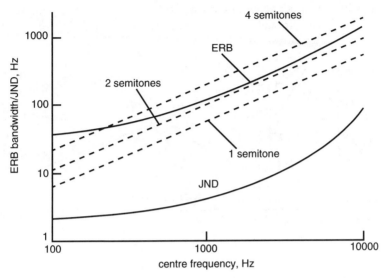

Fig. 2.8   Variation of equivalent rectangular bandwidth (ERB) with auditory filter centre frequency, just noticeable difference (JND) for frequency and 1, 2, and 4 semitone lines for comparison. (adapted from Howard and Angus [21].

2.3.4   Masking

When two or more component pure tones are heard together an effect known as masking can occur, which results in one tone being more difficult to perceive, the maskee, in the presence of another, the masker. A number of aspects of masking are important in the perception of complex sounds such as speech. The masking effect is always greater for maskers whose frequencies lie below rather than above that of the maskee. If masking is considered on an ERB-based frequency scale, then it is possible to model the masking effect in terms of a modification of the threshold of hearing due to the presence of the masker.

Figure 2.9 shows the basis of this effect for three pure-tone maskers at frequencies of 300 Hz, 350 Hz and 400 Hz and levels at 50 dB(SPL), 70 dB(SPL) and 90 dB(SPL) respectively. Sundberg points out that in the critical band of the masker itself, the threshold is approximately 20 dB below the level of the masker. The roll-off is steep below the frequency of the masker and essentially independent of the level of the masker. For frequencies above the masker, the level rolls off at between 5-13 dB per critical band depending on masker level, the effect being greater as the masker level is increased. In complex tones, components mask each other, which is an important aspect to consider when trying to determine particular components that are vital to the perceptual process. Components in speech that fall between formants will often be completely masked and will not therefore contribute to the perceptual process.

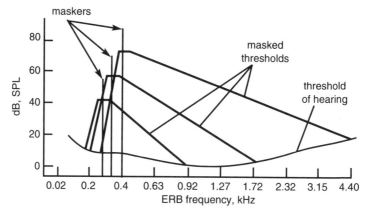

Fig. 2.9   Masking effect for tones at 300 Hz, 350 Hz and 400 Hz at 50 dB(SPL), 70 dB(SPL) and 90 dB(SPL) respectively plotted on an ERB frequency scale (after Sundberg [25]).

2.4   SEGMENTAL AND SUPRASEGMENTAL LEVELS

The descriptions given so far relate to individual speech segments spoken in isolation, but speech does not consist of strings of individual segments, Rather, it is articulated in a continuous manner, and the few examples drawn here serve simply to indicate the importance of these effects in speech production.

The articulation, and therefore the acoustic manifestation, of individual segments may be modified to some extent by the articulation of neighbouring segments. Individual sounds are uttered rapidly during running speech, and in many cases there may be little or no obvious attainment of an acoustic steady-state. Therefore individual segments can become acoustically less distinct, depending on the local context.

2.4.1   Co-articulation effects

Attempts to place acoustic boundaries in order to annotate speech in terms of individual phonemes can often be inappropriate. This is due to coarticulation, where articulatory features associated with a particular sound spread across to its neighbours (e.g. Sharf and Ohde [26]). For example, a vowel followed by a nasal consonant (/m/, /n/ or /N/) tends to be produced with some degree of nasal cavity coupling, since the lowering of the velum for the following nasal consonant is anticipated during the vowel, an effect known as nasalization. The place of articulation of a velar plosive varies in terms of the exact position where contact is made, depending on whether the following vowel is front or back. The articulation of a consonant that is followed by a lip-rounded vowel will often be modified to anticipate the lip rounding — consider the /r/ sounds in /rid/ and /rud/.

Another example is assimilation, in which the place of articulation of a sound is changed to that of the following sound to simplify the articulatory gestures required to utter the phrase. Consider, for example, the phrase ten pit ponies. In deliberate speech this might be uttered as /tEn pIt p@UnIz/, but in running speech could become /tEm pip p@UnIz/ where the final consonant of each word is assimilated to take the place of the following consonant (/n/ becomes /m/ and /t/ becomes /p/), but the voice and manner are retained. It can also happen during compound words such as headmaster (/hEbm{st@/), and uncharacteristic (/uNk{r@kt@rIstIk/). In English, assimilation typically only affects the alveolar sounds /t/, /d/, and /n/, which are assimilated to their bilabial or velar counterparts /p/, /b/, /m/ or /k/, /g/, /N/ respectively. The alveolar fricatives /s/ and /z/ can be assimilated when they are followed by /S/, /Z/ or /j/ (e.g. this ship can become /DIS SIp/).

A further example is elision where sounds in words or at word boundaries are not pronounced or are elided. Wells and Colson [14] note the difference between historical elision in words such as Christmas where the /t/ is elided nowadays to give /krIsm@s/ rather than /krIstm@s/ and contextual elision, such as last time, where either the elided /lAs taIm/ or non-elided form /lAst taIm/ may be heard depending on the speaking rate and formality of the situation.

Effects that are spread over syllables or phrases, are the suprasegmental features of speech production. For example, in conversational as opposed to more formal speech, the rate of delivery tends to be faster and articulation tends to be less extreme, which results in vowels being closer to the neutral /@/ with formants that are pulled closer to the centre of the vowel quadrilateral — an effect known as vowel reduction. Plosives at the end of words tend not to be released, speech is softer, a narrower dynamic range of loudness is employed and the pitch range tends to be smaller and closer to the speaker's mean value.

Variations in pitch and the placement of stress are two of the most important and complicated suprasegmental features of speech and a full discussion can be found elsewhere [5, 27-32]. f_0 should be represented logarithmically in experimental work in order that it can be analysed in a manner that is perceptually salient, and either equal tempered semitones (twelfths of an octave) or cents (hundredths of a semitone) are appropriate and commonly used units. Formulae for converting from frequency ratios to semitones or cents and vice versa are given in Howard and Angus [21]. The average f_0 range for men women and children is about 13 semitones for conversational speech, but it can vary depending on the situation from less than one octave in quiet intimate speech to well over two octaves for professional singers. Mean f_0 values tend in all cases to be lower than the centre of the logarithmic frequency range.

2.5   SPEECH ANALYSIS

The acoustic pressure waveform of speech varies continuously with time as the message is articulated. Changes in articulatory settings manifest themselves as time variations in the frequency response of the sound modifiers (see Fig. 2.1), which are reflected in the speech pressure waveform from the vocal tract. Tracking the time-varying nature of the frequency content of the speech pressure waveform is a key area of analysis required in speech research. A vital consideration in this analysis is the trade-off in accuracy between frequency and time resolution due to the detailed nature of the dynamic variations in speech. This primarily enables information to be gained about the changing settings of the sound modifiers. The other key area about which information is often required is whether the sound source is voiced or voiceless (mixed excitation is usually considered voiced) and, if voiced, its f_0 value.

This section introduces these two areas of analysis and discusses the principles and limitations involved. Firstly, f_0 analysis is considered followed by methods for spectral analysis of the dynamic speech signal.

2.5.1   Fundamental frequency estimation

The pitch of a sound depends on how our hearing system functions and is therefore based on a subjective judgement by a human listener on a scale from low to high. Such a psychoacoustic measurement cannot currently be made algorithmically without the involvement of a human listener. The measurement of the f_0 of vocal fold vibration is an objective measurement which can be made algorithmically and therefore the term fundamental frequency estimation is to be preferred to the term pitch extraction commonly used in the literature. One reason why we adopt estimation rather than extraction is that although changes in pitch are perceived when f_0 is varied, small changes in pitch can also be perceived when the intensity (loudness) or the spectral content (timbre) of the sound are varied while f_0 is kept constant (see Howard and Angus [21]).

The choice of an f_0 measurement technique should be made with direct reference to the particular demands of the intended application [33, 34] in terms of the expected speaker population to be analysed (adult or child, male or female, pathological or non-pathological), the likely competition from acoustic background or foreground noise (others working in the same room, external noises, domestic noises, machine noises, classroom, clinic, children), the material to be analysed (read speech, conversational speech, shouting, sustained vowels, singing), the effect of the speaker to analysis system signal transmission path (room acoustics, microphone placement, telephone line, preamplification), and the measurement errors which can be tolerated (f_0 doubling, f_0 halving, f_0 smoothing, f_0 jitter).

The operation of f_0 estimation algorithms can be considered in terms of:

- the input pressure waveform (time domain);

- the spectrum of the input signal (frequency domain);

- a combination of time and frequency domains (hybrid domain);

- direct measurement of larynx activity.

Most of the errors associated with f_0 estimation techniques are due to:

- the quasi-periodicity of voice speech signals;

- formant movements;

- the difficulties in locating accurately the onsets and offsets of voiced segments.

A highly comprehensive review is given in Hess [33].

The waveform of a periodic sound is, in reality, quasi-periodic. Time domain techniques detect features which occur once per cycle during voiced sounds, such as the major positive or negative peak, positive-going or negative-going zero crossings, the slope changes associated with major peaks or any feature on the waveform which can be readily identified as occurring once per cycle. The input signal might be preprocessed to eliminate most of the effects of the formants and/or the voiceless energy, before applying the period detection processing. Problems with time domain f_0 estimation occur when the shape of the speech pressure waveform changes due to the rapid changes in the formant frequencies. Rabiner et al [35] note that the measured fundamental period can differ, depending on which waveform feature is being tracked, especially when the formants change rapidly. They also note that measurements based on zero crossings produce a small variation, or jitter, in the f_0 output when there is noise present in the input signal due, for example, to a breathy voice quality.

Frequency domain techniques take advantage of the fact that a periodic signal has a harmonic spectrum while a non-periodic signal has a continuous spectrum. Frequency domain f_0 estimation tends to be based on finding and tracking either:

- f_0 itself;

- the frequency difference between any two adjacent harmonics;

- the f_0 which best matches the complete set of harmonics present.

The first method can have problems when the fundamental is either not present, due to a communications channel such as the telephone, or when it is very weak compared to the other harmonics — particularly when the first formant frequency is high in a vowel such as /A/. The second method might exhibit errors when individual harmonics are weak due to the relative formant positions or variations in the cycle-by-cycle nature of vocal-fold excitation. The third method could have difficulty finding a best fit f_0 particularly when short cuts are employed to save processing time. In general, frequency domain techniques cannot be used when an accurate cycle-by-cycle estimate of f_0 is required because windowing of at least one f_0 cycle is needed, but they do tend to be more resilient in the presence of noise. Hybrid techniques make use of a combination of time and frequency domain features such that some merits of one domain might override some drawback of another [33]. Direct measurement of larynx activity includes the use of a throat microphone and electrical impedance measurements systems, such as the electrolaryngograph and electroglottograph [3]. These are widely used as reference f_0 estimation systems in the speech science laboratory [34, 36] due to their immunity to acoustic noise such as banging doors, equipment noise and other people talking in the vicinity, and local environmental acoustics such as reverberation. Such devices also generally give a more rigorous estimation than many microphone-based devices of the instants of onset and offset of voiced segments.

2.5.2   Spectrographic analysis

Since the 1940s, speech analysis has been carried out by means of a device known as a speech spectrograph which provides a spectrogram plot of the energy of the frequency components in the signal. For further details on speech spectrography see Fry [2], Baken [3], Baken and Danilov [4], Rosen and Howell [20], Koenig et al [37], or Potter *et al* [38]. Traditionally, time and frequency are plotted horizontally and vertically respectively, with increasing energy being indicated as a greater darkness of marking on a grey scale which usually incorporates 16 levels. It should be noted that the use of what might be considered to be visually more appealing colour scale can be misleading, since the visually abrupt change between two adjacent colours, such as red to orange, is strongly suggestive of a definite boundary between the energy levels represented. This would not be an appropriate conclusion to reach during detailed interpretation of spectrograms and could lead to misleading results.

The acoustic cues believed to be important in speech production and perception are described in terms of the variation of spectral energy with time during speech. The most rapid time-scale over which such changes typically occur is during the release stages of plosives, which is of the order of 5-10 ms. Tracking individual harmonics of male speech would require a frequency resolution of less than the minimum expected f_0 for males — approximately 50 Hz. Consequently, there is a direct trade-off to be considered in spectrography between frequency and time resolution, and this can be controlled by altering the bandwidth of the spectrograph's analysis filter. Usually this is indicated as wide or narrow, where these terms relate to whether the filter bandwidth is wider or narrower than the f_0 of the speech being analysed.

Traditionally used bandwidth settings are 300 Hz and 45 Hz respectively, but it should be noted from the average f_0 values given Table 1.1 that these are only really suited to the analysis of adult male speech. A filter whose bandwidth is sufficiently narrow to enable individual harmonics to be analysed (i.e. $< f_0$ of the speech sample) will have a time response that is longer than 5-10 ms and rapid spectral changes during speech will be blurred in the resulting spectrogram. This effect is illustrated in terms of frequency resolution in Fig. 2.10 and time resolution in Fig. 2.11. In the frequency domain, it is clear that the individual harmonics are resolved by a narrow-band but not by a wide-band analysis, and that formant positions are indicated more clearly by wide-band analysis. In the time domain, the ringing of a narrowband filter lasts longer and the peak amplitude is lower than that of a wide-band filter, in response to a single, short impulsive event (such as a plosive burst). Thus the wide-band filter responds more rapidly and with a greater amplitude to individual events in time, providing good time resolution. A narrowband filter gives good frequency resolution.

Figure 2.12 shows narrow, and wide-band spectrograms for 'Speech matters to you', uttered by a healthy adult male. The approximate positions of the sounds

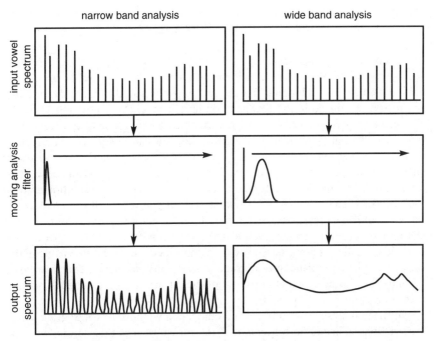

Fig. 2.10   Frequency analysis of the idealized spectrum of the vowel in bee using a narrowband (left) and a wideband (right) filter.

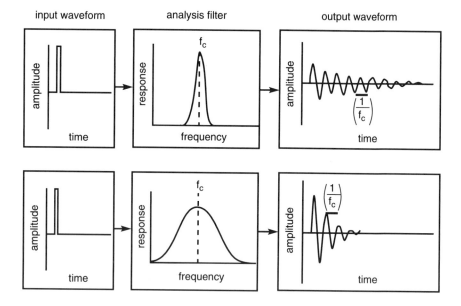

Fig. 2.11   Time resolution of an analysis filter (centre frequency f_c) with narrow (upper) and wide (lower) bandwidth settings for an input short pressure pulse.

in the sentence are indicated by the text above the spectrograms. The good frequency resolution of the narrowband spectrogram manifests itself as horizontal lines, which are the individual frequency components in the signal. During the voiced speech sounds in the utterance (indicated by underlining) /sp<u>i</u>tS m{t@z t@ ju/, these horizontal lines are the harmonics of the fundamental frequency and, since these are integer multiples of f_0, they are spaced equally in frequency because spectrograms are traditionally plotted with linear frequency scales. On the wide-band spectrogram, individual harmonics are not resolved since the filter bandwidth is greater than the f_0 of the speech. The good time resolution associated with wide band spectrograms manifests itself during voiced speech as vertical lines, known as striations, each representing an individual vocal fold closure. During the voiceless sounds, such as the consonants /s/ and /tS/ in speech, the acoustic excitation is noise-based and no harmonics or striations are visible in the narrowband or wide-band spectrograms respectively. However, the underlying structure remains horizontal (narrow) and vertical (wide) during voiceless sounds in the representation of the noise.

Formant analysis is usually carried out from a wide-band spectrogram, since the striations indicate the full extent of the formant bands themselves, whereas the harmonics on the narrow band spectrogram only indicate those that are within the formants (see Fig. 2.12). Formant transitions can be seen on both spectrograms, particularly during the diphthong /Iu/. The dominant high frequency energy in the voiceless fricative sounds which start and end the first word

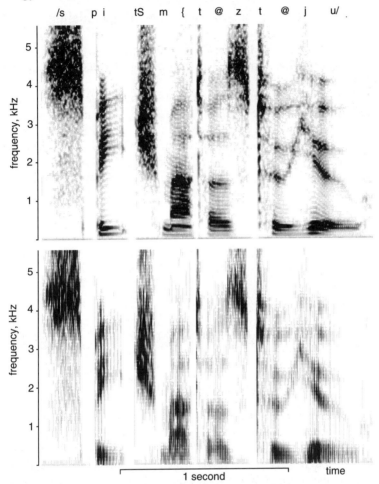

Fig. 2.12   Time-aligned narrowband (NB-upper) and wideband (WB-lower) spectrograms of 'speech matters to you' spoken by a healthy adult male. Analysis filter bandwidths are 45 Hz and 300 Hz for narrow and wide bandwidths respectively.

'speech', are clear in both spectrograms and their durations would be most accurately measured from the wide-band spectrogram due to its shorter time response. The release of /t/ in 'matters' is cleanly defined in time on the wide-band spectrogram, but appears more spread out in time on the narrowband spectrogram. It is possible from only the wide-band spectrogram to assess its energy distribution. The formant transitions in the vowels during /t@ ju/ are again clearly defined in the wideband spectrogram, and on the narrowband spectrogram care should be taken in separating the visual contribution of harmonic movement as f_0 changes.

2.5.3   Spectrography by human hearing modelling

The human hearing system can be considered as a filter bank whose bandwidths change with increasing filter centre frequency as described by the ERB (equation (2.2) and Fig. 2.10).

The GammaTone filter response is often used when modelling the human peripheral hearing system since it describes the shape of the impulse response function of the auditory system as estimated by the reverse correlation function of neural firing times [39]. Figure 2.13 shows the output in spectrographic form from the real-time hearing model described by Swan et al [40], which consists of a bank of 64 ERB-spaced GammaTone filters for the utterance 'speech matters to you' spoken by a healthy adult male for which conventional spectrograms are plotted in Fig. 2.12.

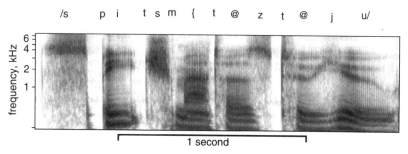

Fig. 2.13   Human hearing model spectrogram of 'speech matters to you' spoken by a healthy adult male (as used in Fig 2.12).

The GammaTone spectrogram exhibits fine time resolution at high frequencies and fine frequency resolution at low frequencies. It should be noted that the frequency axis is based on the ERB scale. Thus high frequencies tend to be compressed visually with respect to low frequencies giving a more appropriate perceptual weight to the different frequency regions. The lower formants are spread out and high frequency information is compacted when compared to conventional spectrograms. The unvoiced fricatives in /spitS/ are clearly represented in terms of their relative spectral balance. The lowest four or five harmonics can be clearly seen during voiced segments and striations are apparent for this speaker above approximately 1 kHz. The last fricative of /m{t@z/ is devoiced which can be seen as the harmonics at its start gradually disappear. The formant transitions are clearly illustrated, especially during /t@ ju/. Howard [9] suggests that such spectrograms may find application in forensic phonetic work.

2.5.4   Wavelet analysis

Another way of achieving a frequency analysis which has a broader bandwidth at higher frequencies is wavelet analysis [41]. In this type of analysis the speech signal is correlated with a set of orthogonal basis functions which represent the impulse responses of a set of increasing bandwidth filters. The resulting computation structure is very similar to the tree-structured quadrature-mirror filterbank used in speech coding.

In fact, the quadrature-mirror filterbank is a form of wavelet transform with the output samples of the filters representing the transform coefficients. Due to the variable bandwidth, which is proportional to frequency, the basic functions are simply versions of each other scaled in time.

One of the important characteristics of wavelet transforms, in addition to their variable bandwidth characteristics, is that they are simultaneously concentrated in time and frequency and this allows them to have the desirable characteristics of good time and frequency resolution at the same time. Press et al [42] provide descriptions and code for programming.

2.5.5   Cepstral analysis

One of the problems of simple spectral analysis is that the resulting output has elements of both the vocal tract (formants) and its excitation (harmonics). This mixture is often confusing and inappropriate for further analysis, such as speech recognition. Ideally, some method of separating out the effects of the vocal tract and the excitation would be appropriate.

Unfortunately these two aspects are convolved together and so cannot be separated by simple filtering. One approach to analysing speech which can resolve this is the cepstrum [43], which allows separation of the effects of the vocal tract and the excitation. This has applications both in pitch detection and vocal tract analysis.

The method relies on applying nonlinear operations to map the operation of convolution into a summation so that signals which are convolved together become signals which are added together and so can be readily separated, providing they do not overlap in this domain.

This was achieved via two mappings:

- convolution in the time domain is equal to multiplication in the frequency domain;

- the sum of the logarithms of two numbers is equal to the logarithm of their product.

Thus by Fourier transforming a signal, representing two convolved signals, and then taking the logarithm, results in a transform which represents the sum of the two convolved signals. This additively combined result can then be transformed back to the time domain and processed to separate the signal into excitation and vocal tract components.

Clearly in order to do this the components must be separable and this requires that there be a strong periodicity in the frequency domain, which is different to that imposed by any formant structure. In practice this means that high-pitched voices and nasals, or any other sound with low harmonic content, are hard to separate.

Figures 2.14a-2.14c show these operations being carried out on the /@/ in the spectrogram example earlier; it can be seen that the log spectrum has a strong ripple in frequency, due to the pitch, and this ripple becomes a peak at about 9.3 ms in the cepstrum. This peak can be used as the basis of a powerful pitch detection system. Cepstral processing is not without its problems as the logarithmic operation results in an unbounded dynamic range and a means of extracting the principle value of the phase if the full complex representation is required. For more details, see Oppenheim [44].

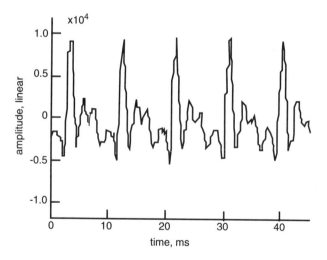

Fig. 2.14a   Input speech waveform /@/.

2.5.6   Linear prediction

An alternative way of analysing and processing speech is through modelling. Here a parametric model of the system is proposed and algorithms are developed

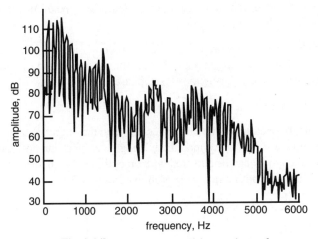

Fig. 2.14b   Log spectrum of the speech waveform.

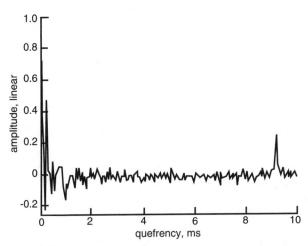

Fig. 2.14c   Cepstrum of the speech waveform.

to determine the parameters of the model and hence of the system. Clearly, the accuracy of these techniques depends on both the accuracy of the model and the algorithms used. Figures 2.1 and 2.4 show a popular model of speech known as the source-filter model. In this model, the effect of the articulators and the voice source spectrum are modelled as linear filters which are excited by an appropriate white source (random noise, impulse train, or a mixture of the two), depending on the voicing of the speech sound being modelled. This type of model is useful, but makes two simplifying assumptions:

- that the source and filter are independent and do not interact with each other — this is not strictly true, since the filter damping is higher during the glottal open phase and lower during the closed phase (see Fig. 2.2);

- that the filters are linear.

In principle, by measuring the difference between the output of the model filter structure and the signal being modelled, using an appropriate cost metric, the parameters of the model can be adjusted with an optimization algorithm to match the real signal in some optimum sense, as shown in Fig. 2.15.

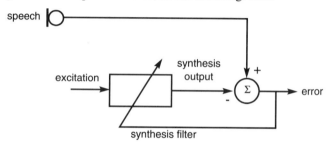

Fig. 2.15   Deriving the filter model for speech with analysis by synthesis.

This technique of parametric modelling can be very effective but suffers from a number of problems:

- computational cost — in general the optimization required is nonlinear and so requires considerable computational effort;

- lack of global optimality — because the performance metric used to judge the quality of the model output in comparison with the real signal can have many local minima it is difficult to know if the optimization algorithm has found the global optimum; this often means that heuristic approaches must be used to resolve these problems;

- difficulty in extracting appropriate performance parameters — in many cases deriving an appropriate metric from real signals is difficult, with a short-term Fourier transform of the closed phase of the pitch period often being used, which gives rise to noise sensitivity problems.

Notwithstanding these problems, analysis by synthesis is an effective way of achieving good copy synthesis of speech. This is probably because the technique is not limited in the filter structure that can be used for the modelling filter, which often allows more effective modelling of sounds such as nasals. With the advent of ever faster processing the computational overhead becomes less of a barrier to this technique, although the difficulties with local optima and extracting suitable performance metrics remain.

Because of the problems associated with general analysis-by-synthesis a particular version of parametric modelling which answers some of these problems has become popular. This technique, known as linear prediction [45], accepts restrictions on the filter structure in order to remove the problems of nonlinear optimization, local minima, and extracting performance parameters.

The basic principle behind linear prediction is that, at a given time instant, the speech signal will consist of two elements, as shown in Fig. 2.16.

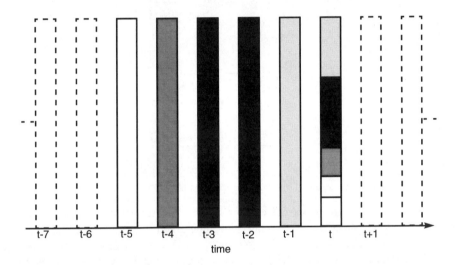

Fig. 2.16   Elements in a sample at a given time due to previous samples from a recursive filter.

- Energy which is due to previous signals that have passed through the vocal tract — in principle the contribution of this to the overall signal can be calculated from a knowledge of the signal that has already passed through the vocal tract, i.e. this contribution is predictable. This is shown by the different hatchings in the sample at time t in Fig. 2.16, which correspond to the similarly hatched earlier samples.

- Energy which is not due to previous signals that have passed through the vocal tract, but is due to the excitation of the vocal tract by the source — the contribution of this to the overall signal cannot be calculated from a knowledge of the signal that has already passed through the vocal tract and therefore this contribution is not predictable, shown in white on Fig 2.16.

This results in the major restriction on linear prediction in that it can only exactly model systems in which the filter only affects samples following the excitation. This type of filter, shown in Fig. 2.17, is a recursive filter with purely feedback terms. It is known as an all-pole filter, as it can only implement the denominator of a general filter-transfer function.

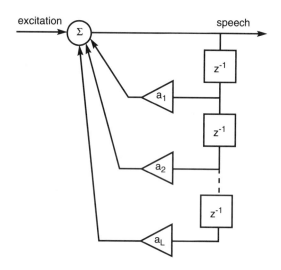

Fig. 2.17 Recursive model filter structure required for linear prediction to be effective.

Unfortunately, this does not cover all possible filter-transfer functions. All pole filters will have an exponentially decaying effect on samples which follow a given sample, due to their feedback structure, and so can be predicted. All-zero finite impulse response (FIR) filters on the other hand have a contribution which can extend both forwards and backwards from a given sample, as shown in Fig. 2.18. Furthermore, the contribution is not constrained to be exponentially decaying in amplitude as a function of time. These effects make it impossible for linear prediction to model an all-zero filter exactly. This would seem to mean that the procedure is doomed to failure on real speech signals which contain correlations due to both poles and zeros (in the case of nasals and laterals). Fortunately, a transfer function with zeros can be approximated by a higher-order all-pole filter which matches the correlations introduced by the zeros and thus the principle is generally applicable to speech signals. Note that even if the model filter is all zero the problem cannot be solved, because, in order to establish the level of correlation in the samples, a knowledge of the excitation of the system is required. This may be established via a hypodermic probe microphone inserted into the oral cavity just above the glottis, although this is not generally an appropriate or acceptable procedure.

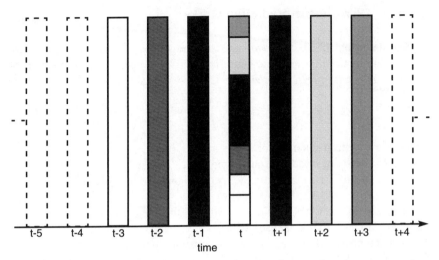

Fig. 2.18   Elements in a sample at a given time due to previous samples using a non-recursive (FIR) filter.

The structure for applying linear prediction is shown in Fig. 2.19 where the speech signal is compared with a prediction based on delayed versions of itself. The mean squared error of the difference between the predicted and the actual signal is used to drive an adaptation algorithm which sets the weights on the prediction filter to their optimum value.

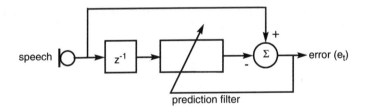

Fig. 2.19   The structure of a linear prediction filter.

It is useful to look at the mathematical derivation of the adaption algorithm to gain an insight into the reason for the utility and popularity of this technique, as well as an appreciation of its limitations. Mathematically we are interested in minimizing the mean squared error of the prediction filter with respect to the coefficients of that filter.

2.5.6.1   Why linear prediction works

To find the mean square error we need to average over all time. In practice, we have to limit this time to some finite number of samples around the time of interest. That is, the mean squared error is approximated by:

$$E_t \approx \frac{1}{n}\sum_n e_t^2(n) \qquad \qquad \dots (2.3)$$

where E_t = the mean square error at time t,
$\quad$ n = the set of samples around time t,
$\quad$ e_t = the difference between the speech signal and the prediction.

This approximation is the first limitation of the linear prediction process. It is necessary because the parameters of the speech signal vary with time, i.e. they are non-stationary. However, one of the assumptions of the linear prediction process is that the system is stationary. Note this assumption is also true in many other forms of speech analysis, e.g. spectral analysis. By using a short enough time period for the measurement of the mean squared error, of the order of 10 ms to 20 ms (80 to 160 samples at 8 kHz sampling rate), the speech can be considered to be locally stationary and thus be suitable for analysis. However, the cost for doing this is that the result of the error estimate is now only an approximation to the true value and will be affected by degradations to the speech signal due to noise or other interference. These effects become more serious as the number of samples used to estimate the mean square error becomes smaller. If the predicted signal is a weighted sum of delayed versions of the input speech signal it can be expressed as:

$$Y_t = \sum_{k=1}^{L} a_k S_t(n-k) \qquad \qquad \dots (2.4)$$

where Y_t = the prediction at time t,
$\quad$ a_k = one of L weighting coefficients,
$\quad$ $S_t(n-k)$ = the input signal delayed by k samples.

This can be represented more compactly in vector form as:

$$Y_t = \begin{bmatrix} a_1 a_2 \dots a_L \end{bmatrix} \begin{bmatrix} S_t(n-1) \\ S_t(n-2) \\ \vdots \\ S_t(n-L) \end{bmatrix} = a^T S \qquad \qquad \dots (2.5)$$

or as

$$Y_t = \left[S_t(n-1)S_t(n-2)...S_t(n-L) \right] \begin{bmatrix} a_1 \\ a_2 \\ \vdots \\ a_L \end{bmatrix} = S^T a \qquad \text{... (2.6)}$$

The squared error is therefore given by:

$$e_t^2(n) = (S_t(n) - a^T S)^2 \qquad \text{... (2.7)}$$

which when expanded gives:

$$e_t^2(n) = a^T S S^T a - 2S_t(n)S^T a + S_t^2(n) \qquad \text{... (2.8)}$$

now:

$$SS^T = \begin{bmatrix} S_t(n-1) \\ S_t(n-2) \\ \vdots \\ S_t(n-L) \end{bmatrix} [S_t(n-1)S_t(n-2)...S_t(n-L)] \qquad \text{... (2.9)}$$

which is equivalent to:

$$SS^T = \begin{bmatrix} S_t(n-1)S_t(n-1) & S_t(n-1)S_t(n-2)...S_t(n-1)S_t(n-L) \\ S_t(n-2)S_t(n-1) & S_t(n-2)S_t(n-2)...S_t(n-2)S_t(n-L) \\ \vdots & \vdots \quad \vdots \quad \vdots \\ S_t(n-L)S_t(n-1) & S_t(n-L)S_t(n-2)...S_t(n-L)S_t(n-L) \end{bmatrix} \quad \text{...(2.10)}$$

and

$$S_t(n)S = S_t(n) \begin{bmatrix} S_t(n-1) \\ S_t(n-2) \\ \vdots \\ S_t(n-L) \end{bmatrix} = \begin{bmatrix} S_t(n)S_t(n-1) \\ S_t(n)S_t(n-2) \\ \vdots \\ S_t(n)S_t(n-L) \end{bmatrix} \qquad \text{... (2.11)}$$

As we are interested in the mean squared error we need to average over n samples and therefore it is helpful to define the following matrix:

$$R = \frac{1}{n}\sum_n SS^T =$$

$$\frac{1}{n}\begin{bmatrix} \sum_n S_t(n-1)S_t(n-1) & \sum_n S_t(n-1)S_t(n-2)...\sum_n S_t(n-1)S_t(n-L) \\ \sum_n S_t(n-2)S_t(n-1) & \sum_n S_t(n-2)S_t(n-2)...\sum_n S_t(n-2)S_t(n-L) \\ \vdots & \vdots \quad \vdots \\ \sum_n S_t(n-L)S_t(n-1) & \sum_n S_t(n-L)S_t(n-2)...\sum_n S_t(n-L)S_t(n-L) \end{bmatrix}$$

$$... (2.12)$$

and the following vector:

$$P = \frac{1}{n}\sum_n S_t(n) S = \frac{1}{n}\begin{bmatrix} \sum_n S_t(n)S_t(n-1) \\ \sum_n S_t(n)S_t(n-2) \\ \vdots \\ \sum_n S_t(n)S_t(n-L) \end{bmatrix}$$

$$... (2.13)$$

This allows us to express the mean squared error of the prediction filter as:

$$E_t = a^T Ra - 2P^T a + \frac{1}{n}\sum_n S_t^2(n) \qquad\qquad ... (2.14)$$

This equation is of the form $ax^2 - bx + c$ and so is quadratic in a. This means that there is one and only one global minimum for the mean square error and therefore one optimum set of coefficients for the filter can be readily found. This uniqueness of solution and lack of local minima is one of the main reasons that linear prediction is so often used. The minima can be readily found by differentiating the equation for the mean squared error and equating it to zero, to give:

$$Ra = P \qquad\qquad ... (2.15)$$

This equation represents a set of linear simultaneous equations in a with coefficients that are given by correlations of the signal with itself. It can be solved directly using any of the usual techniques for solving simultaneous equations, or iteratively via continuous gradient methods such as the LMS algorithm [46]. It is of interest to note that the behaviour of solutions is entirely determined by the conditioning of the matrix R which is a function of the auto-correlations of the input signal.

2.5.6.2   Linear prediction methods

One of the main issues in the direct solution is the means of obtaining the coefficients of the simultaneous equations. These must be obtained by measurement from the signal itself and fall into two basic approaches.

- **Autocorrelation method** — in this method the first L lags of the autocorrelation are formed from the signal using the N samples available. About 20 ms of signal is required which must be windowed. This method guarantees stable filters and can be solved more efficiently than standard matrix methods.

- **Covariance method** — in this method the first L lags of the cross-correlation are formed from the signal using the N samples available. This method requires less than 10 ms of signal but cannot guarantee stable filters. It also has a less efficient solution than the autocorrelation method.

The linear prediction method is really a form of least-squares modelling and these methods are only some of the possible solutions. Although not expressed explicitly, iterative techniques also make autocorrelation or covariance assumptions about the signal. In most cases, the stationarity assumption is made. However, in some of the fast iterative adaption schemes the non-stationarity of the covariance scheme is exploited, with the attendant stability issues.

All linear prediction methods are affected, to a greater or lesser degree, by the excitation of the filter. This is because the method assumes that the excitation is white noise and attempts to find the best solution of the filter which would give the desired waveform given that input. In the case of voiced excitation, the method still works because the excitation spectrum is assumed to be white rather than -12 dB per octave (see Fig. 2.4). However, the presence of strong spectral peaks due to the harmonics of the excitation causes a bias in the resulting filter towards the spectral peaks in the signal, due to the mean squared error criterion. This is not too much of a problem with male speech, where the harmonics are close together, but can cause problems with female's and child's speech, due to the higher pitch.

Another limitation on the accuracy of linear prediction techniques is the dynamic effect of the glottal opening on the filter coefficients. Remember that one of the assumptions behind the source-filter model is the independence of source and filter. Unfortunately, this is only partially correct because the effect of the glottis opening and closing is to change the boundary conditions of the filter. When the glottis is closed there is an acoustically hard boundary with a comparatively low loss and so the filter is lightly damped and can ring for a longer time. This results in narrow formant bandwidths (on the order of 50-80 Hz). However, when the glottis is open the system is damped due to the absorbing effect of the bronchia and the formant bandwidths are much higher. Linear prediction adapts to a compromise position, in which the bandwidth is intermediate between these two values. For speech waveform coding this often does not matter because the subsequent coding of the error signal can disguise the errors. However, in the case of speech synthesis and low bit rate synthesis-based coding, the higher bandwidths so estimated can result in a buzzy quality to the speech. Closed phase analysis is often proposed as a means of alleviating this problem, but it suffers from difficulties associated with identifying the closed phase and with the small amount of data available to form the coefficient solution.

2.5.6.3   Different representations for the linear prediction filter

The results of the linear prediction method can be represented in a very wide variety of ways. These different representations are equivalent in that they all perform the same filtering operation on the excitation signal, and hence the same inverse filtering operation on the speech signal. Some popular representations are the 'a' coefficients, which directly represent the equivalent recursive filter, the 'k' coefficients which represent the reflections in a transmission line model, the frequency and bandwidth of a cascaded set of second order filters, and many more. Because these are all equivalent representations it is possible to convert between any representation, although this may require some considerable computation. These different representations are used because their behaviour when quantized, or interpolated, is different. Thus, when considering a coding scheme it may be important to have a representation such as formant frequencies, in which the effect of an error due to coefficient quantization is subjectively proportional, rather than one where it is not, such as the 'a' coefficients. A similar argument applies when using linear prediction speech synthesis, as one would need to interpolate between different filters to model speech transitions. The 'a' coefficients are not good for interpolating between different filters in this task, because the intermediate filters can be completely different and even unstable, whereas a cascaded set of second-order sections will perform acceptably.

2.5.6.4   Linear prediction and maximum entropy spectral estimation

Linear prediction can also be used as a form of spectral estimation, because the Fourier transform of the 'a' coefficients is the frequency response of the equivalent filter. As these filter coefficients will give an output with a white excitation which has the same spectral magnitude as the original signal, it is a least squares estimate of the frequency spectrum of the waveform. The resolution of this spectrum can be greater than that achieved by simply taking the Fourier transform of the measured autocorrelation coefficients on which the linear prediction solution is based. Thus linear prediction can offer a powerful approach to spectral estimation and forms the basis of maximum entropy methods [47]. Figure 2.20 shows the spectrum that results from taking the Fourier transform of 25 coefficients derived from the /@/ in the spectrogram example shown in Figs. 2.12 and 2.13. It shows that linear prediction can form the basis for spectral smoothing, formant extraction, or even the initial estimate for more sophisticated analysis-by-synthesis techniques.

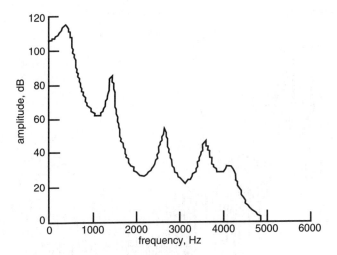

Fig. 2.20   A speech spectrum estimate using linear prediction.

2.5.7   An example of signal processing chain MFCCs

We have looked at the human speech system, showing the importance of formants and have seen that our hearing system has a resolution which is approximately logarithmic in both frequency and amplitude. One representation of speech, currently very popular in speech recognizers, is the mel frequency cepstral coefficient (MFCC) representation. This attempts to analyse speech in a manner which reflects the way we hear. It firstly splits the speech spectrum into bands, which are constant bandwidth below 1 kHz and constant percentage bandwidth above this frequency. Typical bandwidths at low frequencies would be around 100 Hz and these would increase to about 500 Hz at 4 kHz. Note that this variation is similar to the ERB scale mentioned earlier. The varying filter bandwidth approximates the frequency resolution of the ear, and by detecting the logarithm of the energy within the band we can approximate its amplitude resolution. This non-uniform bandwidth logarithmic representation is then cosine-transformed to form a set of orthogonal MFCCs. At the end of this process we could argue that we have a frequency domain representation of the speech signal in which the mean squared spectral error could be psychoacoustically relevant. Figure 2.21 shows an example FFT spectrum overlaid with the corresponding mel spectrum and the spectral effect of only

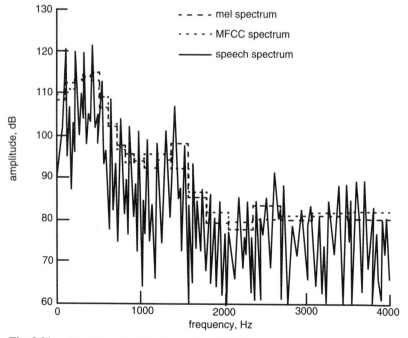

Fig. 2.21   The effect of limiting the number of MFCCs on the accuracy of spectral estimation.

using the first eight MFCCs on our example spectrum. It shows that the variation around F2 and higher is reduced or even suppressed, which may cause problems further down the recognition process. Therefore the effect of a signal processing chain should be carefully analysed to ensure that it is not losing vital information. One useful way is to attempt to hear the result, via resynthesis, or, failing that, do as shown in Fig. 2.21 and reverse the process back to an observable metric such as a spectrum.

2.6   CONCLUSIONS

For those who work in processing, coding, synthesis or automatic recognition of speech, there is no substitute for a proper grounding in the complexities of human speech production and perception, as well as in its physical acoustics. The fields of phonetics, linguistics, acoustics and psychoacoustics all contribute to this knowledge base. This chapter has introduced many of the key points of importance and it offers a list of important seminal textbooks and references which will put tomorrow's engineers in a strong position to pursue speech processing work.

The scientific literature is liberally scattered with misconceptions about speech, often as a result of treating the acoustic pressure waveform in too simplistic a manner. Nature's simplicity is seldom that obvious. It is vital therefore, when designing a chain of signal processing for a speech communications system, to ensure that any information that is lost or discarded in the process is not important — to either the source or the listener. Failure to do this might allow some short-term progress, but the endevour will ultimately be doomed if vital information is missing — whether through carelessness or ignorance.

REFERENCES

1.   Borden G J and Harris K S: 'Speech science primer', Williams and Wilkins, Baltimore (1980).

2.   Fry D B: 'The physics of speech', Cambridge University Press, Cambridge (1979).

3.   Baken R J: 'Clinical measurement of speech and voice', Taylor and Francis, London (1987).

4.   Baken R J and Danilov R F: 'Readings in clinical spectrography of speech', Singular Publishing Group, San Diego (1991).

5.   Kent R D and Read C: 'The acoustic analysis of speech', Singular Publishing Group, San Diego (1992).

6.   Fant C G M: 'Acoustic theory of speech production', Mouton, The Hague (1960).

7.   Van den Berg J W: 'Myoelastic-aerodynamic theory of voice production', Journal of Speech and Hearing Research, 1, pp 227-244 (1958).

8.   Titze I: 'Principles of voice production', Prentice Hall, New Jersey (1994).

9.   Howard D M: ''Variation of electrolaryngographically derived closed quotient for trained and untrained adult singers', Journal of Voice, 9, No 2, pp 163-172 (1995).

10. Wells J C: 'Computer coded phonemic notation of individual languages of the European Community', Journal of the International Phonetic Association, 19, pp 32-34 (1989).

11. Abercrombie D: 'Elements of general phonetics', Edinburgh University Press (1967).

12. Ladefoged P: 'A course in phonetics', Harcourt Brace Jovanovich, New York (1975).

13. Gimson A C: 'The linguistic relevance of stress in English', in Jones W E and Laver J (Eds): 'Phonetics and linguistics', Longmans, London (1966).

14. Wells J C and Colson G: 'Practical phonetics', Pitman, London (1978).

15. O'Connor J D: 'Phonetics', Penguin Books, London (1973).

16. Catford J C: 'Fundamental problems in phonetics', Edinburgh University Press, Edinburgh (1977).

17. Peterson G E and Barney H E: 'Control methods used in the study of vowels', Journal of the Acoustical Society of America, 24, pp 175-184 (1952).

18. Pickles J O: 'An introduction to the physiology of hearing', Academic Press, London (1982).

19. Moore B C J: 'An introduction to the psychology of hearing', Academic Press, London (1982).

20. Rosen S and Howell P: 'Signals and systems for speech and hearing', Academic Press, London (1991).

21. Howard D M and Angus J A S: 'Music technology: acoustics and psychoacoustics', Focal Press, London (1996).

22. von Békésy G: 'Experiments in hearing', McGraw Hill, New York (1960).

23. Moore B C J and Glasberg B P: 'Suggested formulae for calculating auditory-filter bandwidths and excitation patterns', Journal of the Acoustical Society of America, 74, No 3, pp 750-753 (1983).

24. Moore B C J: 'Frequency selectivity in hearing', Academic Press, London (1986).

25. Sunberg J: 'The science of musical sounds', Academic Press (1991).

26. Sharf D J and Ohde R N: 'Physiologic, acoustic and perceptual aspects of coarticulation: implications for the remediation of articulatory disorders', in Lass N J (Ed): 'Speech and language: advances in basic research and practice', Academic Press, New York (1981).

27. O'Connor J D and Arnold G F: 'Intonation of colloquial English', Longmans, London (1961).

28. Gimson A C: 'The linguistic relevance of stress in English', in Jones W E and Laver J (Eds): 'Phonetics and linguistics', Longmans, London (1966).

29. Halliday M A K: 'Intonation and grammar in British English', Mouton, The Hague (1967).

30. Crystal D: 'Prosodic systems and intonation in English', Cambridge University Press, Cambridge (1969).

31. Lehiste I: 'Suprasegmentals', MIT Press, Cambridge Massachusetts (1970).

32. Witten I H: 'Principles of computer speech', Academic Press, London (1982).

33. Hess W: 'Pitch determination of speech signals', Springer Verlag, Berlin (1983).

34. Howard D M: 'Peak-picking fundamental period estimation for hearing prostheses', Journal of the Acoustical Society of America, 86, No 3, pp 902-910 (1989).

35. Rabiner L R, Cheng M J, Rosenberg A E and McGonegal C A: 'A comparative performance study of several pitch detectors', IEEE Transactions on Acoustics Speech and Signal Processing', ASSP-24, No 5, pp 399-413 (1976).

36. Hess W and Indefrey H: 'Accurate pitch determination of speech signals by means of a laryngograph', Proceedings of the IEE International Conference on Acoustics, Speech and Signal Processing, ICASSP-84, pp 1-4 (1984).

37. Koenig W, Dunn H K and Lacy L Y: 'The sound spectrograph', Journal of the Acoustical Society of America, 18, pp 19-49 (1946).

38. Potter R K, Kopp G A and Green H: 'Visible speech', Van Nostrand, New York (1947).

39.   Patterson R D: 'Auditory filter shapes derived with noise stimuli', Journal of the Acoustical Society of America, 59, pp 640-654 (1976).

40.   Swan C, Tyrell A M and Howard D M: 'Real-time transputer simulation of the human peripheral hearing system', Microprocessors and Microsystems, 18, No 4, pp 215-221 (1994).

41.   Rioul O and Vetterli M: 'Wavelets and signal processing', IEEE Signal Processing Magazine, 8, No 4, pp 14-38 (1991).

42.   Press et al: 'Numerical recipes in C', in: 'The art of scientific computing', 2nd Edition, Cambridge University Press, pp 591-606 (1995).

43.   Noll A M: 'Cepstrum pitch determination', Journal of the Acoustical Society of America, 41, pp 293-309 (1967).

44.   Oppenheim A V and Schafer R W: 'Digital signal processing', Prentice Hall, New Jersey (1975).

45.   Markel J D and Gray A H: 'Linear prediction of speech', Springer-Verlag, Berlin (1976).

46.   Widrow B and Stearns S D: 'Adaptive signal processing', Prentice Hall, Englewood Cliffs, New Jersey (1985).

47.   Schroeder M R: 'Linear prediction, entropy and signal analysis', IEE ASSP Magazine, 1, No 3, pp 3-11 (1984).

3

LOW RATE SPEECH CODING FOR TELECOMMUNICATIONS

W T K Wong, R M Mack, B M G Cheetham and X Q Sun

3.1 INTRODUCTION

This chapter highlights the major speech coding systems used in the existing networks, and summarizes the activities in speech coding standards. The evolution of the key speech coding techniques is reviewed. Finally, a new class of speech coding technique, known as speech interpolation coding, is discussed. Recently, this type of speech coding technique has attracted much attention, and is believed to have the potential to provide good quality speech coding suitable for telephone use at or below 4 kbit/s.

Low rate speech coding or compression attempts to provide toll-quality speech at a minimum bit rate for digital transmission or storage. The advantages to be gained are bandwidth efficiency, cost reduction, security, robustness to transmission errors, and flexibility in the digital world. The trade-offs, dependent on the particular coding technique and the application, are coding delay and distortion, and increased equipment costs.

Throughout the history of digital communications, dating back to the early 1960s and the selection of *A*-law coding for the first 24-channel pulse code modulation (PCM) systems [1], BT has maintained an active role in speech coding development and standardization. Research into more complex lower bit rate coding schemes was initially inhibited by practical implementation considerations imposed by the semiconductor technology of the day. As a consequence, research into sophisticated low bit rate algorithms did not gather momentum until the late 1970s, and it was not until 1984 that the first worldwide lower bit rate coding standard was achieved. This standard was the International Telegraph and Telephone Consultative Committee (CCITT) G.721 Recommendation for 32 kbit/s adaptive differential PCM (ADPCM), and BT played an important role in the standardization process.

Since that time, the major advances made in microelectronics and digital signal processor technology have spurred research into increasingly complex speech
coding methods. Sophisticated speech coding techniques are now commercially
viable, and can be implemented at low cost for more affordable mobile, computer
and audio-visual terminals. BT has continued to support codec standardization
processes via organizations such as the CCITT, the International Telecommunications Union (ITU-T), the European Telecommunications Standards Institute
(ETSI), and the International Maritime Satellite Organization (Inmarsat), and has
achieved notable successes with in-house speech coding technology. Among the
latter are the Skyphone™ codec [2] developed at BT Laboratories (BTL), which
became the codec chosen by Inmarsat and the Airlines Electronic Engineering
Committee (AEEC) as the most suitable codec for aeronautical satellite
communications at 9.6 kbit/s [3]. More recently, a low bit rate coder designed at
BTL has found application in CallMinder™ [4], a new network-based call
answering and message storage service. BTL has also developed the voice
activity detector (VAD) [5, 6] adopted for use with the global system for mobile
(GSM) communications speech coder, and this VAD is attracting interest for use
with other mobile radio systems. An efficient VAD is crucial to overall system
performance, and reliable operation under high background noise conditions is a
key feature of the BTL design.

There was a period in the early 1980s when the need for further speech coding development was brought into question. With the rapid migration to optical
fibre transmission that was occurring in world networks, would the potentially
vast bandwidth offered by full exploitation of this technology make speech compression unnecessary? This perspective was flawed in that it overlooked the huge
growth in demand that would occur for mobile telecommunications. Here, the
limited available radio bandwidth makes speech compression essential for the
provision of viable services. Furthermore, although the introduction of optical
fibre transmission has proceeded apace, bandwidth constraints remain a significant issue in areas of the fixed network. This is particularly true for international
links, whether they are satellite or cable, where speech compression in the form
of digital circuit multiplication equipment (DCME) [7] is increasingly employed.
Competition from other network operators has also been an important driver for
speech coding technology, since more efficient use of the transmission medium
permits lower tariffs, leading to increased market share. Another area where
speech coding plays a vital role is in the new multi-media services [8]. In these
applications, the reduced data rate required for the transmission of the audio
component maximizes the bandwidth available to the visual element.

The position today is that a number of low rate speech coding techniques
have already been adopted as international standards for various network applications. For many new network developments such as global virtual private networks, third generation mobile, cellular satellite and even the asynchronous

transfer mode (ATM) [9] networks, it is no longer a question of whether speech compression should be used, but which speech coding technology provides the target speech quality at lowest cost.

3.2   PERFORMANCE ISSUES

The use of any speech processing system over the public switched telephone network (PSTN) is normally bounded by a set of stringent network planning rules in order to maintain a high quality of service and not to degrade existing applications. The principal aspects of performance against which a speech coder is assessed are [10]:

- speech quality;
- delay;
- transparency to other voiceband signals;
- ability to support existing services using speech recognition or verification.

In addition, parameters such as speech quality and transparency are affected by a number of influences external to the coder, so it is important to take account of the following:

- dynamic range;
- robustness to channel errors;
- robustness to circuit noise;
- robustness to ambient background noise.

In general, the existing speech coding standards will have taken into consideration all of the above performance issues.

Table 3.1 lists well-known telephone-bandwidth international speech coding standards. Many of the proprietary codecs used for voice messaging or Internet telephony are derivatives of these ITU-T standards. The bit rate, algorithmic delay, subjective performance, and class of coding scheme are given for comparison. The bit rates indicated are the net speech coding rates, excluding the error control coding used on the transmission channels. Algorithmic delay is purely the coding delay introduced by the fundamental speech encoding and decoding operations, excluding the processing and transmission delay. The performance of a speech processing system is normally evaluated using mean opinion score (MOS), which is a formal subjective measure of received speech quality (see Chapter 12).

Table 3.1 Comparison of the well-known telephone-band speech-coding standards.

Standard	Year	Coding Type	Bit rate (kbit/s)	MOS*	Algorithm delay (ms)
ITU-G.711	1972	PCM	64	4.3	0.125
ITU-G.721	1984	ADPCM	32	4.0	0.125
ITU-G.726	1991	VBR-ADPCM	16, 24, 32 and 40	2.0, 3.2, 4.0 and 4.2	0.125
ITU-G.727	1991	Embedded-ADPCM	16, 25, 32 and 40	—	0.125
ITU-G.728	1992	LD-CELP	16	4.0	0.625
Inmarsat-B		APC	9.6/16	communication	20
GSM Full-rate	1989	LTP-RPE	13	3.7	20
GSM-EFR	1995	ACELP	13	4.0	20
Skyphone	1989	BT-MPLPC	8.9	3.5	28
DAMPS Full-rate IS54	1991	VSELP	7.95	3.6	20
ITU-G.729	1995	CSA-CELP	8	4.0	15
IS-96	1991	Qualcomm CELP	1, 2, 4, and 8	3.5	—
JDC Japanese Full-rate		VSELP	6.7	communication	20
GSM Half-rate	1994	VSELP	5.6	3.5	24.375
ITU-G.723	1995	A/MP-MLQ CELP	5.27/6.3	communication	37.5
American DOD FS1016	1990	CELP	4.8	3.0	45
TETRA	1994	ACELP	4.56	communication	35
ITU	(1998)	(To be defined)	4	(toll)	(25)
Inmarsat-M	1990	IMBE	4.15	3.4	78.75
JDC Japanese Half-rate	1993	PSI-CELP	3.45	communication	40
American DOD FS1015	1984	LPC-10	2.4	synthetic	22.5 (minimum)

* **WARNING:** The Mean Opinion Score (MOS) coding performance figures given above are obtained from different formal subjective tests using different test material, and they are evaluated under the most favourable listening conditions including the optimum speech input level, optimum listening level, no added background noise, no bit errors and no transcoding. The given MOS figures are therefore useful as a guide, but should not be taken as a definitive indication of codec performance. This measure is on a scale from 1 to 5, interpreted as follows: 1 = bad, 2 = poor, 3 = fair, 4 = good, and 5 = excellent.

Generally, coding quality with MOS higher than 4 is considered as toll quality, between 3.5 and 4 as communication quality, between 3 and 3.5 as professional quality, and below 3 as synthetic quality [11].

3.3   SPEECH CODING IN THE NETWORK

3.3.1   National and international networks

The well-known 64-kbit/s PCM coding method of the ITU G.711 Recommendation, was standardized in 1972, but it was 1990 before the UK national network finally became fully converted to this digital speech transmission technique. PCM coding provides toll-quality speech transmission, and is transparent to all the standard voiceband data transmission techniques.

As the national networks are normally owned and operated by the country's PTT organization, low operating costs using simple PCM coding are more important than high bandwidth efficiency using more complex speech coding methods. In contrast, the international networks, such as networks for international direct dial (IDD) services, operate over undersea cables and satellites. These are costly shared resources, and therefore result in a more expensive service. A way to provide more efficient digital transmission in this situation is to combine the 64 kbit/s PCM coding with a technique known as digital speech interpolation (DSI) [12]. As normal conversational speech is active only about 40% of the time in either direction of transmission, a large group of simultaneous calls can be statistically multiplexed. The DSI suppresses the silent periods between the spoken words and syllables using reliable silence elimination and regeneration process respectively at the transmitting and receiving terminals. While a talker is silent, the transmission channel is made available to the other active calls in progress. At the receive end of the link, 'comfort' noise is usually substituted during the periods of disconnection, so that the listener is less aware of the breaks in transmission. By sharing channels in this way, the capacity of the international links can be approximately doubled. However, if the system is not properly dimensioned, annoying speech-clipping effects may occur during busy periods, due to speech bursts competing for the shared transmission resource. This is known as competitive clipping.

A more commonly used technique over international networks is to use DCME [7], which combines DSI and ADPCM speech coding to provide a compression gain of up to 5:1 for speech transmission. The ITU G.763 Recommendation is the international standard for DCME, and combines DSI with G.726 variable bit rate ADPCM. Recommendation G.763 specifies that the DCME should always use 40-kbit/s ADPCM to transmit voiceband data, but speech is transmitted at 32 kbit/s, dropping to 24 or even 16 kbit/s when necessary. This variable bit rate feature for speech transmission aims to reduce the amount of competitive speech clipping under heavy traffic loading conditions, but the speech quality is inevitably degraded due to the reduced mean bit rate of the coded speech. It is possible to exercise control over the transmission performance of the system by setting a minimum mean speech coding bit rate objective. Typically, this would be about 3.6 bits/sample to ensure toll-quality transmission. If the mean bit rate falls below this value, new calls are prevented from gaining access to the system until the heavy loading subsides. This feature is known as dynamic load control (DLC). A block diagram illustrating the main DCME processes is shown in Fig. 3.1.

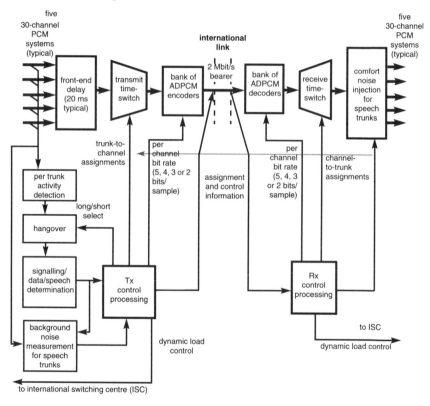

Fig. 3.1   Block diagram to illustrate DCME processes.

An ITU 16-kbit/s speech coder, known as the G.728 low-delay code-excited linear predictive (LD-CELP) coder, became a standard in 1992, and demonstrated coded speech quality comparable to G.726 ADPCM at 32 kbit/s. A variable bit rate version of the LD-CELP coder is now being studied by the ITU. If this is used to replace the ADPCM coding in DCME, speech compression gain as high as 9:1 could theoretically be achieved.

Packetized networks, such as Frame Relay and ATM [10], are considered more flexible network technologies than the existing circuit-switched PSTN networks. A new ITU standard, the G.765 packet circuit multiplication equipment (PCME), has all the basic functions of a DCME and more advanced features, such as higher bit rate transmission (up to 150 Mbit/s) and the capability to support bandwidth-on-demand services. As this is a new standard, the use of PCME is not as popular as DCME, and it is viewed as a more advanced DCME for packet-switched networks. A well-known problem in packet transmission is packet loss during traffic congestion. To counteract this, PCME uses the ITU G.727 embedded ADPCM coding, which is similar to G.726 variable bit rate coding, but with the crucial difference that the variable bit rate aspect can be controlled within the transmission path without informing the encoder. With embedded coding, the transmitted values consist of 'core' bits essential for the signal to be decoded, plus 'enhancement' bits which contribute to the quality of the decoded signal. For each encoded sample, the enhancement bits may selectively be dropped to relieve congestion. In PCME, a block of speech samples is assigned to a packet, but, before transmission, the PCME rearranges the bits within the packets according to their significance. When heavy traffic causes congestion, it is initially only the least significant part of a packet that is dropped, with progressively more significant parts being dropped as congestion increases. Thus the embedded coding property leads to a graceful decline in speech quality as traffic congestion rises [13].

Many manufacturers claim that DCME and PCME offer toll-quality transmission at high compression gain. However, the cost saving is route-dependent and not directly proportional to the maximum compression gain of the equipment. The use of such equipment requires careful consideration of many factors, including cost of the equipment, cost of the international circuits, the traffic profile of the route, the relative proportions of speech and data traffic, system dimensioning, and the regulatory issues between the countries involved.

3.3.2   Mobile and satellite networks

It is already evident that digital public land mobile radio (PLMR) [14] systems are more robust, more flexible and offer a wider range of telecommunications services than analogue mobile systems. The development of digital PLMR has been the primary driving force in the past decade for the advances in speech coding technology that have been achieved. International mobile radio standards organizations, such as ETSI in Europe, and the Telecommunications Industry Association (TIA) in the United States, conduct comparative studies to choose the best speech coding systems available for their networks. Due to the demands for bandwidth efficiency and short time-scales, and the limitations imposed by available technology, the standards organizations are usually prepared to accept a small compromise in the quality of the speech coding system in order to maximize the channel capacity. This near toll-quality speech is often referred to as communication quality.

The existing speech coding systems used in digital PLMRs are the 13-kbit/s long-term-predictive, regular-pulse-excited (LTP-RPE) codec in GSM, the 7.95-kbit/s vector-sum excited linear predictive (VSELP) codec in the digital American mobile public system (DAMPS), and the 6.7-kbit/s VSELP codec in the Japanese digital communications (JDC) system. These codec bit rates apply to the basic, full-rate systems. Half-rate systems are achieved by halving the transmitting bit rate of the radio link, and so doubling the channel capacity. Half-rate systems for GSM and JDC have been completed, and the associated speech codecs are a 5.6-kbit/s VSELP codec for half-rate GSM and a 3.45-kbit/s pitch-synchronous innovation (PSI-CELP) codec for half-rate JDC. The subjective performance of these codecs (not the overall system performance) is presented in Table 3.1.

The above-mentioned mobile radio standards are not intended to provide services for maritime and aeronautical applications, nor even for remote or sparsely populated areas. Instead, Inmarsat has set up several standards and services to provide coverage in these areas via satellite. The well-known speech coding standards are the 9.6/16-kbit/s adaptive predictive coding (APC) for Inmarsat-B, the 9.6-kbit/s multipulse linear predictive coding (MPLPC) for Sky-phone (otherwise known as Inmarsat-aero), and the 6.4-kbit/s improved multiband excited (IMBE) coding for Inmarsat-M applications.

The second generation digital cordless telephone systems, such as CT-2 and the digital European cordless telephone (DECT) system, place the emphasis on low delay, good quality and inexpensive equipment rather than spectrum efficiency. ADPCM at 32 kbit/s, conforming to the G.726 standard, is the speech coding method adopted for these systems. A hybrid dual-mode system, DECT/GSM, to combine cordless and mobile applications into one handset is currently under study. The DECT system would be used in the indoor environment, automatically switching to the GSM system in an open area.

3.3.3   Coding design in mobile systems

It is becoming clear that the transmission performance of mobile systems is not comparable to the fixed wireline systems. Currently, ETSI has a new activity to develop a significantly better speech coder system for the existing full-rate GSM system. A new toll-quality 8-kbit/s coder, the G.729 conjugate structure algebraic (CSA)-CELP coder, was standardized by the ITU in late 1995. A primary application for this coder is to provide robust, toll-quality transmission for the future third-generation mobile systems.

Applying speech coding over any radio network is generally much more difficult than on the fixed wireline network, since mobile radio users are usually in a noisy environment where the signal-to-background noise ratio could be as low as 10 dB. The speech codec must also face the harsh radio-fading effect which, in the GSM radio channel, can lead to a mean bit error rate as high as 10%. The design of a speech coding system must be robust to these conditions.

Figure 3.2 shows the functional block diagram of a speech coding system typical of that used in digital mobile radio. The front-end of the speech encoding unit contains:

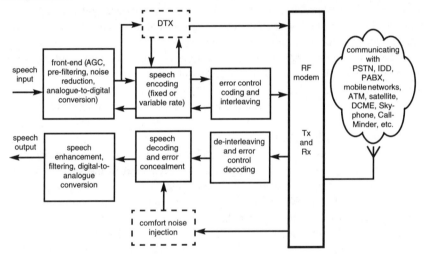

Fig. 3.2   Block diagram of a typical mobile radio transceiver.

- a handset that provides good acoustic noise rejection;

- automatic gain control (AGC) to adjust the input level to the optimum operating range;

- analogue to digital conversion;

- an optional digital noise cancellation unit to further reduce the noise effect before encoding;

- pre-filtering or a signal modification process that aims to maximize the speech coding efficiency.

The expected communication quality can be achieved and maintained only if these functions are properly designed and implemented [15]. The coding performance could be further improved if all these functions are controlled by the speech encoding unit in a closed-loop manner based on some form of optimization process [16].

After the pre-conditioning process, the speech signal is compressed by either a fixed or variable bit rate speech coding algorithm. Fixed bit rate coding is normally adopted if the speech coding and error control coding (channel coding) are designed separately. A more robust design approach, for a given gross channel bit rate, is to jointly optimize the speech and channel coding. The different component parts of the speech signal are encoded at their minimum acceptable rates, leading to a varying data rate (variable rate speech coding) [17]. The compressed speech is then error protected by different error-control methods which take into account the error sensitivity of the speech component and the remaining bits available for error protection [18-20]. The combined speech and channel coding can also be transmitted at variable rates; examples of this approach can be seen in the Qualcomm IS-96 speech coding standard.

Transmission errors will degrade the received speech quality, but the nature of the perceived distortion is dependent on the type of speech coder used. In general, as the speech coder bit rate reduces, the information content per bit becomes increasingly more significant. For the lowest bit rate speech coders, even the corruption of a single data bit could cause a severe speech degradation or result in loss of intelligibility. Fortunately, the error-control coding helps to maintain acceptable transmission performance under poor radio channel conditions by greatly reducing the number of digital errors presented to the speech decoder. Error-control coding, however, adds redundancy and interleaving delay to the protected data. Typically, the inclusion of error-control coding doubles the bandwidth relative to that required for the unprotected compressed speech, and the additional interleaving delay is significantly higher than the algorithmic delay of the speech codec.

At the receiver, the speech decoder performs the inverse of the encoding process to reconstruct the speech signal. Error-control coding has its limitations, so that under severe fading conditions residual errors will occur in the received data, causing transient distortions such as clicks and pops in the decoded speech, and badly degrading the quality of service. Error-concealment methods are used to suppress these transient distortions once channel errors are detected, and the corrupted speech signal is smoothly muted under long-term fading conditions. To improve robustness to residual errors, current research also attempts to use speech parameters with inherent error-detection capability [20] or to optimize the bit representation for the coded speech parameters to reduce the effect of bit errors [21].

An optional function, known as discontinuous transmission (DTX) [5] (see also Chapter 4) has been standardized for GSM to reduce co-channel interference and power consumption at the mobile terminal. This is achieved by transmitting only during the active periods of speech. The operation has similarities to DSI as described in section 3.3.1. The DTX feature has attracted considerable interest from other international mobile radio standards organizations. However, within GSM, the use of DTX is left to the discretion of the mobile radio network operator.

3.3.4   Military developments

Very low bit rate speech coding at 2.4 kbit/s is in common use for military communications. Here, the objective is to provide robust transmission for various radio links even when subject to severe radio jamming. The coding scheme must also provide highly intelligible speech under unusual background noise conditions such as tank, helicopter, and machine-gun noise. As other aspects of speech quality are of less importance to military communications, even a synthetic quality coder might be acceptable if the objectives of high intelligibility and robustness to transmission conditions are achieved.

The American Department of Defense (DOD) plays an important role in the advancement of speech coding technology. Two well-known military standards are the LPC-10e 2.4-kbit/s vocoder based on linear predictive coding, and the FS1016 4.8-kbit/s CELP coder. The LPC-10e vocoder provides synthetic but intelligible speech quality at a very economical bit rate. The CELP coder became a standard in 1990, and the algorithm software is freely available on the Internet [22]. Several manufacturers have implemented the standard on various digital signal processing (DSP) devices. Although the speech quality is not high, this codec has attracted much attention due to the potential for further improvements to the algorithm.

Currently, the DOD is the major driving force in developing a new 2.4-kbit/s speech coding standard, and recent results indicate that some 2.4-kbit/s coding methods already exceed the FS1016 4.8-kbit/s coding performance [23]. It is expected that even more significant improvement could be made by the time of the standardization in mid-1996.

3.4   FUTURE DEMANDS

For mobile networks, state-of-the-art speech coding will continue to be crucial, but the objectives will become even more demanding as these networks grow and customer expectations increase. In the near future, the objective for the enhanced full-rate GSM system is to provide transmission performance at least as good as the fixed wireline network. This will also be the objective for third-generation mobile systems [15].

If a good-quality mobile service could be provided at sufficiently low cost, the number of domestic mobile customers would be expected to increase dramatically. The present norm of one fixed telephone per household might well become one mobile phone for each member of the family! This will not be achieved without extremely efficient use of radio bandwidth, leading to a requirement for speech coders at even lower bit rates than those used today. The ITU has an objective to standardize a toll-quality 4-kbit/s speech coding system in 1998 [10]. The target application areas for this codec are mobile, satellite, and very low rate visual telephony. For a number of satellite applications such as Inmarsat-P [24], and Iridium [25], there are also requirements to develop and use 2.4-kbit/s speech coding.

The future will also see an expansion of telephone voice services outside the traditional person-to-person conversation call, and low rate speech coding will play a crucial role in the economic provision of many of these services. The recent introduction of 'CallMinder' [4] into the national network is an example of the type of interactive voice messaging service destined to become common-place. CallMinder is a sophisticated call-answering and message-storage facility, network-based so that no additional equipment is required at the customer's home or office. Its advanced capabilities allow two incoming calls to be handled at once, even if the customer is already on the phone, and for each customer, up to thirty separate messages may be stored. The role that low bit rate speech coding plays in this service is clear, since to store numerous voice messages using 64-kbit/s PCM would make the system prohibitively expensive; 1 Mbyte of storage would afford only about two minutes of recorded speech.

Archiving conversational speech is important for legal purposes in many spheres of work, such as business meetings, telephone banking, betting shops, police-questioning, insurance companies, and the legal profession. Speech compression will be increasingly exploited for this purpose. With high-quality 2.4-kbit/s speech coding operating with a VAD, about 2.5 hours of coded speech can be stored on a low-cost 1.4-Mbyte high-density floppy disk. The storage costs are significantly cheaper than for analogue tape recording. As digital speech also allows the use of additional control software, selected speech passages can be very quickly retrieved from the storage system for listening, or for transfer to a document by an audio-typist. Sometimes, a very large speech database is required to be transferred between different systems; with a compressed speech database, communications costs are cheaper and the transfer time is shorter.

Multimedia is another area destined to become a part of everyday life. The combined use of low bit rate speech and visual coding will mean that multimedia services can be made available over the PSTN. Emerging packet speech transmission technology allows full-duplex conversation over the Internet. There are already a number of products — usually referred to as Internet phone or Iphone [26] — on the market, and they all use low bit rate speech coding to maximize their transmission performance over the Internet medium. However, the Internet was not originally engineered to cope with the requirements of voice traffic. The transmission delay is variable, and unacceptably high when the Internet link is heavily loaded. For current Internet voice products, the quality of service cannot be guaranteed. It will be interesting to see whether these systems ever become more than the equivalent of a 'CB radio' channel for the Internet amateur.

3.5   THE TECHNOLOGY

Over the past three decades, many different types of speech coder have been developed. For PSTN applications, speech quality and delay are the key performance issues, but these must be balanced against the cost of implementation. The simplest form of coding in the PSTN is the toll-quality G.711 64-kbit/s PCM coding standardized in 1972. Today, an 8-kbit/s CSA-CELP coder has been shown to provide near toll-quality coding, and is due to become the ITU G.729 standard in 1995 — speech compression by a factor of eight achieved in little more than two decades. Although the G.729 codec is much more complex than the G.711 codec, the advances in microelectronics allow it to be implemented at low cost.

There are four main classes of speech coder, these being waveform coding, vocoding, hybrid coding, and frequency domain coding. These classifications are fully described in Boyd [27]. Within each class there are many types and variants, each having their own attributes. This section does not attempt to describe every known speech coding method, but concentrates on highlighting the key techniques used in speech coding standards.

3.5.1   Memoryless coding

This is the simplest form of coding, and G.711 64-kbit/s PCM coding is an example of this method. Each input signal sample is rounded (or quantized) to the nearest representative level, independent of the past or future nature of the signal. The quantized levels may either have a uniform separation, or a non-uniform (typically pseudo-logarithmic) arrangement as in the case of G.711 PCM. For a uniform quantizer, the coding distortion is directly proportional to the variance (or the energy) of the input signal. If the input signal to the uniform quantizer is adjusted to the full operating range, the signal-to-quantization noise ratio, SNR, is estimated as $6R$, where R is the number of coding bits per sample. However, for a non-uniform quantizer, the lower signal levels are encoded with finer precision, leading to a more consistent coding performance over the dynamic range. The SNR of this non-uniform quantizer can be approximated as [28]:

$$SNR_{PCM} = 6R - 10 \text{ (dB)}$$

For the 8-bit non-uniform quantizer of G.711, this corresponds to an SNR of about 38 dB for either *A*-law or μ-law PCM coding, and its subjective performance is considered to be good, achieving a MOS rating normally above 4.

3.5.2 Adaptive differential predictive coding

The distortion produced by a quantizer is proportional to the signal variance, and its subjective performance degrades rapidly as the bit rate is reduced. It has been observed that speech is a quasi-stationary signal in the short term, so that the difference amplitude between adjacent speech samples has much lower variance than the sample amplitude itself. This characteristic of a speech waveform is exploited in differential PCM (DPCM) and its derivatives such as ADPCM.

The DPCM coder uses a predictor to estimate the amplitude of the next input speech sample from the history of past samples. It is the difference between this estimate and the actual sample amplitude which is coded. With an appropriate predictor, this difference signal will be small and therefore have lower variance than the actual signal. The better the predictor, the smaller the variance. This variance-reduced signal, usually known as the prediction error or residual signal, can then be quantized at a lower PCM coding rate. The amount of the variance or energy reduction is known as prediction gain, G_p. The overall SNR of a predictive PCM coding system [28] is given by:

$$\text{SNR} = SNR_{PCM} + G_p \text{ (dB)}$$

Two approaches are possible to the design of an adaptive predictor. The prediction can either be derived using the history of the quantized signal, which is known as backward prediction, or using a segment of the uncoded input speech, which is known as forward prediction. A higher prediction gain can normally be achieved by the forward prediction method, but the input signal must be buffered and the predictor parameters are transmitted as side information. The buffering delay introduced for forward prediction is usually the dominant part of the algorithmic delay introduced by the coder.

In ADPCM, which is the speech coding technique used in the ITU G.726 and G.727 standards, the predictor is made adaptive to the continuously changing characteristics of the speech waveform, leading to a more accurate prediction and, hence, increased prediction gain. To satisfy the low delay requirement, the G.726 and G.727 [29] coders use a 2-pole and 6-zero adaptive backward prediction method to encode the signal on a sample-by-sample basis. These coders also use adaptive quantizers which attempt to match the quantizer range to the short-term power of the signal being coded. As the signal level reduces, the quantizer range contracts, and vice versa. Because of this, the separation between the discrete quantizer levels is reduced for low-level signals, allowing these signals to be encoded with finer precision. At 32 kbit/s, the SNR for the decoded speech signal is about 22 dB which is about 16 dB lower than for 64-kbit/s PCM coding, but the subjective MOS performance is normally about 4, only slightly lower than for 64 kbit/s PCM. This highlights the difficulty of using objective measures to quantify the performance of a speech coder.

In recent speech coder designs, the performance of the adaptive prediction has been further improved by exploiting both the short-term and the long-term nature of the speech signal [30]. Many techniques have been proposed to derive these predictors, but linear predictive analysis is the most commonly used in speech coding. The short-term predictor normally aims to model the vocal tract function, and the long-term predictor to model the pitch periodicity of the speech signal. For 4-kHz bandwidth speech, a 10th order (or 10 taps) short-term predictor and a 1-tap long-term predictor are usually used and, typically, they together provide a prediction gain of between 8 and 15 dB. As this combination of predictors has proved to be robust and practical to implement, they have been used in many digital mobile radio coding standards.

However, the achievable prediction gain depends on the types of predictor used and the nature of the speech signal to be coded. The prediction gain does not increase linearly as the number of taps is increased, and tends to level off at high orders of prediction.

3.5.3   Analysis-by-synthesis vector quantization

Instead of coding the speech sample or the prediction error on a sample-by-sample basis such as in the G.711 PCM and the G.726 ADPCM coders, higher coding efficiency can be achieved if a set of related parameters, or a vector of elements, is encoded to a representative vector extracted from a pre-stored codebook of optimized vectors. This is known as vector quantization (VQ) [31]. Even better coding performance can be achieved if VQ is combined with an analysis-by-synthesis (AbS) method [32, 33]. The AbS-VQ method forms the basis for coding the prediction error in code-excited linear predictive (CELP) coding. The basic difference between the direct VQ and the AbS VQ method is in the definition of the quantization distortion measure used in the VQ codebook search.

For example, the 16-kbit/s ITU G.728 LD-CELP coder uses a 50th order backward linear prediction method, and a 10-bit vector codebook to encode a vector of 5-sample prediction error. The direct VQ approach would select the stored vector that gives minimum quantization distortion on the 5-sample prediction error, while the AbS VQ method used in G.728, and illustrated in Fig. 3.3, searches for the vector that gives minimum quantization distortion between the input and the coded speech. As the coded speech is computed from the prediction error through a filtering process and this has to be performed on every codebook vector to be tested, the AbS VQ method is much more complex than the direct VQ method.

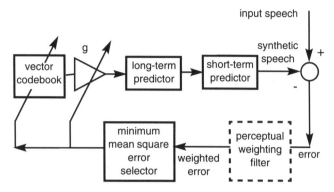

Fig. 3.3   AbS-VQ in CELP coding.

The general drawbacks of VQ are the large memory needed to store an optimized vector codebook at the encoder and at the decoder, and the high computational complexity to determine the best vector via an exhaustive codebook search. For these reasons, systems using VQ coding are more expensive to implement than those using PCM or ADPCM coding methods.

For a speech signal, the measured SNR of the 16-kbit/s LD-CELP coder is about 18 dB, which is about 20 dB lower than for the G.711 64-kbit/s PCM coder. The ITU 8-kbit/s CSA-CELP coder uses forward prediction and the AbS VQ method to encode the prediction error. The measured SNR is about 13 dB.

3.5.4   Noise shaping and masking

Inevitably, even with the advanced speech coding technology described above, the signal-to-quantization noise ratio decreases as the coding rate decreases. It has been observed [34] that the perceived quantization noise is affected by the spectral energy of the speech. A speech signal usually exhibits a large variation in the frequency domain, as illustrated in Fig. 3.4, but the frequency spectrum of quantization noise is normally evenly spread across the frequency band. This frequency varying nature of the SNR suggests that the perceived noise in the high spectral energy regions, such as the speech formant regions, is not as significant as in the low spectral energy regions, such as the trough regions.

Based on this principle, a technique known as perceptual weighting or noise shaping, aims to reduce the perceived noise by shaping the noise in the frequency domain so as to improve the SNR in the trough region at the expense of reducing the SNR in the formant region. In fact, a perceptual weighting filter (PWF) is used in all CELP coding standards. The PWF is incorporated with the AbS VQ in the encoding process, as shown in Fig. 3.3, so that the quantized prediction error vector selected from the codebook would minimize the perceptually weighted error between the input and the coded speech. An example of the frequency response of a PWF is shown in Fig. 3.4.

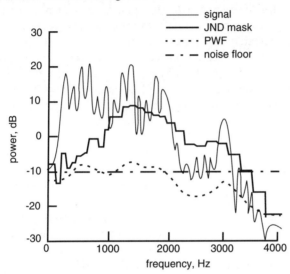

Fig. 3.4   Noise shaping for speech coding.

At the decoder, the perceived quantization noise may be further reduced by a post-filtering process [35], also known as 'speech enhancement' as in the case of the Inmarsat-M IMBE codec. The technique exploits the noise-masking effect in human perception. One signal is said to be masked when its audibility is reduced by the presence of another, louder signal at the same or near frequency. An example of masking is the failure to hear a telephone ringing when listening to loud music. The basic principle of post-filtering used in CELP coding is to artificially strengthen the formants and the peaks of the pitch harmonics so that quantization noise between the peaks is partially or totally masked. This technique becomes increasingly important at the lower speech coding rates.

However, since post-filtering modifies the coded signal, consideration must be given to the possibility of codecs being operated in tandem. If a post-filter is not carefully designed or disabled in the tandeming situation, the coded signal may be modified several times leading to degraded performance.

3.5.5   Latest developments in low rate coding

Combining the latest coding technology and good engineering, the 8-kbit/s G.729 CELP coder has achieved near toll-quality comparable to the G.726 32-kbit/s ADPCM coder. Based on current evidence, it seems unlikely that CELP coding could achieve near toll-quality coding at 4 kbit/s or below to satisfy the future demand mentioned in section 3.4. The primary limitation is the performance when coding voiced speech such as the vowel sounds. Three major themes in speech coding research work are currently being actively pursued, namely, efficient VQ, auditory modelling, and coding methods for voiced speech.

The use of VQ is computationally expensive and tends to be signal dependent, but it is a crucial technique for very low bit rate coding. The future research objectives are to develop low-complexity VQ techniques which are robust to different types of background noise, frequency/attenuation distortion of the input signal, and channel errors.

Noise shaping and masking, as discussed in section 3.5.4, have been successfully used in speech coding. The trend is to exploit more general masking properties of the human hearing process to derive a dynamic, frequency-dependent threshold, below which noise will be perceptually inaudible. This is sometimes referred to as the just noticeable distortion (JND) threshold, and it is already widely used in wideband audio coding [36]. An example of this threshold and its corresponding signal spectrum are given in Fig. 3.4, and clearly, it has a very different characteristic to the more 'speech-specific' PWF threshold. In principle, as long as the coding noise is maintained entirely below the JND threshold — even with noise higher than the signal in some spectral regions — transparent coded speech performance could be achieved. The use of JND or its combination with perceptual weighting and post-filtering would make a significant contribution to achieving toll-quality coding at very low bit rates.

Toll-quality coding of voiced speech has been a difficult objective for low bit rate speech coding. Recent developments show that more effective coding of these sounds can be achieved via interpolation of representative waveforms. As voiced speech is a fairly stationary signal, it consists of near replicas of itself when observed in a short time window. When coding speech below 4 kbit/s, a 'representative' waveform segment can be extracted regularly and efficiently coded; the signal between the coded segments can then be regenerated via interpolation. With an appropriate speech interpolation process, this method can result in lower overall quantization noise than competing techniques. Several interpolation methods have been proposed in recent years. This chapter classifies this type of technique as speech interpolation coding, and the more popular approaches are summarized in the next section.

3.6   SPEECH INTERPOLATION CODING

The use of time-domain interpolation is a common feature of many of the speech coding techniques now being proposed for very low bit rates. The principle is to characterize speech by analysing short representative segments extracted at suitable time intervals. The characteristics of each analysis segment are assumed to be 'instantaneous' measurements which describe the speech waveform in the vicinity of a particular point in time referred to as an update point. These characteristics are efficiently represented by encoded parameters. At the decoder, frames of speech are synthesized with smoothly changing parameters obtained by interpolating between the encoded parameters. Synthesis frames begin and end at update points and, in general, the lengths of analysis segments and synthesis frames will be different. The interpolation process is intended to approximate the changes that occur in the original speech, from one update point to the next.

Among the techniques actively studied in recent years for very low bit rate speech coding are the following which use time-domain interpolation:

- sinusoidal transform coding (STC) [37, 38];

- multiband excitation coding (MBE) [39, 40];

- prototype waveform interpolation (PWI) [41].

3.6.1   Analysis and coding

Each of the three techniques mentioned above may be implemented with update points at regular intervals of about 20 ms. At each update point, an estimate of the pitch-frequency (the fundamental frequency of the vocal cord vibration) is required for the speech waveform in the vicinity of the update point when it may be considered voiced. The pitch frequency is considered variable, and the estimate obtained is taken as an 'instantaneous frequency' at the update point. Some quite sophisticated pitch-frequency estimation techniques are required as this is a critical aspect of all three techniques.

Sinusoidal transform coding (STC) is based on a method of representing speech as the sum of time-varying sinusoids whose amplitudes, frequencies and phases vary with time and are specified at regular update points. The method was first applied to low bit rate speech coding [37] in 1985. At the analysis stage, segments of the original speech, each of duration about two and a half pitch-periods and centred on an update point, are spectrally analysed via a fast Fourier transform (FFT). For voiced segments, the power spectra will have peaks, in principle, at harmonics of the pitch-frequency as illustrated in Fig. 3.5. Unvoiced segments will have randomly distributed peaks. The frequencies of the peaks are identified by applying a simple peak-picking algorithm and the corresponding magnitudes and phases are then obtained. These measurements become the parameters of the STC representation. It has been found [38] that speech synthesized from these parameters can be made essentially indistinguishable from the original speech when the parameters are unquantized..

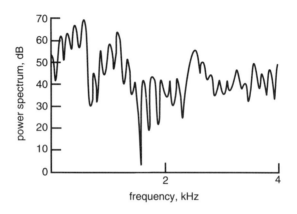

Fig. 3.5   Power spectrum modelled by STC.

For low bit rate coding, it is not possible to encode, with sufficient accuracy, the STC model parameters directly. An indirect method is adopted whereby a spectral envelope represented by a small number of parameters, a 'sign (ambiguity) bit', a pitch-frequency measurement and a 'voicing probability' frequency are encoded at each update point. The sign bit is needed, as $s(t)$ and $-s(t)$ have the same spectral envelope. The voicing probability frequency divides the speech spectrum into two bands; a lower band considered voiced and an upper band considered unvoiced. This frequency, and the sign-bit are determined by an 'analysis-by-synthesis' procedure to determine its voicing decisions. A means of deriving the sinusoidal model parameters from this information at the decoder will be discussed in the next section.

Multiband excitation (MBE) coding was originally proposed [39] for 9.6-kbit/s coding and later improved and adapted to 4.15 kbit/s, referred to as the IMBE codec. The IMBE codec, combined with error protection to yield an overall 6.4 kbit/s, is used for Inmarsat-M satellite mobile communications [40]. MBE coding involves extracting parameters at the update points from relatively large analysis segments of fixed length which overlap as illustrated in Fig. 3.6.

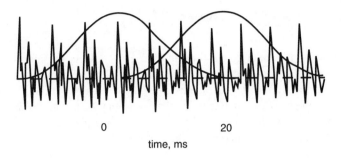

0   20

time, ms

Fig. 3.6   Overlapped windows used by IMBE.

A two-stage pitch-period extraction procedure is applied, the first stage achieving an accuracy of one half of a sampling interval. The second stage refines the result of the first stage to an accuracy of one quarter of a sampling interval. From the first pitch-period estimate, non-overlapping frequency bands are defined, each of bandwidth equal to a specified number of pitch-frequency harmonics — typically three. The number of bands is therefore pitch-period dependent and may lie between 3 and 12. A separate voiced/unvoiced decision is next made for each of these bands by measuring the similarity between the original speech in the band and the closest approximation that can be synthesized by harmonics of the more accurately estimated pitch-frequency. When the similarity is close, the band is declared voiced — otherwise it is classified as unvoiced. Three 'spectral amplitudes' are determined, corresponding either to the pitch-frequency harmonics within the band or, for an unvoiced band, to samples of the speech spectral envelope. The input speech is now assumed to be characterized at each update point by a pitch-period estimate from which the pitch frequency and the number of frequency bands may be determined, and for each of these bands, a voiced/unvoiced decision and three spectral amplitudes. This information is then efficiently coded for transmission.

Prototype waveform interpolation (PWI) was proposed [41] in 1991 for speech coding at 3 to 4 kbit/s. A prototype waveform is a segment of voiced speech of length equal to one pitch period, starting at any point within a pitch cycle. The original idea of PWI was to encode separately the lengths and shapes of such waveforms extracted, as illustrated in Fig. 3.7, at update points which lie within voiced portions of the speech. It was proposed that a voiced/unvoiced decision should be made for each update point, and that unvoiced segments should be encoded by switching to a form of CELP. Each prototype waveform is characterized by passing it through an LPC inverse filter to generate a spectrally flattened residual. The residual is then analysed to produce a Fourier series:

$$e(t) = \left[A_k \cos \left(\frac{2\pi kt}{P} \right) + B_k \sin \left(\frac{2\pi kt}{P} \right) \right]$$

whose period P is equal to the estimated pitch-period. Over the analysis interval, $e(t)$ is equal to the prototype waveform residual. 'Time alignment' is applied to $e(t)$ to make the A_k and B_k coefficients as close as possible to the A_k and B_k coefficients of the previous update point. The resulting A_k, B_k and linear predictive coefficients (LPC), the pitch period P, and a one-bit voicing decision are quantized and efficiently encoded for voiced speech.

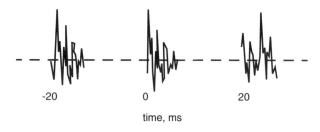

-20   0   20

time, ms

Fig. 3.7   Prototype waveforms for PWI

Quantization distorts not only the shape of each prototype waveform, but also the way this shape changes from one update point to the next, i.e. the waveform 'dynamics'. The result may be too much, too little and/or inappropriate periodicity in the decoded speech. An important innovation of PWI is the use of a 'signal-to-change ratio' (SCR) in the quantization procedure. This ratio measures the rate at which the periodicity is changing in the original speech, and attempts are made to quantize the parameters in such a way that this rate of change is preserved at the decoder, even when the decoded waveforms are considerably different from the original.

Recently [42], the concept of a prototype waveform has been generalized to include arbitrary length segments of unvoiced speech. The term 'characteristic waveform' is now used. Instead of extracting a single characteristic waveform at each update point, a sequence of about ten are extracted at regular intervals between consecutive update points. When these waveforms are time aligned, as illustrated in Fig. 3.8, the changes that occur to their corresponding Fourier series coefficients are indicative of the nature of the speech. Rapid changes occur for unvoiced speech, slow changes for voiced speech. High- and low-pass digital filtering is applied to separate the effect of these changes, and the resulting filtered Fourier series coefficients characterize a 'slowly evolving waveform' (SEW) and a 'rapidly evolving waveform' (REW) which sum to form the true characteristic waveform coefficients. The SEW is down-sampled to one waveform per update point, rapid fluctuations having been eliminated by the low-pass filtering. The REW cannot be accurately represented at low bit rate. Fortunately, it may be replaced at the decoder by a random waveform with similar spectral shape. The parameters for this generalization of PWI, i.e. LPC coefficients, pitch period and characteristics of the SEW and REW, may be encoded at 2.4 kbit/s.

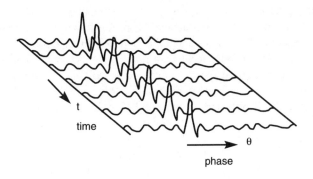

Fig. 3.8   Evolution of prototype waveform.

Comparison of analysis and coding techniques — as described above, all three techniques may be considered as sinusoidal coding techniques, where fundamental differences lie in the lengths of the analysis segments. For MBE, the relatively large analysis segment gives good spectral resolution and noise immunity. However, the effects of non-stationarities, such as that due to pitch-frequency variation, will be apparent in the resulting spectra. These effects are accommodated to a considerable extent by the MBE coding technique. A PWI analysis segment, being of length equal to a single pitch period, may be expected to give a more accurate representation of voiced speech at an update point and be less affected by pitch-frequency variation and other non-stationarities. The variable length of the analysis segment adopted by STC is not a critical factor and is chosen to allow an accurate spectral envelope to be determined without serious distortion caused by non-stationarity. Vector quantization is used liberally by all three techniques, e.g. for LPC coefficients, spectral amplitudes, SEW and REW. The principles of all three techniques are still the subject of active research.

3.6.2   Decoding and synthesis

All three techniques synthesize speech at the decoder as sums of sinusoids whose amplitudes, instantaneous frequencies and phases are changing on a sample-by-sample basis and are obtained by interpolating between values obtained from decoded parameters at each update point.

A straightforward linear interpolation scheme is used for the amplitudes, which must change monotonically across the synthesis frame.

The term 'instantaneous frequency' is used for a frequency modulated sine wave to describe its apparent frequency at a given instant in time. Voiced speech conforms more naturally to this sinusoidal model than unvoiced, and for all three techniques, the instantaneous frequencies at each update point are simply the harmonics of the decoded pitch frequency. There are important differences in the way each of the three techniques adapt the model to unvoiced frames.

STC — the STC decoder receives a pitch frequency measurement, a set of parameters representing the spectral envelope, a 'voicing probability' frequency and a sign bit. Below the voicing probability frequency, the amplitudes of the sinusoidal model are obtained at each update point by sampling the decoded envelope at the pitch-frequency harmonics. The phase of each sinusoid is not available directly from the decoded parameters, but is deduced from the spectral envelope on the assumption that it is the gain response of a minimum phase transfer function. STC models speech above the voicing probability frequency by sinusoids closely spaced in frequency (say 100 Hz apart) synthesized with random phase. Again the amplitudes of these sinusoids are obtained by sampling the decoded envelope. The sign bit applied to each sinusoid will eliminate the possibility of random 180° phase reversals from one update point to the next.

As the phases of the sinusoids are specified or estimated at the update points, the instantaneous phase of each sinusoid is computed using cubic interpolation for a more faithful reconstruction of the original speech waveform.

IMBE — although earlier versions of MBE encoded phase information, the improved version (IMBE) encodes only spectral amplitude measurements for each band. These determine the amplitudes of (normally) three sinusoids for each voiced band, and the power and spectral shape of the required band-limited random signal generated for each unvoiced band. The number of bands and the frequencies of the sinusoids are deduced from the decoded pitch period.

IMBE makes no attempt to model the true phase relationships between sinusoids at the update points; instead the phases at the update points are simply chosen to maintain continuity with the waveform synthesized in the previous frame. This is achieved by linear interpolation from frame to frame of the corresponding pitch frequency harmonics. It can be shown that linear interpolation of a harmonic frequency is equivalent to a quadratic interpolation of its phase, but this form of interpolation introduces arbitrary phase shifts. IMBE only applies this method for relatively stationary voiced harmonics. It is known that disregarding the true phase relationships in this way entails some loss of naturalness, and the original waveshape is not necessarily preserved.

A less complex but a reasonable approximation to the effects of interpolation is obtained by an 'overlap and add' technique, used for harmonics with large variation in pitch frequency and for transitions in either direction between voiced and unvoiced conditions. Centred on each update point, frames are initially reconstructed without interpolation using sinusoids and/or random signals derived from the coded parameters. Between two update points, the current and previous reconstructed frames are merged to form the final synthesized speech frame. To merge the frames, the current frame is multiplied by a gradually increasing window, and is then added to the previous frame multiplied by a gradually decreasing window.

PWI — the original version of PWI is restricted to voiced portions of speech, unvoiced portions requiring a switch to CELP. As the sine and cosine amplitudes of the Fourier series are encoded, the correct phase relationships between pitch frequency harmonics may be preserved. However, pitch synchronism with the original speech is not maintained because the prototype waveforms are time aligned by the encoder for efficient quantization. Further time alignment is needed at the decoder to maintain the continuity of the synthesized waveform. This is achieved by equating the phases at the update point to the instantaneous phases attained at the end of the previous synthesis frame.

PWI employs linear interpolation of the pitch period, which is broadly equivalent to quadratic interpolation of the phase information except that the use of sine as well as cosine terms in the Fourier series preserves the original phase relationships between harmonics. A synthesized waveform similar to the original, but with an imperceptible time shift, which varies from frame to frame, is characteristic of PWI-encoded waveforms.

3.6.3   The way forward

Three basic forms of speech interpolation coding have been briefly summarized. These coding methods have enabled great advances to be made in the field of low bit rate coding for telephone bandwidth speech, and are still an active research topic. Although the development of the 2.4-kbit/s speech coder for the American DOD standard has shown very promising results with these coding methods, it is not yet certain whether these methods could achieve toll-quality coding for the future ITU 4-kbit/s coding standards. A number of questions need to be resolved.

Do the simple rule-based interpolation methods (linear, quadratic or cubic) impose a barrier to toll-quality coding? Could these methods adequately track the speech dynamic behaviour even within a short speech segment (typically about 20 ms)? Would these artificial rule-based interpolation methods tend to introduce more synthetic decoded speech under codec tandeming conditions?

Although some speech signals look fairly stationary, some frequency regions could have larger variations at certain 'time zones'. It may be desirable to adaptively allocate encoded bits to adjust the interpolation strategy for the frequency components at the time of large variations. Could better coding efficiency be gained with such an adaptive time-frequency bit allocation scheme?

Some 'representative' speech waveforms may be formed as a combination of two or more complex waveforms; they may be subject to different time evolution from the previous representative waveform. Does toll-quality coding require careful control of different interpolation functions over these complex waveforms?

At present it seems possible that there is much potential for refinement and cross-fertilization of the principles behind the three coding techniques reported here. New and practical ideas are being reported all the time. However, there is also the possibility of a totally different approach revolutionizing this field of research yet again.

3.7   NETWORK INTERACTIONS

This chapter has highlighted the general performance issues, the bit rates, system complexities and key techniques of speech coding systems used, or destined to be used, in the telephone network. Low bit rate speech coding has undoubtedly proved beneficial in many network applications, but it also imposes limitations. The fundamental reasons for these limitations, and the techniques used to evaluate and resolve the consequent network issues, are discussed below.

The telephone's history was dominated by essentially linear systems (see Chapter 18). Amplitude quantization, as in PCM coding, is commonly believed to be the only nonlinear process used in speech coding. For low bit rate coding as discussed in section 3.6, quantization can, in fact, come in various different forms or in combinations of these. The STC coder quantizes the frequency spectrum into a set of spectral peaks — frequency domain quantization. The STC and PWI coders quantize a long speech segment as a representative speech waveform — time domain quantation. Noise masking coders quantize speech into perceptually important signal components — 'hearing' quantization. Time domain interpolation also introduces 'artificial' time variation into the decoded signal. These different types of time variant nonlinear processes are important for achieving high-quality speech coding at low bit rates, but this means that white granular quantization noise is no longer the only speech coding distortion that exists in the network.

As the bit rate decreases, the design of speech coders will become more 'speech specific', and they will be less likely to offer transparent transmission for voiceband data such as modem and facsimile data, or even in-band network signalling tones, such as dual-tone multifrequency (DTMF). These disadvantages can be resolved using reliable signal classification methods to identify speech, modem data, facsimile and network tones. The different signal types can then be transmitted using different speech coding rates or techniques such as modem or facsimile demodulation [43].

As a result of these nonlinear factors, when low bit rate coders are used, problems may arise due to adverse interactions [15, 44, 45] with other elements of the telephone network. For example, echo canceller performance may be impaired, and high bit rate voiceband data transmission may become unworkable over the link. If codecs of the same or different type are connected in tandem, unacceptable distortion may result. It is therefore necessary to give consideration to possibly problematic configurations, so that appropriate advice can be given to network planners.

BT Laboratories have facilities to evaluate an extensive range of network configurations by both objective and subjective test methods. A portable, modular DSP platform developed at the Laboratories and known as the network emulator [44] is routinely used to facilitate these investigations. Software emulations exist for all of the common network elements, and selected emulations are downloaded to the DSP hardware to provide a real time emulation of the chosen network configuration. When it is desired to evaluate the performance of a particular codec in a network scenario, this is usually achieved by converting the codec algorithm into a software module for the network emulator. It may then be tested in any network configuration. Modules exist for many of the codec standards, e.g. G.711, G.726, G.728, GSM, and Skyphone.

A recent addition to the library of software modules is that of DCME to CCITT Recommendation G.763. This module allows any DCME trunk-to-channel ratio and loading to be emulated, and will continually output important system statistics such as mean bit rate, and the amount of speech clipping. Since DCME involves the combination of voice activity detection, DSI, comfort-noise injection, and variable bit rate ADPCM, there is considerable scope for adverse interactions to occur when other low rate codecs form part of the link.

3.8   CONCLUSIONS

Contrary to popular belief, G.711 64-kbit/s PCM coding is no longer the only digital transmission method in the switched telephone network. It still reigns supreme in the PSTN trunk network due to the full-scale deployment of optical fibre, and the relatively cheap bandwidth this affords. However, low rate speech coding is in common use in international, mobile, satellite and private networks, and these form a major part of the overall network. In digital mobile radio systems, low rate coding offers greatly improved channel capacity — indeed, these systems would not be viable without it. In fixed networks including satellite, international and private networks, low rate coding is in high demand because of the substantial cost savings. Low rate coding is also essential to allow multimedia services over conventional telephone lines.

Utilizing the latest speech coding technology, the new ITU G.729 8-kbit/s coder achieves toll-quality speech coding performance. Recent research results show that speech interpolation combined with noise masking speech coding techniques for achieving toll-quality performance at 4 kbit/s. The ITU already has a plan in place to achieve this goal by 1998.

As coding performance improves at lower bit rates, speech codecs will become widespread in the switched network. With a variety of speech coders used in different network scenarios, the likelihood of tandemed codecs will increase, and the need to guard against adverse network interactions will become increasingly relevant. If a connection comprises more than one speech codec, the transmission delay and distortion will increase significantly so that echo control may be required. This is a particular case where careful planning is vital to prevent serious degradation occurring. For the network operators, a good understanding of network interactions with low rate coding systems will become an important issue both to guarantee a quality service to customers and to ensure that none of the existing services are degraded.

REFERENCES

1.   Boag J F: 'The end of the first pulse-code modulation era in the UK', The Post Office Electrical Engineers Journal, 71, Part 1, pp 2-4 (April 1978).

2.   Boyd I and Southcott C B: 'A speech codec for the Skyphone service', BT Technol J, 6, No 2, pp 50-59 (April 1988).

3.   AEEC: 'Characteristics 741, Part 2', (1990).

4.   Rose K R and Hughes P M: 'Speech systems', BT Technol J, 13, No 2, pp 51-63 (April 1995).

5.   Freeman D K, Cosier G, Southcott C B and Boyd I: 'The voice activity detector for the pan-European digital cellular mobile telephone service', Proc ICASSP'89, pp 369-372 (1989).

6.   Barrett P A: 'Information tone handling in the half-rate GSM voice activity detector', Proc ICC'95, pp 72-76 (1995).

7.   Kessler E A: 'Digital circuit multiplication equipment and systems — an overview', British Telecommunications Eng J, 11, Part 2, pp 106-111 (July 1992).

8.   Mills G and Griffiths D: 'PC-based visual communications services', British Telecommunications Eng J, 12, Part 4, pp 266-271 (January 1994).

9.   Gallagher I and Wardlaw M: 'Asynchronous transfer mode', Supplement to the British Telecommunications Eng J: 'A Structured Information Programme', Chapter 6.3 (1994).

10. Dimolitsas S, Ravishankar C and Schroder G: 'Current objectives in 4 kbit/s wireline-quality speech coding standardization', IEEE Signal Processing Lett, 1, No. 11, pp 157-159 (November 1994).

11. Delprat M, Urie A and Evci C: 'Speech coding requirements from the perspective of the future mobile systems', Proc IEEE Workshop on Speech Coding for Telecommunications, pp 89-90 (October 1993).

12. Campanella S J: 'Digital speech interpolation', COMSAT Technical Review, 6, No 1, pp 127-158 (Spring 1976).

13. Karanam V R, Sriram K and Bowker D O: 'Performance evaluation of variable bit rate voice in packet networks', AT&T Technical Journal, pp 41-56 (1988).

14. Hall K: 'Global system for mobile communications', British Telecommunications Eng J, 13, pp 281-286 (January 1995).

15. Goetz I: 'Ensuring mobile communication transmission quality in the year 2000', Electronics and Communications Engineering Journal, pp 141-146 (June 1993).

16. Aarskog A I and Guren H C: 'Predictive coding of speech using microphone/speaker adaptation and vector quantization', IEEE Trans on Speech and Audio Processing, 2, No 2, pp 266-273 (April 1994).

17. Cellario L et al: 'A VR-CELP codec implementation for CDMA mobile communications', Proc ICASSP'94, p I-281 (1994).

18. Goodman D J: 'Combined source and channel coding for matching the speech transmission rate to the quality of the channel', Rec IEEE Global Telecom Conf, pp 316-321 (November 1982).

19. Cox R V et al: 'Sub-band speech coding and matched convolutional channel coding for mobile radio channels', IEEE Trans on Signal Processing, 39, No 8, pp 1717-1731 (August 1991).

20. Wong W T K: 'A speech coder design for land mobile radio communications', PhD thesis, University of Liverpool (June 1989).

21. Kleijn W B: 'Source-dependent channel coding for CELP', Proc ICASSP'90, 1, pp 1-4 (1990).

22. Internet site: ftp:://svr-ftp.eng.cam.ac.uk/pub/comp.speech/sources.

23. Kohler M A, Supplee L M and Tremain T E: 'Progress towards a new government standard 2400 bit/s voice coder', Proc ICASSP'95, pp 488—491 (1995).

24. Chambers P: 'Personal mobile satellite communications', IEE Colloquium, Digest No 12, pp 2/1-2/9 (January 1993).

25. Leopold R J: 'The Iridium communication systems', Proc ICC'92, 2, pp 451-455 (1992).

26. Internet site: http://www.vocaltec.com.

27. Boyd I: 'Speech coding for telecommunications', in Westall F A and Ip S F A (Eds): 'Digital signal processing in telecommunications', Chapman & Hall, pp 300-325 (1993).

28. Jayant N S and Noll P: 'Digital coding of waveforms', Prentice-Hall, (1984).

29. Sherif M H: 'Overview and performance of CCITT/ANSI embedded ADPCM algorithms', IEEE Trans on Comm, 41, No 2, pp 391-399 (February 1993).

30. Atal B S: 'Predictive coding of speech at low bit rates', IEEE Trans on Comm, COM-30, No 4, pp 600-614 (April 1982).

31. Makhoul J, Roucos S and Gish H: 'Vector quantization in speech coding', Proc of the IEEE, 73, No 11, pp 1551-1588 (November 1985).

32. Atal B S and Schroeder M R: 'Stochastic coding of speech signals at very low bit rates', Proc ICC'84, Part 2, pp 1610-1613 (May 1984).

33. Kroon P and Deprettere E F: 'A class of analysis-by-synthesis predictive coders for high quality speech coding at rates between 4.8 and 16 kbit/s', IEEE SAC, 6, No 2, pp 353-363 (February 1988).

34. Schroeder M R, Atal B S and Hall J L: 'Optimising digital speech coders by exploiting masking properties of the human ear', JASA, 66, pp 1647-1652 (1979).

35. Chen J H and Gersho A: 'Adaptive postfiltering for quality enhancement of coded speech', IEEE Trans on Speech and Audio Processing, 3, No 1, pp 59-71 (January 1995).

36. Jayant N, Johnston J and Safranek R: 'Signal compression based on models of human perception', Proc of IEEE, 81, No 10, pp 1385-1422 (October 1993).

37. McAulay R J and Quatieri T F: 'Mid-rate coding based on a sinusoidal representation of speech', Proc ICASSP'85, pp 945-948 (1995).

38. McAulay R J and Quatieri T F: 'Low-rate speech coding based on the sinusoidal model', in Furui S and Sondhi M M (Eds): 'Advances in speech signal processing', Marcel Dekker Inc (1992).

39. Griffin D W and Lim J S: 'A high quality 9.6 kbit/s speech coding system', Proc ICASSP'86, pp 125-128 (1986).

40. Hardwich J C and Lim J S: 'The application of the IMBE speech coder to mobile communications', Proc ICASSP'91, pp 249-252 (1991).

41. Kleijn W B: 'Continuous representation in linear predictive coding', Proc ICASSP'91, 1, pp 201-204 (1991).

42. Kleijn W B and Haagen J: 'A speech coder based on decomposition of characteristic waveforms', Proc ICASSP'95, pp 508-511 (1995).

43. Lewis A V, Gosling C D, Evans K G, Davis A G and Wong W T K: 'Aeronautical facsimile — over the oceans by satellite', BT Technol J, 12, No 1, pp 83-97 (January 1994).

44. Guard D R and Goetz I: 'The DSP network emulator for subjective assessment', in Westall F A and Ip S F A (Eds): 'Digital signal processing in telecommunications', Chapman & Hall, pp 352-372 (1993).

45. Dimolitsas S et al: 'Voice quality of interconnected PCS, Japanese Cellular, and Public Switched Telephone Networks', Proc ICASSP'95, pp 273-276 (1995).

4

SPEECH TRANSMISSION OVER DIGITAL MOBILE RADIO CHANNELS

P A Barrett, R M Voelcker and A V Lewis

4.1 INTRODUCTION

The last few years have witnessed a remarkable growth in the mobile telephony market, and digital systems now support a significant portion of the customer base. With the apparent ubiquity of digital technology, its application to mobile radio might be taken for granted. In practice, speech transmission over a digital mobile radio channel requires a difficult trade-off between a number of competing factors. This chapter identifies the considerations in selecting a speech codec for mobile telephony applications, outlines techniques for robust and efficient speech transmission over a digital mobile radio channel and discusses how the resulting performance can be assessed.

Throughout this chapter, the Global System for Mobile Communications (GSM) half-rate speech channel will be used as an illustrative example. At the time of writing, GSM is the most widely adopted digital mobile radio standard, and by the end of the century there will be an estimated 25 million GSM customers in Western Europe alone. GSM has also been adopted in Eastern Europe, the Far East, Australia and Africa. In these regions, GSM is licenced to operate in the 900 MHz and 1800 MHz radio bands. In North America, GSM is operated in the 1900 MHz band and uses an improved speech codec, which will soon be available to operators using the other frequency bands. Of course, GSM is not the only digital mobile radio standard and further examples are provided in Chapter 3.

Figure 4.1 highlights two of the key components for speech transmission over a digital mobile radio channel — a speech codec and a channel codec. The function of the speech codec is to represent the speech signal using as few bits as possible. The performance of the speech codec determines an upper limit for the speech quality. However, poor radio reception introduces errors and the channel codec further encodes the speech data so that these errors can be detected and corrected. The degradation in speech quality under different error conditions is largely determined by the channel codec.

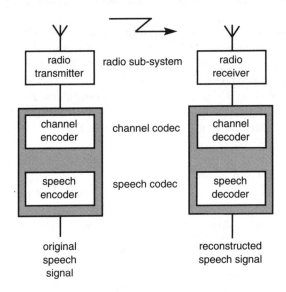

Fig. 4.1   Key components for speech transmission over a digital mobile radio channel.

Section 4.2 discusses the performance trade-offs that must be made in a practical digital mobile radio system. Next, section 4.3 considers the requirements of a speech codec for mobile radio applications, while techniques for robust and efficient transmission are introduced in sections 4.4 and 4.5. As an illustrative example, section 4.6 describes some aspects of the half-rate GSM speech channel. In section 4.7 discussion covers how the performance of such a system can be measured. Finally, sections 4.8 and 4.9 consider future directions and conclusions.

4.2   PERFORMANCE DIMENSIONS

The purpose of a communications system is to enable two or more people to communicate. This may sound obvious, but what is the definition of communication? In a military radio application the primary requirement may simply be the transfer of spoken information; radio operators can be trained in the use of the system and reliability is likely to be more important than naturalness. By contrast, in telephony the customer requires a system that is simple to use, offers natural communication, and conveys as many nuances of the original speech as possible. It is therefore important to balance the various dimensions of system performance to suit the needs of the application in question. The most important performance dimensions of a digital speech transmission system are speech quality, robustness to errors, delay, complexity and bit rate [1].

If the ability to communicate at all times is paramount, the majority of bits available in the radio channel will be allocated to the channel coding. However, this will inevitably degrade the basic speech quality since there will be less bits available to the speech codec. Conversely, if good speech quality is more important than providing service under difficult radio conditions, the majority of bits are likely to be allocated to the speech codec — at the expense of robustness to channel errors. Military applications and public mobile telephony provide good examples of these two cases.

Delay is also an important factor in determining the naturalness of a communications system. From a customer's point of view, the important figure is the delay between the production of a sound by the talker and the arrival of that sound at the ear of the remote listener. Delays that exceed a few hundred milliseconds rapidly lead to difficult and unnatural conversation. In cases of extreme delay, conversation may break down altogether. The ITU recommendations on delay are defined in recommendation G.114.

Codec complexity can have a large impact on mobile terminal design. In digital mobile terminals, the power consumption of the radio sub-system is often small compared to that of the signal processing devices. A more complex codec may therefore reduce talk time or necessitate larger and heavier batteries. The number of silicon devices required to implement a codec can also affect the physical dimensions of the terminal. Ideally, it should be possible to implement a speech or channel codec on a single device. However, even in mass-production quantities, state-of-the-art devices can be prohibitively expensive.

In the development of speech codecs for the digital public switched telephone network (PSTN), the emphasis has been on maintaining good speech quality and low delay. The need for a cost-effective network has also meant low complexity codecs, with the result that the majority of speech traffic is currently carried at 64 kbit/s, although lower rates are sometimes used (see Chapter 3).

The data rate available to a user in a mobile radio system depends on a large number of factors including:

- the radio bandwidth allocated to the service, which is determined by regional and international agreement;

- the type of system, for example, cellular radio or satellite;

- the operating environment and associated radio propagation problems, such as multipath and shadowing;

- sources of radio interference, for example, other users of the system;

- limitations on radio transmission power, which could be regulatory or technical.

These and many other factors spanning physics, engineering and economics, combine to severely constrain the bit rate available to individual users in a practical radio system.

Obviously there are limits to which any one performance factor can be compromised. However, the restricted data rate and high occurrence of errors found in many mobile radio systems often dictate lower speech quality, greater complexity and longer delay compared to a PSTN link.

4.3   SPEECH CODEC CONSIDERATIONS

A survey of speech codecs used in mobile radio systems is provided in Chapter 3. This section will consider the basic requirements of a speech codec suitable for use in a digital mobile radio system.

Codecs suitable for general audio often exploit the limitations of the human auditory system to provide near-transparent compression of arbitrary signals. Speech codecs achieve further compression by making judicious assumptions about the source signal to be encoded. A source model commonly used in speech codecs is the linear predictive model of speech production [2]. While the use of a source model reduces the bit rate of a codec, it also reduces the range of signals that can be faithfully transmitted. Problems arise when a speech codec is required to transmit signals that are not well described by the source model, and, as a consequence, transmission quality may be significantly degraded. In general, the lower the bit rate of a speech codec, the more dependent it will be on the assumptions of the source model.

4.3.1   Talker and language dependency

The use of source models can lead to codec performance that varies between different talkers and languages. The character of sounds produced by an individual talker is determined by a number of factors, including the talker's native language, regional accent and physical attributes. In the latter case, the articulators used in speech production not only vary between individuals, but also as a function of gender and age. The physical mechanisms used to generate speech sounds can also differ between languages — as do the nuances that carry the meaning.

In a mobile radio system, the speech codec must operate in both the send and receive directions, and in a system connected to the PSTN, calls can be made to and from anywhere in the world. The speech codec should therefore be capable of faithfully encoding all human utterances. In practice, however, design considerations often result in speech codec performance that is biased towards a particular language, gender or age group.

4.3.2   Signals other than speech

The typical acoustic environment of a mobile radio user differs significantly from that of a PSTN user. The sound pressure level (SPL) of ambient noise in an office or home environment is generally in the range of 35 to 55 dB(A). However, the SPL experienced by a mobile radio user, for example in a moving vehicle, in an aircraft, or in a busy street, may exceed 85 dB(A).

A high ambient noise level creates two problems. Firstly, it increases the listening effort required by the mobile radio user, since the signal produced by the handset ear piece must compete with the ambient noise. Secondly, it introduces noise into the speech received by the handset microphone. A speech codec selected for mobile radio applications should therefore exhibit acceptable performance for signals containing high acoustic noise levels. It should be noted that PSTN speech codecs may also have to carry mobile radio traffic and should also be capable of handling ambient noise.

A partial solution to the problem of ambient acoustic noise is to simply reduce the amount entering the codec. Careful acoustic design of a handset can dramatically reduce its sensitivity to ambient noise compared to the user's speech. Even better noise rejection figures are possible using noise cancelling microphones [3]. Figure 4.2 shows the noise rejection achievable with a well-designed noise-cancelling handset. While rejection is modest above 2 kHz, the degree achieved at low frequencies is very valuable. For example, much of the ambient acoustic energy in a moving vehicle is present below 200 Hz. An alternative approach is to use signal processing techniques to achieve noise suppression [4]. However, substantial reductions in noise level without perceptible distortion of the speech signal are difficult to achieve.

Further signals that may have to be transmitted include call progress tones (such as ring tone and busy tone), music on hold and dual-tone multifrequency (DTMF) signals. The latter represent a particular problem since they must be reli-

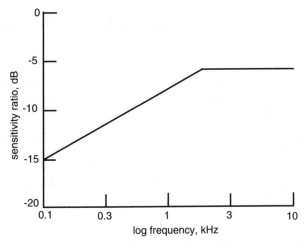

Fig. 4.2   The ratio of the noise sensitivity to speech sensitivity for a well-designed noise-cancelling handset.

ably interpreted by a DTMF detection algorithm. In practice, many low bit rate speech codecs do not transmit DTMF tones reliably.

4.3.3   Codec tandeming issues

The term tandeming is used to describe the interconnection of multiple speech codecs. Tandeming can introduce substantial distortion into a link and is an important consideration in speech codec design.

Imagine the following scenario — a customer on an aircraft over the Pacific Ocean uses Skyphone™ to call the UK. The call is routed to the UK through digital circuit multiplication equipment (DCME) and is answered by a voice mail service. Finally, the message is retrieved by a GSM customer. Figure 4.3 shows the speech codecs required to make such a call. As the number of mobile telephone users grows, scenarios such as this will occur with increasing frequency.

A problem for network planners is how to estimate the distortion introduced by a link such as the one in Fig. 4.3. One method would be to measure the distortion introduced by each codec in isolation and sum the individual figures. This technique is sometimes used in PSTN circuits, where the performance of each codec is described in quantization distortion units (qdu). Unfortunately, this approach is not valid for the type of distortion introduced by low bit rate speech codecs, which cannot be treated as linear network elements (see Chapters 14 and 18).

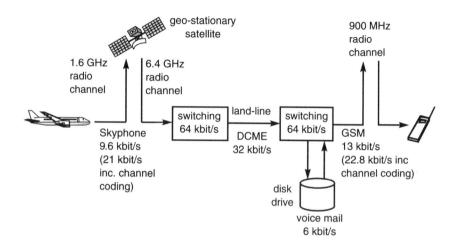

Fig. 4.3   In this scenario, the voice mail message left by the Skyphone customer passes through seven speech codecs before being heard by the GSM customer.

As mentioned earlier, many speech codecs rely on describing the input signal in terms of a source model. If the input to a codec is corrupted by the presence of ambient acoustic noise, or distortion introduced by previous codecs, the performance of the codec may be significantly worse than for clean speech. As a result, speech codecs in a tandem configuration may not perform as expected. In a digital mobile radio system connected to the PSTN, heterogeneous tandeming must be considered as well as homogeneous tandeming. The performance of a link containing heterogeneous speech codecs can even depend on the order in which the codecs appear.

A second difficulty in estimating the subjective performance of speech codecs in tandem may occur if they exploit the properties of the human auditory system. Many speech codecs shape the frequency spectrum of their distortion so that it is rendered less audible. Unfortunately, each additional codec introduces further distortion, and there is a limit to how much distortion can be hidden from the listener using this technique. Consequently, a codec with very good subjective performance for a single encoding may demonstrate significantly worse subjective performance after two or three encodings.

4.3.4   Sensitivity to bit errors

So far, it has been assumed that the impact of channel errors on speech quality is determined solely by the channel codec. In practice, however, not all channel errors can be detected and corrected, and it is important that the subjective impact of residual bit errors is minimized. Robustness to residual errors is therefore a further consideration in the selection of a speech codec for mobile radio channels, and section 4.6.3 discusses how the half-rate GSM speech codec is designed to reduce the impact of such errors.

4.4   COPING WITH CHANNEL ERRORS

The bit error rate of a mobile radio channel can be substantially higher than would be tolerated in a PSTN channel. For example, in the PSTN bit error rates do not usually exceed 0.001%, while 0.1% is considered an exceptionally poor channel. However, in a radio system, the error rates experienced in areas of marginal reception can be as high as 10%. To compound the problem, moving terminals are subject to variable radio path loss. Periods of very low radio signal strength are referred to as fades and introduce error-bursts. Slow-moving terminals experience the longest fades, which may last for many hundreds of milliseconds.

4.4.1   Error correction and detection

To minimize the effect of channel errors, forward error correction (FEC) is applied to the speech data before it is transmitted. FEC increases the number of bits that must be transmitted but enables occasional errors to be corrected. Cyclic redundancy check (CRC) parity bits may also be added to the speech data prior to FEC to enable the detection of uncorrected errors.

Basic error protection can be provided by simple block coding. However, mobile radio channel codecs often employ convolutional encoding [5] combined with soft-decision Viterbi decoding [6]. These techniques are significantly more complex than block coding, but offer a higher degree of error protection.

Not all of the bits that are generated by a speech encoder make an equal contribution to the reconstructed speech quality. Therefore, to reduce the overhead of channel coding, FEC is often only applied to the speech bits that are subjectively most important.

4.4.2   Interleaving and frequency hopping

Although FEC is capable of correcting sparsely distributed bit errors, it cannot correct sustained periods of errors. Since the error characteristics of mobile radio channels are inherently bursty due to radio fades, the encoded data bits are spread across a number of transmission frames. This process is called interleaving and an example is shown in Fig. 4.4. The number of transmission frames over which data is spread is referred to as the interleaving depth. However, interleaving introduces significant delay into the channel, and is generally only suitable for improving FEC performance over short radio fades.

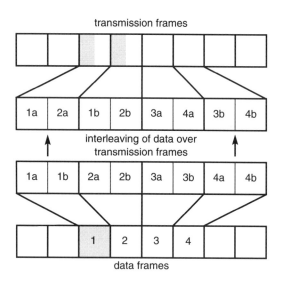

Fig. 4.4   A simple example of interleaving; in this case the interleaving depth is 2.

To enable FEC to work effectively during long radio fades, a technique called frequency hopping can be employed. For a specific physical location, radio fades will only occur at particular carrier frequencies. By changing the frequency of the radio carrier at regular intervals, the effect of fading can be limited to short time periods, thus enabling interleaving and FEC to work more effectively. Frequency hopping benefits slow-moving users most, for example a person walking along a street, because they experience the longest radio fades.

4.4.3   Error concealment

In poor radio conditions the channel decoder may be unable to correct all the bit errors in the speech data, and the speech decoder will enter an error concealment mode. Uncorrected bit errors may be detected either from parity bits added to the speech data, or in the case of a Viterbi decoder, from the soft decision value or the distance metric [7].

It has already been shown that many speech codecs use a parametric model to represent the speech to be encoded. Such codecs usually encode frames of speech, which are typically 10 to 30 ms in duration. In many source models, for example, the widely used linear predictive model, the parameter values change slowly over a number of frames. Speech codecs exploit this gradual variation of parameter values for error concealment purposes. When corrupted speech frames are detected, the speech decoder can use the parameter values from the last good frame to reconstruct an approximation of the missing speech. The same sound cannot be generated *ad infinitum*, and, after a small number of consecutive corrupted frames the output level of the speech decoder is gradually reduced until it has been completely muted.

4.5   DISCONTINUOUS TRANSMISSION

The previous section introduced techniques for robust transmission over a radio channel. This section outlines a technique that is intended to increase the efficiency of the system — discontinuous transmission (DTX).

DTX is a mechanism that limits a user's radio transmissions to the periods when they are speaking. The benefits of DTX are twofold:

- it reduces the mean activity of the radio channel, reducing interference to distant users operating on the same carrier frequency;

- it lessens the power consumption of hand-held terminals, through reduced radio transmission and associated processing tasks.

The key component of a DTX system is the voice activity detector (VAD). The VAD identifies which frames contain speech and should therefore be transmitted. Occasionally, the VAD may fail to detect some of the speech, most commonly in high levels of acoustic noise. The resultant gaps in the transmitted speech can reduce its intelligibility, and are referred to as clipping. In non-stationary acoustic noise, the VAD may also incorrectly classify periods of noise as speech. This increases the proportion of frames that are transmitted and thus reduces the benefit of the DTX system. VAD design is therefore a compromise between reducing clipping and maintaining a useful DTX gain.

When the VAD indicates that speech is not present, noise is generated at the receiver to provide continuity between speech bursts. This is called comfort noise and is intended to simulate the ambient acoustic noise present in the original speech signal. Under very noisy conditions the insertion of comfort noise greatly improves speech intelligibility. To ensure good continuity in the reconstructed signal, the energy of the comfort noise is matched to that of the ambient noise at the transmitter; continuity can be further improved by matching the frequency spectrum. In some schemes, the noise model at the receiver is regularly updated at the expense of a marginal increase in radio activity.

4.6   AN EXAMPLE DIGITAL MOBILE RADIO SYSTEM — HALF-RATE GSM

The preceding sections have introduced some of the techniques and considerations that go into the design of a digital mobile radio speech channel. This section describes a specific example in more detail — the half-rate GSM speech channel, which has recently been standardized by the European Telecommunications Standards Institute (ETSI). GSM is a good example of a system where the need to maximize the number of users, while maintaining acceptable speech quality, complexity and delay, has pushed speech coding technology to its limits.

The current GSM system uses a 13 kbit/s speech codec, which, after error correction coding, requires a 22.8 kbit/s traffic channel. To double the maximum number of users, a new set of recommendations has been defined which specify a 5.6 kbit/s speech codec and an 11.4 kbit/s channel. The two schemes are respectively referred to as full-rate and half-rate GSM [8-10].

The half-rate GSM speech codec belongs to the code-excited linear prediction (CELP) class of codecs [11]. Before considering some of the features of the speech codec in detail, it is useful to review the fundamentals of speech production and CELP coding.

4.6.1   Speech production

The characteristics of a particular speech sound are determined by the shape of the vocal tract and the excitation mechanism. The vocal tract is considered to extend from the vocal cord to the lips and the nose. The shape of the vocal tract determines the gross frequency spectrum of a speech sound, although the effect of a sound radiating from the lips also modifies its frequency spectrum. The resonances introduced by the vocal tract are called formants, and in many languages their frequency carries information.

Three basic types of excitation are evident in speech production — voiced, fricative and plosive. Voiced speech is generated by the periodic opening and closing of the vocal cords. The frequency at which the cords vibrate is referred to as the pitch of the speech. Fricative speech is generated by forcing air past a constriction in the vocal tract and has the characteristics of noise. Plosives are generated by creating a closure in the vocal tract, increasing the air pressure behind it and then releasing the pressure suddenly. The / i / in s*eem*, the / f / in *f*ace and the / p / in *p*uff are examples of voiced, fricative and plosive sounds respectively. Further sounds can be generated from a mixture of excitation mechanisms, for example the /z/ in ma*z*e (mixed voiced and fricative). A more detailed treatment of speech production is given in Chapter 2.

4.6.2   Fundamentals of CELP coding

CELP coding is based on the linear predictive model of speech production [2], in which speech is generated by filtering an excitation signal. The filter coefficients are referred to as linear predictor (LP) coefficients, and are calculated by assuming that the excitation signal has a flat frequency spectrum. In practice, this is not always true, and the LP filter represents the combined frequency characteristics of the vocal tract, lip radiation and the excitation spectrum.

The linear predictive model of speech is reflected in the structure of the basic CELP decoder shown in Fig. 4.5. The reconstructed speech signal $s'(t)$ is produced by filtering an excitation signal $e'(t)$. The synthesis filter uses LP coefficients calculated from the original speech. The excitation signal is constructed from the sum of two codewords, which are selected from the adaptive and fixed codebooks. The adaptive codebook contains delayed copies of the excitation signal $e'(t)$. This enables a periodic excitation signal to be generated during voiced speech. The fixed codebook contains random entries, which contribute mainly during fricatives, plosives and transitions. The contribution from each code book is determined by its respective gain factor. The excitation information is typically updated more frequently than the LP coefficients.

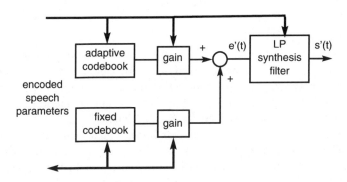

Fig. 4.5   CELP decoder.

The CELP encoder is shown in Fig. 4.6. The LP coefficients are calculated from the current frame of speech, quantized and transmitted to the decoder. The optimum codebook entries are determined by the analysis-by-synthesis method, where speech $s''(t)$ is synthesized from each codeword in turn. The index of the codeword that minimizes the mean squared error between the synthesized speech and the input speech is transmitted to the decoder. The optimum codeword gain is also calculated and is transmitted after being quantized. The adaptive codebook search is performed first and its contribution incorporated into the fixed codebook search. It is searching the codebooks that makes CELP coding inherently complex.

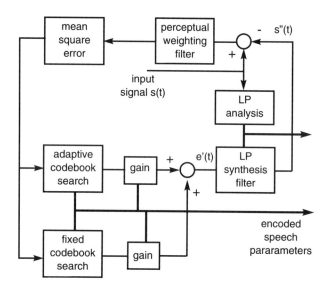

Fig. 4.6   CELP encoder.

The subjective performance of the CELP codec can be improved by weighting the error calculation in the codebook searches. This weighting favours codewords that minimize the distortion at frequencies where it will be most perceptible. The error signal is weighted by the perceptual weighting filter, whose coefficients are derived from the LP coefficients.

The structure of the CELP decoder determines which sounds can be faithfully reconstructed. It has been shown that the linear predictive model has much in common with the physical mechanisms of speech production. However, the CELP decoder is less capable of faithfully reconstructing many non-speech sounds. Consequently, the performance in the presence of ambient acoustic noise and under tandem conditions may be significantly worse than for a single encoding of clean speech.

4.6.3   The half-rate GSM speech codec

Having introduced the basics of speech production and CELP coding, the design of the half-rate GSM speech codec will now be considered. For a complete description of the codec, the reader is referred to the ETSI Recommendation [12]. This section will outline some of the techniques the half-rate GSM speech codec uses to achieve speech quality at 5.6 kbit/s that is comparable to that of the 13-kbit/s full-rate speech codec.

The half-rate GSM speech codec is a vector sum excited linear prediction (VSELP) codec [13]. In a VSELP codec, the codewords in the fixed codebook are constructed by summing a set of basis vectors, which makes the codebook more robust to bit errors than a random codebook. In a random codebook, a single bit error in the index will return a codeword with no relation to the desired codeword. However, in the vector sum approach, a bit error in the index will only corrupt one of many contributions to the desired codeword. The vector sum technique also reduces the complexity of the codebook search.

The half-rate codec processes speech in 20 ms frames, which are divided into 5 ms sub-frames. The LP coefficients and frame energy are calculated and transmitted once every frame, while the codebook parameters are calculated and transmitted every sub-frame.

The characteristics of speech differ for different degrees of voicing. To enable good quality modelling of speech at a low bit rate, the half-rate codec has four modes of operation — one unvoiced mode and three voiced modes. The mode is determined by the success of the adaptive codebook search. For each mode, separate quantizers are defined for the frame energy and codebook gains. While faithfully representing the four classes of speech, this enables the number of quantizer bits to be kept to a minimum. Since unvoiced speech is not periodic, the adaptive codebook contribution is replaced by a second VSELP codebook in the unvoiced mode.

Further reductions in bit rate are achieved by employing vector quantization (VQ) to quantize the LP coefficient vector and the two codebook gains. The VQ process searches a codebook to find the codeword that best matches the vector to be quantized, and transmits the index of the selected codeword. VQ codebooks are optimized using training data that is representative of the data to be quantized [14]. For a given number of bits, VQ introduces less distortion than individually quantizing each vector element. The disadvantage of VQ is the complexity of the codebook search. To reduce the complexity of the half-rate codec, the LP coefficients are quantized using a sub-optimal two-stage search.

To smooth the spectrum of the reconstructed speech, the LP coefficients may be interpolated over the sub-frames. However, this implies that the LP analysis window, which is used to calculate the LP coefficients, should be centred on the end of a frame, rather than its middle. In practice, the LP analysis window only extends 4.4 ms into the subsequent frame, increasing the algorithmic delay of the codec from 20 ms to 24.4 ms.

The half-rate speech decoder is a basic CELP decoder with the addition of a post-filter. The post-filter modifies the spectrum of the reconstructed speech signal to reduce the audibility of the codec distortion. This is achieved by emphasizing the formant frequencies and, during voiced speech, the harmonic frequencies. The post-filter is a significant factor in improving the overall subjective performance of the codec.

4.6.4   Error correction coding

In half-rate GSM, the error correction coding increases the number of bits representing each 20 ms frame from 112 to 228 bits. The encoded speech bits are divided into class 1 (95 bits) and class 2 (17 bits) according to their impact on the subjective quality of the decoded speech. Three parity bits are generated for the most important class 1 bits. FEC is only applied to the class 1 bits and the parity bits. A further 6 tail bits must also be encoded to flush the memory of the convolutional encoder.

The ratio of the number of bits before error correction coding to the number of bits after coding is called the rate of a code. The class 1 bits (including parity and tail bits) are protected by a 104/211 rate-punctured convolutional code. The punctured code is derived by encoding the speech data with a 1/3 rate code and discarding some of the encoded bits. Punctured codes enable different portions of the data to be encoded at different rates [5]. In this case, 1/2 rate coding is applied to the class 1 bits and tail bits and 1/3 rate coding to the parity bits.

The number of errors that can be corrected by a convolutional code is determined by its constraint length.

Due to the lower interleaving depth of the half-rate speech channel, the convolutional code has a constraint length of 6 compared to 4 in the full-rate speech channel. This increase in constraint length raises the complexity of the Viterbi decoder by a factor of 4.

4.6.5   TDMA structure and interleaving

Each GSM radio channel supports 8 full-rate users or 16 half-rate users. Time division multiple access (TDMA) is used to share the radio channel capacity among the different users. Figure 4.7 shows the breakdown of a 120 ms TDMA multiframe. The structure of a TDMA frame is divided into eight time-slots. The physical contents of a time-slot are called a transmission burst, and contain 114 data bits. In the half-rate case, each time-slot is shared between two users, who use alternate bursts.

After error correction coding, the speech data is re-ordered, interleaved and mapped on to the transmission bursts. The encoded bits are re-ordered to remove the correlation between adjacent bit values. The 228 re-ordered bits are then divided into two groups and interleaved over the 114-bit transmission bursts.

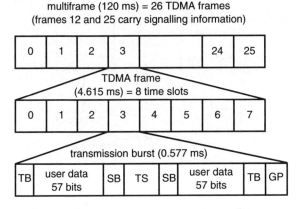

multiframe (120 ms) = 26 TDMA frames
(frames 12 and 25 carry signalling information)

TDMA frame
(4.615 ms) = 8 time slots

transmission burst (0.577 ms)

TB: tail bits (3 bit)
SB: stealing bit (1 bit)
TS: training sequence (26 bit)
GP: guard period (8.25 bit)

Fig. 4.7   TDMA structure for a GSM channel.

4.6.6   Discontinuous transmission

The GSM recommendations define an optional DTX mode of operation for the speech traffic channel [15]. Figure 4.8 shows a block diagram of the full-rate GSM VAD structure [16]. The input signal frame is first filtered by an inverse linear predictive model of the acoustic noise. The energy of the filtered signal frame is then compared to an adaptive threshold to form an intermediate decision. A hangover mechanism is used to calculate the final VAD decision, which extends the duration of periods classified as speech. The noise model and adaptive threshold are updated if the secondary VAD indicates the absence of speech. This occurs if the signal is non-periodic and spectrally stationary. Hangover is added to the secondary VAD decision to reduce the probability of adaptation during speech.

At the end of a speech burst a further seven frames are transmitted. During this period, the acoustic noise is characterized at the encoder by averaging the spectral information and energy of the input signal. A silence descriptor (SID) frame, which contains the noise information, is then transmitted. The decoder uses the SID information to generate comfort noise between the speech bursts.

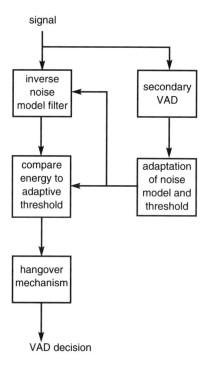

Fig. 4.8 Structure of the GSM VAD.

SID frames are transmitted every 240 ms to update the comfort noise parameters at the decoder. These parameters are interpolated over the SID update period to avoid discontinuities in the comfort noise.

4.6.7 Delay

The GSM recommendations define a round trip delay budget that must be met by manufacturers. Table 4.1 provides a coarse breakdown of the delay budget for the half-rate GSM channel.

The figures quoted in Table 4.1 are round-trip figures, which are the sum of the uplink and downlink delays. The implementation budget is the time allocated to perform the speech and channel coding tasks, and effectively determines the minimum processing power of the devices used if the delay requirement is to be met. The 'other' figure is the sum of the remaining system delays and budgets.

Table 4.1 Breakdown of the half-rate GSM speech channel delay budget.

Delay element	Delay, ms
Algorithmic delay	48.8
Interleaving delay	65.8
Implementation budget*	29.6
Other*	44.3
Total round trip delay*	188.5
* These figures are subject to review by ETSI.	

4.6.8   Complexity

The computational description of the half-rate system is based on the arithmetic operations defined for the full-rate system. The arithmetic is fixed-point and uses 16- and 32-bit operations. The exact complexity of an individual implementation will depend on the signal-processing hardware used. However, for the purposes of complexity evaluation, a set of basic computational operations is defined, each operation being allocated a weight equal to the number of instructions required to perform it on a typical DSP device.

The theoretical worst-case complexity of the half-rate codec has been evaluated at 21.2 weighted million operations per second (wMOPS). This is 4.5 times the worst-case complexity of the full-rate codec. In practice, the computation required by the half-rate codec to process different signal frames may vary by up to 20%. These figures include channel coding, but not DTX functions.

4.7   SPEECH QUALITY ASSESSMENT

The wide range of competing factors to consider and balance in the design of a mobile radio speech channel has been described. There is a similar challenge in determining suitable processess to verify that the speech performance meets the specified criteria. In common with past practice, the speech quality of the half-rate GSM system was measured by subjective assessment. However, objective assessment techniques, based on models of human perception, are also being developed [17]. In this section, some of the main issues relating to subjective speech quality assessment are outlined.

4.7.1   Methods of assessment

The simplest speech quality assessments are informal. They are quick and simple to perform and are commonly used, for example, to provide basic information on the effects of design changes during speech codec development. Such assessments often do not reflect real network scenarios and only provide limited information. As such, they are not sufficient to give a deep understanding of the performance issues that are important to the users of a system and to service providers.

Formal assessment methodologies have been developed to address such requirements, with much greater emphasis placed on controlling the process. Some of the factors considered include:

- acoustic environment;

- terminal equipment;

- the order in which impairments (such as channel errors or tandeming) are introduced;

- the selection of participants for the test;

- the questions asked of the participants.

International collaborative work has led to the production of the ITU-T P-series of recommendations, which form the basis for many of these tests.

Formal assessment methods are in a continual state of flux. This is caused by the development of new systems and services, which attempt to meet the increasing demand for improved and more accessible communications. Many of these developments affect the way in which speech performance is assessed, forcing existing methods to evolve and new methods to be developed. For example, the use of mobile radio systems has introduced impairments caused by radio channel errors and a wide range of different acoustic environments such as train stations, the interiors of moving vehicles and busy streets. These were not considerations in the development of the fixed telephone network, and traditional methodologies have been forced to evolve, in order to provide relevant information. This process of evolution has led to the development of a wide range of formal methods (see Chapter 8), each designed to address a specific issue.

4.7.2   Selecting appropriate methods

Almost all modern speech quality assessments use opinion-based methods, where the participants are asked directly for their opinion on some aspect of the quality of the system they are assessing. To determine the most suitable methodology, different aspects of the system being investigated have to be considered, for example, the types of impairment that are likely, how and where the system is to be used, and how it will interwork with other telecommunications networks. These factors will all have some effect on the quality of received speech, the ability to speak without difficulty and the ability to converse naturally. It would be an impossible task to cover all of these aspects in any single performance assessment, since the cost and time-scale would be prohibitive. Hence, a balance has to be reached through understanding the relative importance of different factors, the relevance of existing information, and practical considerations such as time and cost.

In the GSM standardization process, there were several different phases of assessment. The majority of these phases were used to select the most suitable speech codec, based on its performance for the more important impairments. Once a codec had been chosen, the emphasis changed to characterizing its performance over a wider range of operating condition. This was to provide more detailed information for potential service providers.

In the half-rate GSM standardization process, the emphasis in assessment was placed on the quality of received speech, with the aim of attaining a system with a performance level comparable to that of the existing full-rate GSM system. This enabled the exclusive use of listening-only tests, where there is no need for the participants to converse with one another.

Experience at BT Laboratories and other laboratories around the world has led to the use of two primary types of listening-only methodologies — those based on listening quality and those based on listening effort. In listening quality assessments, participants listen to speech and then give their opinion as to whether the quality was excellent, good, fair, poor, or bad. In listening effort assessments, the participants are asked how much effort is required to understand the meanings of the sentences they have just heard. The listening effort method provides more discrimination at the lower end of acceptable performance for telephony and hence is typically used to investigate performance for the more critical impairments, whereas the listening quality method provides valuable information over typical operational conditions. Both of these methods were used in the half-rate GSM performance assessment process.

4.7.3   Drawing conclusions

Once an assessment has been completed, the results are analysed using statistical methods to determine what conclusions can be drawn and how best to express them. In the GSM process, the results of this are summarized by Usai et al [18] and in an ETSI Technical Report [19]. The latter serves as an informative document, which supports the recommendations with background information on the performance of the half-rate system. The main conclusion from this information was that the performance of the half-rate speech codec is comparable to that of the full-rate speech codec, except when used in mobile-to-mobile situations and in certain noisy environments.

4.8   FUTURE DIRECTIONS

As the start of this chapter it was shown how the performance requirements of a system depend heavily upon its application. As the mobile telephony market has grown, the proportion of non-business users has increased dramatically. A typical user may no longer be a business person willing to trade quality of service for the benefits of mobility, but a member of the general public who will expect a quality of service similar to that provided by the PSTN. Consider the difference between instructing a broker to sell shares and whispering sweet nothings to a loved one. It is arguable that the latter requires a degree of naturalness not offered by current mobile telephony systems.

It has already been noted that the performance of speech codecs is often biased towards a particular gender, age or language group. As the market-place for mobile communications grows to encompass a greater proportion of the population, both in socio-economic and ethnic terms, this trend may become increasingly unacceptable. Does the concept of an egalitarian society extend to the performance of mobile communications systems?

The mobile communications systems of the future will need to provide communication that is, at the very least, as natural as the current PSTN. This means lower distortion, lower delay, and possibly wider bandwidth. These considerations appear to indicate that a higher data rate must be offered per user. However, in direct conflict with this is the requirement to handle an increasing volume of traffic. For the moment at least, the public's desire for mobile communications would appear to be insatiable, and speech quality improvements will have to be made within existing, and possibly lower, data rates.

The adoption of a small number of global codec standards could reduce the need for unnecessary speech coding stages. However, recent advances in speech coding and device technology have resulted in a proliferation of speech coding standards. With many new markets for speech transmission emerging, perhaps the communications industry should heed the following ironic observation:

> 'The nice thing about standards is that there are so many of them to choose from.'
>
> A. S. Tanenbaum

On a final note, the use of more immersive environments for quality assessment must be considered in the future, so the effects of more than just the acoustic surroundings can be understood. This will be particularly relevant to the design and assessment of mobile radio systems. For example, the effects of the radio shadows created by buildings could be examined by giving participants the opportunity to move around in a virtual environment. Such techniques will enable controlled studies of system performance in more realistic contexts.

4.9   CONCLUSIONS

As the market for mobile communications grows and evolves, customers are increasingly likely to demand a better quality of service from both manufacturers and operators. The designers of speech and channel codecs are already facing many of the fundamental problems presented by the mobile telephony markets of the future. One of the main challenges is to design speech channels that offer near-transparent quality, low delay, and consistent performance over a wide range of error conditions. In conflict with this are the physical, engineering and economic factors that limit the bit rate available to individual users. One of the immediate challenges is to design sub-4-kbit/s speech codecs that are speaker- and language-independent, and, perhaps most difficult of all, virtually unaffected by the presence of ambient acoustic noise and tandem conditions.

In this chapter the authors have described the major considerations involved in speech transmission over digital mobile radio channels. Both the design and assessment of such systems present considerable engineering challenges, requiring a careful trade-off between many competing factors. An appropriate balance is often very difficult to achieve, and must evolve with changing customer expectations. However, achieving this balance is vital to customer satisfaction.

REFERENCES

1.   Jayant N et al: 'Signal compression based on models of human perception', Proc IEEE, 81, No 10, pp 1385-1422 (October 1993).

2.   Markel J D and Gray A H Jr: 'Linear prediction of speech', Springer-Verlag, New York, Heidelberg, Berlin (1980).

3.   Hollier M P: 'Sound fields around the head/handset system and their exploitation for pressure gradient noise cancellation', Proc Inst Acoustics, 12, Part 10, pp 523-530 (1990).

4.   Lim J S and Oppenheim A V: 'Enhancement and bandwidth compression of noisy speech', Proc IEEE, 67, No 12, pp 1586-1604 (December 1979).

5.   Hagenauer J: 'Rate-compatible punctured convolutional codes (RCPC codes) and their applications', IEEE Trans Communication, COM-36, pp 389-400 (1988).

6.   Viterbi A J: 'Error bounds for convolutional codes and an asymptotically optimum decoding algorithm', IEEE Trans Information Theory, IT-15, pp 177-179 (1969).

7.   'Integrated speech and channel coding for personal communications', Dept of Trade and Industry Link Personal Communications Program (1993).

8.   ETSI: 'Physical layer on the radio path; general description (GSM 05.01)', ETS 300 573 (1992).

9.   ETSI: 'Full rate speech Part1: full rate speech processing functions (GSM 06.01)', ETS 300 580-1 (1992).

10. ETSI: 'Half rate speech Part 1: half rate speech processing functions (GSM 06.02)', ETS 300 581-1 (1992).

11. Schroeder M R and Atal B S: 'Code-excited linear prediction (CELP): high quality speech at very low bit rates', Proc ICASSP, pp 937-940 (March 1985).

12. ETSI: 'Half rate speech Part 2: Speech transcoding for half-rate speech traffic channels (GSM 06.20)', ETS 300 581-2 (1992).

13. Gerson I and Jasiuk M: 'Vector sum excited linear prediction (VSELP) at 8 kbit/s', Proc ICASSP, pp 461-464 (April 1990).

14. Linde Y et al: 'An algorithm for vector quantizer design', IEEE Trans Communication, COM-28, pp 84-95 (January 1980).

15. ETSI: 'Half rate speech Part 5: discontinuous transmission (DTX) for half-rate speech traffic channels (GSM 06.41)', ETS 300 581-5 (1992).

16. Freeman D K et al: 'The voice activity detector for the pan-European digital cellular mobile telephone service', Proc ICASSP (May 1989).

17. Hollier M P and Hawksford M O: 'A perception based speech quality assessment for telecommunications', IEE Digest No 1995/089 (May 1995).

18. Usai P et al: 'Subjective performance evaluation of the GSM half rate coding algorithm (with voice signals)', Proc ICC (June 1995).

19. ETSI: 'European digital cellular telecommunications system: half rate speech; performance characterization of the GSM half rate speech codec (GSM 06.08)', ETR 229 (1992).

5

THE LAUREATE TEXT-TO-SPEECH SYSTEM — ARCHITECTURE AND APPLICATIONS

J H Page and A P Breen

5.1   INTRODUCTION

Making machines that can talk has been a goal of engineers and scientists for many decades. The speaking clock, which was produced by the Research Branch of the British Post Office in 1936 [1], was one of the first systems to address this dream, and used analogue recording and playback techniques. The speech was stored using an optical method on twelve inch glass discs (Fig. 5.1).

This system was very sophisticated for the 1930s because, as well as getting the time right, it used concatenated speech to avoid recording all the material for the complete 24-hour cycle. With the technology then available, that would have taken nearly 8 miles of sound track! With improvements in recording techniques, network announcements and recorded information lines have become commonplace. However, it was not until the development of digital recording that it became possible to easily manipulate recorded speech on computer systems. Prior to that, the use of continuously running looped tape systems, for recorded announcements, resulted in callers being connected at any point within the message. Today, sophisticated interactive systems, capable of extensive dialogues using fixed recorded messages, are being deployed. However, when there is a need to utter a variable piece of speech, it is very difficult to make it sound natural with simple recordings and concatenation techniques. In such circumstances, the synthesis of natural sounding speech from a textual input becomes a desirable capability, but one that is several orders of magnitude harder to achieve than speaking a prerecorded message.

Fig. 5.1   Picture of 1936 speaking clock.

Several factors complicate the process, but the main one is that the media of speech and text have quite different purposes. Speech is the primary **human communication mechanism**, whereas the main purpose of text is to **record information**. The conversion from one medium to the other is rarely a one-to-one mapping — compare the work required to turn a book into a play.

Early attempts at synthesis were based on mechanical analogues of the human vocal tract (see Chapter 2), to be later replaced by sophisticated electronic models of the acoustic signal [2]. Provided that accurate parametric information on the speech signal was available, such models could produce natural-sounding synthetic speech. Unfortunately, these advances in the synthesis of speech were not matched by a corresponding increase in the understanding of how to automatically generate the appropriate parametric information to drive these synthesis models. As a consequence, the quality of synthetic speech generated from text sounded robotic and unnatural. With advances in computer technology, doubling processing power every 18 months or so, the possibilities for modelling, encoding and storing fragments of the speech signal have led to improvements in quality and a resurgence of interest in text-to-speech generation. However, the depiction of HAL, the talking computer in '*2001: A Space Odyssey*' by Arthur C Clarke, still provides a quality target for text-to-speech synthesis systems.

Today, the quality of such synthetic speech has improved to such an extent that real-world applications are now viable and the subject has moved from being a predominantly academic discipline to one of increasing importance to the tele-communications and computer industries. This shift in emphasis has led to a need for industry to evolve methods of research and development which enable novel ideas to be investigated, tested and developed, while simultaneously ensuring that products can be produced rapidly and cost effectively. The solution to this problem, adopted by BT Laboratories, is described later in this chapter. But first, a description is given of some applications where the technology can produce some real advantages over other forms of speech output.

5.2   APPLICATIONS FOR SPEECH SYNTHESIS

The applications for a synthetic speech system which provides human-quality speech are unlimited. Some of these applications place the synthesis at the point of use, e.g. talking machines for people who have lost their voices, or reading machines for the visually impaired. Many personal computers also have speech synthesis, and this use probably constitutes the single largest market for text to speech today. However, with the convergence of communications and computing technology, the use of text to speech within a communications system produces the greatest benefit. Using an ordinary telephone, of which there are more than 26 million in the UK and over half a billion world-wide, speech synthesis enables everyone to access network-based information sources.

The quality of the voice synthesizer influences the choice of application. Clearly, if the synthesizer output was equal to that of a good-quality recording, then any system requiring voice output would use it, if only for the generation of stored voice prompts. However, since the voice quality produced by text-to-speech synthesizers is, in general, not yet up to the standard of a good-quality recording, applications have to be carefully matched to the quality of the voice. To a certain extent, the choice of application depends on how important it is to the user to get the information, as well as its type and presentation. To illustrate this point, consider talking books for the blind and visually impaired. Most sighted people would choose to read a book themselves given a choice of delivery, but they may also choose to have a book read to them by another person. Blind and visually impaired people are also likely to prefer a human reader, but would probably make use of a speech synthesizer if they needed access to the information, and no recordings were available. However, some people prefer a synthesis system where they can control the speed of the speech, as this enables them to speed it up and absorb the information at a higher rate.

Another factor involved in matching text to speech (TTS) to the application is the quality of textual information. If the text consists of short sentences and phrases with good punctuation, then the quality of the synthesis will be perceived as being high. On the other hand, if the text is poorly conceived and difficult to read out loud for a human reader, then the synthesis quality will be similarly impaired.

5.2.1   When to use synthetic speech

There are a number of situations where synthetic speech is likely to be selected in preference to the simpler technology solutions for providing speech output.

- Where the text is unpredictable and dynamic

    Text-to-speech can be used in situations where the messages are short, typically a few lines, but the information content varies significantly and cannot be constrained to a standard format. Here the text is totally unpredictable and a speech-synthesis system is the only feasible delivery method.

- Where access is required to a large database of information

    For large databases it is not feasible to store the information as recorded speech due to the costs of recording and storage. Large databases are rarely completely stable and this also favours a text-to-speech approach.

- Where the output is relatively stable, but cost of provision and lead time are critical

    An example of this is a telephone network announcement system where most of the announcements remain static, but situations can arise where new messages are necessary, often at short notice, and must be in the same voice. Voice synthesis greatly simplifies the process of generating fixed-voice prompts for new services, as compared with live recordings which take a significant time to prepare.

- Where consistency of voice is a requirement

    Many services require the same voice for all announcements. This is not a problem if the service, once built, requires no future modifications. If, on the other hand, future enhancements are envisaged, there may be a problem with the availability of the original speaker. In this situation it may be better to choose a synthesized voice in the first place.

- Where low bandwidth is required

  By transmitting information in text form and converting to speech at the receiving end, only a very low bandwidth is used.

  The next section gives more detailed descriptions of a number of applications which are well suited to using text-to-speech synthesis.

5.2.2   Example applications

5.2.2.1   Information lines

The importance and conciseness of the information is key to this type of service. With a suitably interactive dialogue (user input via speech recognition or TouchTone® — dual tone multifrequency (DTMF) tones generated from the telephone keypad), the user can pinpoint the information required. This also means that the information can be provided in convenient chunks appropriate to the synthesis process.

There are currently many services available over the telephone, distributing all kinds of information. Those with unchanging messages are best served using recordings. Other services, such as weather forecasts, are updated two to three times a day, and an automatic text-to-speech conversion system is a practical lower-cost option in these cases.

A prototype system, developed at BT Laboratories (BTL), demonstrates access to detailed, localized weather information. This is achieved by providing a system with forecast areas linked to dialling code areas. Access to the weather forecast for the region of interest is simply obtained by keying in the relevant dialling code.

Other types of information, such as stocks and shares prices, vary continuously but have a fixed format. Here speech concatenation techniques can be appropriate. An example of a service giving the current price of some item might be:

'*The price of gold is <n> dollars/ounce*'

where the main phrase is a single recording, but has the variable *n*, which will be one of a set of recorded numbers embedded in it. These concatenation systems can work well provided that the information content is simple. With more complex examples employing a larger number of fields with greater variability, the resulting speech quality becomes disconcerting with disjointed intonation and impairment.

The use of interactive systems, employing either speech recognition or DTMF tone detection for user input, will enable sophisticated information services to be provided. Here it will be possible to extract information from large databases which are too extensive to be provided in a pre-recorded form.

5.2.2.2   Automated catalogue ordering system

The extra information given to the customer, by using text to speech in this application, is likely to lead to a far higher level of customer confidence and satisfaction. It is also a good example of an application where information flows in both directions, as well as being an example of accessing a large database. Automated catalogue ordering systems can be provided with relatively simple technology, using DTMF tone detection to input the number of the item to be ordered and speech output via digit concatenation to confirm the correct receipt of this number. When TTS technology is used, a far more user friendly system can be constructed. With the TTS system the customer would enter the item number by speech or DTMF tone detection as before, but in this case the system responds with a full spoken description of the item ordered. This not only confirms that the correct item number has been recognized, but also that the customer had looked up and entered the correct item number from the catalogue. The TTS-based system would also read out the name and address to whom the order was to be sent, so that, at the end of the transaction, the customer would be confident that the right item had been ordered and that it would be sent to the right address.

5.2.2.3   Remote e-mail reading

Speech access to e-mail from any fixed or mobile telephone can only be achieved by using text-to-speech synthesis. It is particularly useful for people whose jobs take them away from their offices, but who do much of their communication via e-mail.

A prototype telephone access to e-mail systems has been developed at BTL, which provides both an experimental platform and a demonstration system. A user interacts with the current prototype via a combination of DTMF tone detection and speech recognition. An advantage of using DTMF tones is that they are reliably detected in the presence of outgoing speech. Consequently, comprehensive prompts which have been provided for novices can be interrupted to give faster access for the experienced user. The facilities provided by the system are as follows:

- PIN code security;

- review of messages by sender or subject;

- read-out of e-mail with the facility to go back, repeat or go forward;

- facility to delete messages;

- facility to send messages by facsimile to any destination.

Additional facilities which would be provided in a fully featured system include:

- voice-mail reply with an e-mail notification;

- facility to file messages.

A highly desirable feature is the facility to provide a spoken reply to e-mails, with the speech converted to text using speech recognition. However, since current speaker-independent recognition systems can only have a restricted vocabulary (see Chapter 8), a compromise solution of a voice-mail reply along with an e-mail notification of the voice mail, would be a useful future extension.

5.2.2.4   Personal navigation systems

With the advent of inexpensive global positioning system (GPS) receivers [3] and mobile phones, it is possible to provide navigational guidance facilities, with one basic arrangement, for a variety of navigation-related applications. As shown in Fig. 5.2, this consists of a GPS receiver combined with a mobile phone linking back to a central computer located within the telephone network. The mobile phone transmits the current position from the GPS receiver to the central computer which then reads directions via a text-to-speech synthesizer to direct the user to the previously entered destination.

Other navigation systems link a GPS receiver directly to an on-board computer which stores map and routeing information; but the advantage of using a network-based computer for the information storage is that it can be constantly updated with any relevant information. This would be particularly useful in a car navigation system where varying road traffic information could be taken into account when giving directions. Speech-based systems also have obvious safety advantages over in-car electronic map systems which require drivers to take their eyes off the road.

Blind and visually impaired people could also benefit from a similar system, by using the more accurate differential GPS systems [4]. These are sufficiently accurate to guide them along the street avoiding all fixed objects such as lamp posts.

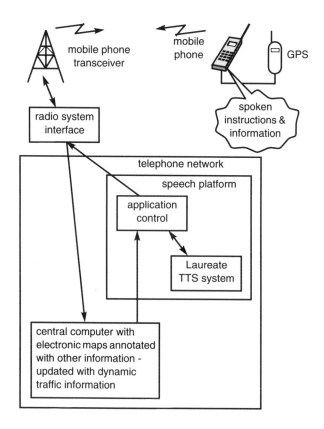

Fig. 5.2   Personal navigation and information system.

Navigation systems of this type are universally applicable, providing people with voice-based personal guidance systems which would be just as useful navigating a route in London as it would when out walking in the countryside. It would even be possible to have guided countryside walks along with commentaries about interesting features along the route. Guides to the local pubs and restaurants could also be included.

5.2.2.5   Natural language or conversational interactive voice response (IVR) systems

One area of current IVR research (expanded on in Chapter 17) is directed towards making systems more conversational, allowing the user to speak to the machine in a more natural manner, unconstrained by today's menu-driven approach. Natural-language systems will almost inevitably require the use of on-line text-to-speech synthesis to utilize the greater variety of response provided.

5.2.2.6   Synthetic persona

This application provides a lifelike synthetic talking face (Fig. 5.3). The head and shoulder images are generated using a polygonal wire-frame model on to which

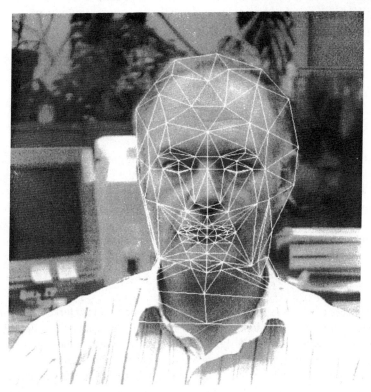

Fig. 5.3   Picture of synthetic persona.

the image of a person is mapped [5]. If the wire frame is rotated or translated and the stored image texture projected into a moving sequence of images, a convincing 3-D effect is produced. This is achieved despite only two images being used initially for the texture mapping.

Moreover, by transforming the positions of sets of the vertices of the wire-frame model resulting in distortion of the surface texture, synthesized facial expressions can be generated. Linking this synthetic facial image of a talker to the synthetic speech of the same talker produces a very compelling and entertaining way of providing computer-based information. The basic source data driving the system is computer-stored text, but appears to be delivered to the user by a human presenter.

This arrangement can be used to provide a more personal, human interface to network services, an example being an automated announcer for video-on-demand services. By providing suitable helper applications to World Wide Web (WWW) browsers, this type of system could also be used for the presentation of information in an interesting way.

Laureate, BT's text-to-speech system, which has an architecture designed to support these communications applications, is described in the next section.

5.3   ARCHITECTURE OF THE LAUREATE TEXT-TO-SPEECH SYSTEM

A simple description of a text-to-speech system and the associated technical challenges introduces this section. Subsequently, the philosophy and structure of the Laureate system architecture is presented. Two companion chapters describe the algorithms employed in text-to-speech systems. The first deals with text and linguistic analysis (Chapter 6) and the second, with prosody and speech generation (Chapter 7).

5.3.1   Technical challenges

The process of text-to-speech conversion can be broken into three main parts, as shown in Fig. 5.4. The first activity is text analysis (more fully described in Chapter 6), where the underlying linguistic structure is determined and a phonemic transcription of the text is produced. Next come the parallel processes of prosody synthesis and speech sound selection. Prosody synthesis generates the intonation and rhythm of the passage to be synthesized, while speech sound selection involves choosing speech fragments from a recorded speech database. Finally the speech waveform is generated by smoothly joining the selected fragments of recorded speech and imposing the prosody on to it using speech modification algorithms. Speech and prosody generation are discussed in detail in Chapter 7.

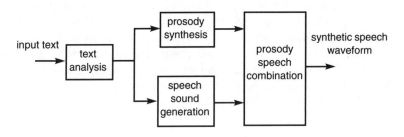

Fig. 5.4   Text-to-speech conversion system.

Stated simply, there are two problems facing text-to-speech system developers — technical constraints and theoretical unknowns. With the recent increase in computing power and the availability of cheap storage, many of the technical constraints have been solved or at least significantly eased. However, there has not been an equivalent increase in our understanding of the production and perception of speech. A number of the latest developments in text-to-speech have tended towards encoding ignorance rather than attempting to model the underlying mechanics. This has led to a divergence in system design, those systems which attempt to understand and model the speech signal (e.g. formant synthesis) and those which attempt to encode significant aspects of the speech signal (e.g. concatenative synthesis). Neither approach has, as yet, been shown to be clearly superior to the other.

By far the most significant advance in the generation of synthetic speech in recent years has been as a result of the impact of low-cost, high-volume storage media, combined with efficient speech-modification algorithms. This combination of technology gave a step increase in the perceived naturalness of synthetic speech, with the added bonus that different speakers could be modelled comparatively easily. For the first time speech researchers were able to assess the importance of the prosodic structure of language without being encumbered by the comparatively poor segmental quality which masked significant effects. It is now clear that high-quality natural-sounding speech synthesis will not be achieved through the simple concatenation of segments of sampled speech data. The acceptable segmental quality of the current generation of speech modification algorithms has forced researchers to address significant problems in the specification of synthetic prosody in speech synthesis.

5.3.2   Laureate system architecture

The Laureate text-to-speech system has been designed specifically with the aim of allowing a number of different models of spoken language to coexist within the same computational framework. The basic design premise of Laureate is shown in Fig. 5.5. The different-sized balloons represent different shades and types of linguistic theory, which in isolation are incompatible with each other. The purpose of the Laureate core and component interface is to provide a consistent and sufficient minimum set of standard linguistic representations which are common across the different models and production methods, so allowing different theories to coexist.

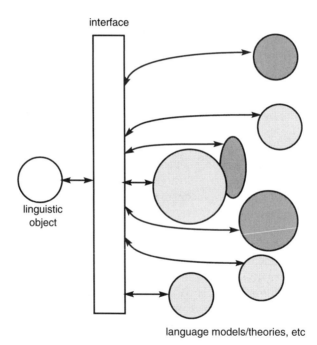

Fig. 5.5   Laureate's basic design premise.

There are a number of potential problems in mixing different theories:

● each theory will require its own internal data structures;

- different theories may have different and even conflicting linguistic definitions;

- different theories will require different amounts of given knowledge.

Figure 5.6 shows diagrammatically how the Laureate architecture addresses these problems. The Laureate system consists of two main parts, a core component and a number of satellite components. The core represents the backbone of the system. All information within the system passes from one component to another via the linguistic object contained within this core. The linguistic object is a dynamic database of information, which has a simple understanding of the relative hierarchical structure of language.

Satellite components can only access the object via a set of formalized linguistic questions or statements. This means that each satellite component is completely isolated from the rest of the system. It may therefore have its own internal representational linguistic structure, which need not necessarily conform to that of any other component within the system. The formalized linguistic interface ensures that data returned to the linguistic object can be used by other components of the system in a consistent manner. Due to the modular nature of the system, components may be placed at different points along the backbone. Thus, components that require large amounts of linguistic knowledge can be placed after components that add such knowledge to the linguistic object.

The linguistic object expands and contracts as text blocks are processed by the system. A text block represents the minimum amount of text which contains sufficient linguistic information for the system to function correctly. Text blocks are determined within the text normalization component of the system and may vary in size from single words to paragraphs. As a text block cascades through the system, each satellite component loads information into the linguistic object. All components, with the exception of the realization component, are only concerned with adding information to the object. Once the realization component has completed its tasks, all information in the linguistic object will be cleared in preparation for the next text block.

The linguistic object is, in effect, a relational database which has no understanding of the information contained within each database record, but only allows certain linguistic relationships between records to exist, for example, 'words contain syllables'. This relationship is reflected in the way word and syllable records relate. As the specific meaning of a piece of information contained in a record may change as satellite components develop, the meaning of record elements is stored externally to the linguistic object so that it is common to all components.

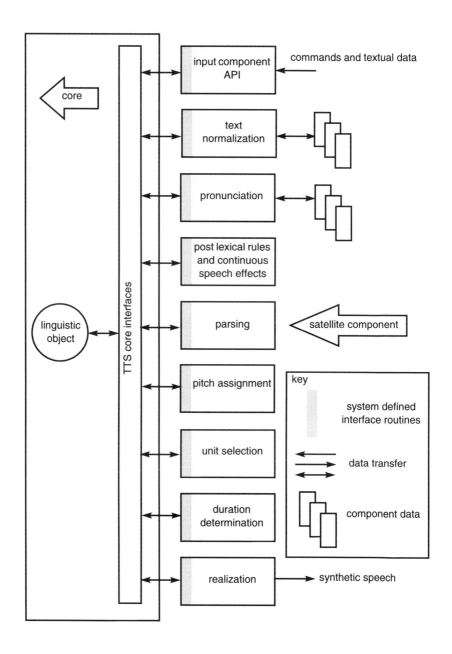

Fig. 5.6 The Laureate architecture.

As a practical example of the above, consider the generation of intonation within the Laureate text-to-speech system. There are a number of different theories on how best to specify and generate an intonation contour. Two prominent models are the Fujisaki model and the Silverman [6] model. Basically, the Fujisaki model consists of two components — an accent component and a phrasal component. Values for both of these components must be specified at the outset. A contour is generated by superimposing accent commands of different length and height, on a basic phrasal component which extends over the total length of the phrase. In contrast, the Silverman model views intonation as a realization of a linear sequence of tonal accents, and so does not contain a phrase level component. A contour is generated by interpolation between specified tones. The two theories clearly make very different assumptions about the processes leading to the generation of intonation, but the Laureate system can accommodate either model as follows:

- two components are commonly used in the generation of intonation within Laureate — the pitch assignment component and the realization component;

- the pitch assignment component generates the accent information needed to generate a contour;

- the realization component generates the actual pitch values as dictated by the accent information found in the linguistic object.

Thus, either theory can be accommodated in the Laureate system by simply including the appropriate satellite components. While the meaning of the information contained in the linguistic object has changed, structurally the systems are identical.

Another advantage of the theory-independent structure of Laureate is that it is also a language-independent structure. This minimizes the changes required to implement synthesis in alternative languages. Simplistically, the structure of Laureate can be viewed as consisting of three parts, as in Fig. 5.7 — the linguistic object (data store), the satellite components (language processing engines) and the language-specific data. The meaning of information stored in the linguistic object is stored as part of the language-specific data. By separating the data from the linguistic object the Laureate architecture provides a structure which can be more easily tailored to different accents and languages. The aim of the linguistic object is to provide a language-independent store, while the aim of the components is to manipulate language-specific data in such a way as to provide new information which can be stored in the linguistic object.

In practice, it is impossible to generate completely language-independent components, but this architecture ensures the maximum reuse of code. For example, if a satellite component needs to be completely replaced this can be done simply without significantly affecting the other components. If extra components are required these may be inserted without modification to existing ones.

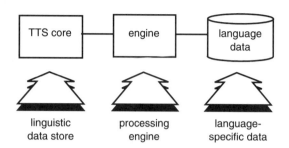

Fig 5.7   Relationships between linguistic data store, processing components and language-specific data.

The Laureate system has been designed to meet the challenge of developing a software suite, which can act as an effective research tool and provide the basis for robust and cost-effective applications. The list given below highlights some of the design characteristics provided within the Laureate architecture to achieve this aim.

- Modular design — this, coupled with the concept of a linguistic object, means that each component can be developed in isolation. Consequently the cost of upgrading the system, once it has been downstreamed to a platform, is greatly reduced, as the cost of downstreaming a new or improved component is predominantly the cost of the component testing.

- Execution sequence flexibility — the ability to change the order of execution, insert or remove components without the need for recompilation provides a flexible structure, which will be able to accommodate new ideas as they appear and extend the lifetime of the code.

- Industry standard software — the code has been written in ANSI standard C, which is available on most hardware platforms. Languages other than 'C' may be included in the system provided they are capable of linking with 'C'. For extra flexibility, each satellite component in the system may be configured without recompilation to run as a separately scheduled task.

- API simplifies integration into applications — the Laureate system includes an API (application programming interface) which eases integration with application software. The application developer may view the Laureate system as a black-box component.

- Multichannel — the code has been designed to support multichannel operation. Multiple instances of the Laureate system may exist simultaneously within the same or different applications. An application may choose which instance of the Laureate system to use at any one time. The Laureate API code will manage all system resources required to support multiple channels as efficiently as possible.

The linkage of a flexible, portable code structure to a clearly motivated and eclectic theoretical design enables the Laureate code to meet the needs of industry, while acting as an effective research and development tool.

5.3.3   Tailoring Laureate for specific applications

5.3.3.1   New voices, accents and languages

The Laureate technology and architecture permits new voices, accents, and languages to be incorporated into the system with minimal changes. New voices are introduced by recording the voice of the chosen speaker reading a specially devised script, which has a good coverage of the sounds of the language. This recorded voice data is phonetically annotated before incorporation into Laureate. For regional accents a new voice is required, but depending on the characteristics of the accent, modifications to the pronunciation (lexicons and rules) and pitch assignment systems may also be necessary. A different language also requires changes in areas such as text normalization and parsing. The core system does not require changes for any accent or language. Some languages and accents will only require different data files. The processing components, which, in general, have been designed to be data-processing engines operating on rules and data, may occasionally require modifications where language-specific features have been unavoidable.

5.3.3.2   Application-specific data

The Laureate text-to-speech system has been designed to enable ease of tailoring to specific applications. To this end, its behaviour can be significantly altered simply by changing data files. Wherever possible these are text based, making it easy to change or enhance the content. The Laureate system has the following main types of data file:

- configuration file;

- abbreviation dictionary;

- acronyms dictionary;

- pronunciation dictionary;

- pronunciation rules;

- speech files.

The configuration file determines which of the other data files are used as well as controlling the behaviour of the pronunciation, parser and intonation components. The speech output file format is also selected in the configuration file.

The abbreviation dictionary can be significant to the behaviour of the system in certain types of application, as it controls the way abbreviations are expanded. Abbreviations can pose problems for a text-to-speech system as they rarely have a unique expansion. The common examples of 'st.' and 'dr.' are good examples as they both commonly have two expansions: Saint and street for 'st.' and doctor and drive for 'dr.' Intelligence is built in to the text normalization component of Laureate to resolve these common ambiguities [7]. However, in many specialist areas abbreviations are reused with their own local meanings in a way that is impossible to disambiguate. In these situations a specialist abbreviations dictionary must be generated just covering the area of the application.

A dictionary is provided for acronyms that are pronounced as a complete word such as 'DEC', and 'ASCII'.

Pronunciation dictionaries can also be tailored to the application. An example of when this is useful is in an interactive voice response (IVR) system, which confirms a name by reading it back using synthesis. In this situation the only option is to have large name dictionaries to get good synthesis performance, as names in the UK are notoriously difficult to pronounce by rules. However, the Laureate system uses a specialist technique for name pronunciation called 'pronunciation by analogy' which can perform significantly better than normal English pronunciation rules.

5.4   LAUREATE PERFORMANCE

Measuring the performance of text-to-speech systems is not a straightforward matter. An obvious measurement is intelligibility, which did provide a useful measure for the early systems. The quality of today's text-to-speech systems are still not as good as real speech, but most systems are intelligible most of the time. Consequently, a measurement is needed that can extract the differences between systems that have a performance between basic intelligibility and high-quality recorded speech.

The techniques used by BTL to measure the performance of speech synthesizers are based on methods that have been devised over many years for the measurement of the performance of telephone system components. These tried and tested techniques have been chosen for their reliability and independence from the ability of the subject rather than the performance of the TTS system under consideration.

One of the factors often overlooked in this type of testing is the sound level. This has a significant effect on the 'listening effort' and therefore measurements must be done over a range of levels for meaningful results to be obtained. The subjects are asked to rate the speech in terms of the effort required to understand the meaning of sentences. They are given a five-point scale, where the points have the following descriptions:

1   complete relaxation possible — no effort required;

2   attention necessary — no appreciable effort required;

3   moderate effort required;

4   considerable effort required;

5   no meaning understood with any feasible effort.

Measurements obtained in this way are reliable and consistent, when used in combination with a reference system, as shown in Chapter 8. The performance of Laureate and many leading commercial speech synthesis systems have been measured using these techniques. At normal listening levels the performance of Laureate is better than any other system tested and this is shown in the graphs of two different performance tests against four other commercial synthesis systems (see Figs. 5.8 and 5.9).

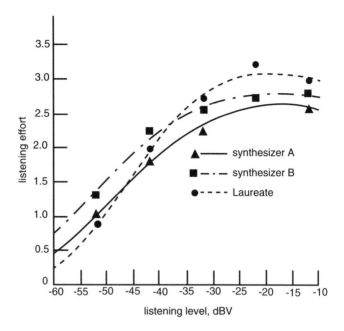

Fig. 5.8   Sample synthesizer performance test results.

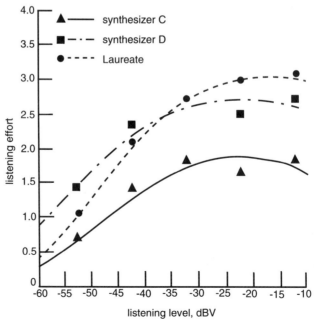

Fig. 5.9   Sample synthesizer performance test results.

5.5 CONCLUSIONS

Text-to-speech synthesis is generating increasing interest due to its ability to transform information from one medium to another. This is of growing importance in the move towards a truly multimedia world, where people will want information presented in the medium of their choice. The possibilities for applications delivering computer-based textual information over telephone networks are almost without limit. This makes text-to-speech synthesis of great importance in an information age, where the major form of communication used by customers is still speech, using the ubiquitous telephone. This chapter has given examples of some applications where text-to-speech synthesis can currently be used to greatest benefit. The unique design philosophy and structure of the Laureate text-to-speech system, which can support both downstreaming and continuing research, has been presented, as well as the ways in which the Laureate synthesis system can be tailored to applications.

The Laureate system has been easily integrated into a number of different prototype systems which include telephone access to e-mail, weather forecast, road traffic congestion and medical information lines, and WWW pages. It has also been ported to a number of Unix™ workstations and personal computers as well as to the BT speech applications platform [8] where other speech processing algorithms can be called on by application programmers building speech-based services.

Two companion chapters describe the modern techniques used to perform the conversion of text into speech. The first deals with text analysis (Chapter 6) and the second, prosody and speech generation (Chapter 7).

Finally, with the accelerating advances in synthesis technology, the predictions of Arthur C Clarke may well be fulfilled, and a speaking machine may yet sound as convincing as HAL by the year 2001.

REFERENCES

1. Magnusson L E, Speight E A and Gill O W: 'The speaking clock', The Post Office Electrical Engineers, J, 29, Part 4 (January 1937).

2. Klatt D: 'Review of text-to-speech conversion for English', J Acoust Soc Am, 82, No 3, pp 737-739 (September 1987).

3. Hurn J: 'GPS — a guide to the next utility', Trimble Navigation (1993).

4. Hurn J P: 'Differential GPS explained — beyond prompt and response', Trimble Navigation (1993).

5.   Welsh W J, Simons A D, Hutchinson R A and Searby S: 'Synthetic face generation for enhancing a user interface', Proceedings of Image Com 90, Bordeaux, France, pp 177-182 (November 1990).

6.   Silverman K: 'The structure and processing of fundamental frequency contours', PhD thesis, University of Cambridge (1987).

7.   Gaved M: 'Pronunciation and text normalization in applied text-to-speech systems', Proc Eurospeech'93, Berlin, pp 897-900 (September 1993).

8.   Rose K R and Hughes P M: 'Speech systems', BT Technol J, 13, No 2, pp 51-63 (April 1995).

6

OVERVIEW OF CURRENT TEXT-TO-SPEECH TECHNIQUES: PART I — TEXT AND LINGUISTIC ANALYSIS

M Edgington, A Lowry, P Jackson, A P Breen and S Minnis

6.1 INTRODUCTION

Speech is the primary communication method of human beings, evolving in *Homo Sapiens* at least 100 000 years ago, making it the most natural and powerful expressive medium between people. In contrast, writing has been part of human culture for only a few thousand years, and originated as a way of permanently recording information, since speech is an inherently transient phenomenon. In the modern world, extremely large amounts of information are stored as text, and an increasing amount of this is stored electronically. A method of converting this electronically stored text, in any major language, into high-quality natural-sounding speech would provide an extremely useful tool for many applications. This is the long-term goal of many speech scientists and engineers, although current text-to-speech (TTS) systems have a long way to go to reach this goal. However, commercial systems, such as BT's Laureate (see Chapter 5), are capable of producing highly intelligible speech which, under ideal conditions, has a naturalness approaching that of a human speaker (see Chapter 8).

The last decade has seen a substantial improvement in the naturalness of synthetic speech, due largely to the availability of high-quality speech concatenation algorithms [1]. The resulting improvement in segmental quality has provided an impetus to the development of more complex and subtle intonation and phrasing models. This paper presents an overview of the algorithmic techniques used by state-of-the-art commercial TTS systems in converting from the input text stream to the final speech signal. Chapter 5 describes major applications of TTS synthesis and details a particular system under development at BT called Laureate. The authors have all been actively involved in the development of the BT Laureate system, and this chapter and Chapter 7 consequently have a slight bias towards the methods used in that system, though they focus on techniques that can be universally applied. A description of earlier work in TTS synthesis can be found in the excellent review paper by Klatt [2] which covers the fundamentals, and developments up to 1987.

The focus of this chapter (and its companion — Chapter 7) is on 'real world' commercial TTS systems, rather than systems designed purely for research. In this arena some issues rarely addressed in research-only systems become significant. In particular, robustness, maintainability and computational complexity of algorithms are important to commercial developers, while the primary motivation for a research-only system is often to provide a test bed for a particular underlying theory. This is not to say that a commercial system will not employ the latest research — for example, BT's Laureate TTS was designed explicitly to accommodate different theoretical models within its processing architecture to encourage its use within the research community.

Multilingual TTS systems are becoming increasingly important, especially within Europe, and although this chapter concentrates on English, most of the techniques described are equally applicable to many other languages.

6.1.1   Overview of TTS conversion process

The process of generating synthetic speech from unrestricted text is essentially a two-stage process, as shown in Fig 6.1.

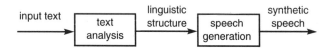

Fig. 6.1   Text-to-speech as an analysis and synthesis process.

* The first stage analyses the input text to determine its underlying linguistic structure, including the sound sequence, or phonemic representation, of every word in the text. This stage is sometimes referred to as 'text-to-phoneme' conversion, but this does not adequately describe the whole process, as it neglects other parts of the linguistic structural analysis.

* The second stage synthesizes a speech waveform from this linguistic structure. This synthesis step can be further subdivided into two steps, that of **prosody generation,** which generates the intonation and rhythm of the

speech, and **speech generation,** which produces the final speech waveform. This whole stage is sometimes referred to as 'phoneme-to-speech' conversion in the literature.

Consequently, the subject is split into two chapters, following the structure of Fig 6.1. Part I, which includes this introduction, describes techniques for text and linguistic analysis, and Part II (Chapter 7) describes techniques for the automatic generation of prosody and the speech signal.

6.2   TEXT ANALYSIS OVERVIEW

The text analysis stage of a TTS system takes the input text and analyses it to generate a linguistic description and pronunciation for the whole text. This information is used by the prosody generation and speech synthesis stages to produce the final speech. Traditionally, much research effort has gone into the speech generation aspects of the TTS process, while text analysis for TTS systems has received less attention. However, as the quality of synthesis for short phrases and sentences has improved, it has become increasingly clear that the text analysis stage is crucial in improving and maintaining TTS performance for longer stretches of text covering many sentences or paragraphs [3].

The text analysis stage can be split into the following four tasks.

Text segmentation and normalization — the input text must be split into a hierarchy of segments: at the highest level are paragraphs, which are split into sentences, which are further divided into words. Decisions have to be made at each stage to determine the exact boundaries. In parallel with this segmentation, the text is normalized, which involves the conversion of abbreviations, digit sequences, and other symbols into a predefined format, and introduces a standard treatment of punctuation. It is essential that the text segmentation and normalization stage works correctly for unrestricted text, as the rest of the TTS relies on the output of this stage.

Morphology and pronunciation — pronunciations must be generated for every word in the text so that they can be spoken. This is partly achieved by using a pronouncing dictionary or lexicon, but unfortunately most languages have a very large and ever-increasing number of words, making it impossible to produce an absolutely complete dictionary. Therefore the dictionary is supple-mented by one or more rule- or statistic-based pronunciation methods. To aid this process, a morphological (or structural) analysis of the words is usually performed. This determines the 'root form' of every word, e.g. love from loves, and thus it allows the dictionary to store just headword entries, rather than all derived forms of a word. At this stage other word-level information is normally generated, such as part-of-speech (or syntactic) and possibly lexical semantic infor-mation. Certain sequences of words may also be merged into short lexical phrases, for example it is easier to treat the phrase 'to and fro' as a single unit, than as a sequence of three separate words.

Tagging and parsing — the morphology and pronunciation tasks produce a sequence of words (or lexical phrases), each with a pronunciation and some syntactic information. For further analysis it is necessary to consider the overall structure of the sentence, which can give a clue to its meaning. The first part of this task is to unambiguously determine the part-of-speech for each word in the sentence, a process known as tagging. This has the added benefit that most homographs (words spelt the same way but with different pronunciations) differ in their part-of-speech. For example, 'live' rhyming with 'give' is used as a verb, while 'live' rhyming with 'hive' is used as an adjective. Thus, in most cases, if the part-of-speech for the word is known, the correct pronunciation should also be known. The second part of the task is to build up a syntactic parse which represents the sentence structure. The syntactic parse represents the structural meaning of the sentence, so it can identify phrases, which phrases are linked, and how they are linked within a hierarchy.

Continuous speech effects — the pronunciation task produces phoneme transcriptions for individual words (or lexical phrases), as if each word were spoken in isolation. However, there are many processes in continuous speech which operate across word boundaries, or modify the isolated word pronunciation in some other way, dependent on the syntactic function of the word in the sentence. The so-called weak forms (consider the difference in pronunciation of 'the' in 'the apple' and 'the car') are an example of the latter class of processes. It is important to model these effects in high-quality TTS systems, otherwise the synthetic speech can sound over-articulated and stilted.

6.2.1   The variability of input text

In a truly unrestricted TTS system, there are very few strong assumptions that can be made about the input text, other than the basic character encoding (e.g. ASCII). A text stream is essentially flat and linear, but is normally used to convey information that has a structure, so a number of 'orthographic conventions' are employed by the writer. Some of these conventions are well-defined, such as marking the end of a sentence with a full stop; others are rather more ill-defined, such as the use of digits rather than words, and the use of intra-sentence punctuation [4]. For example, sentences *a* to *d* below have different punctuation and number formats, yet have exactly the same meaning. This does not mean that punctuation can be ignored, since sentence *e* only differs from *a* in commas, yet has a significantly different meaning. (The first four sentences give the husband's age as additional information, while the last implies polygamy; it is the twenty seven year old husband who has left, not one of the others — the first four sentences have a 'non-defining relative clause', but the last has a 'defining relative clause'.) This distinction can be clearly made when speaking by using a different intonation for each sentence type, and inserting a pause before the relative pronoun 'who'.

a   My husband, who is 27, has left me.

b   My husband, who is twenty-seven, has left me.

c   My husband — who is twenty seven — has left me.

d   My husband (who is 27) has left me.

e   My husband who is 27 has left me.

The traditions of the original media, and the formality of the text can also have an effect on which conventions are used. Figure 6.2 shows three text extracts: from a local newspaper, a personal e-mail, and a novel — the original formatting has been retained as far as possible. Most readers should easily identify which extract comes from which source, from their content, style and format. It is clear that different conventions are used in each extract.

The newspaper text (Fig. 6.2 (a)) uses different fonts to distinguish the headline from the article, and typically there is no punctuation in the headline. The first word of the article is written entirely in capitals, new paragraphs are indicated by first line indentation, and there are often many words broken across lines, such as the word 'summer' in the first paragraph. It should be noted that large local and national newspapers are produced using an editorial style guide, so there is consistency between all articles in a particular paper, but not necessarily between papers.

There is an interesting contrast in the e-mail (Fig. 6.2 (b)) between the strict format of the header and the very informal, poorly punctuated text. Two typical problems in e-mail occur in the extract, the infamous 'smiley' face (used as an irony marker) and the signature footer, which often contains ASCII graphics. The specialized format of e-mail lends itself to the use of a preprocessor to extract the important header information and the message body [8].

The extract from the novel (Fig. 6.2 (c)) is very similar to what would be regarded as 'normal' text, although this extract was chosen to show some of the features introduced by written dialogue. The dialogue seen in this extract is mainly written in complete sentences (unlike real speech) and has a different style to the quotations that appear in journalism.

The role of the text normalization process is thus to impose some kind of order upon the unpredictable input. This then allows the rest of the system to be insulated from the 'messy' world of the original text.

Pot-plant tree has peach of a summer

WHILE many plants are withering in the dry conditions this summer a couple from Snape have finally seen their peach tree emerge in all its glory.

John and Jane Ling planted a peach stone in a pot about eight years ago.

It was later planted in the garden but, as far as fruits were concerned, it was something of a disappointment

(a)

```
From mde@paris Mon Jun 26 09:59 BST 1995
Received: by paris
(1.3.99.11/1.22) id AA0194; Mon, 26 Jun 1995
09:59:54 +0100
From: Mike Edgington <mde@paris>
Return-Path: <mde@paris>
Subject: Gloat
To: pj@paris, devaney@paris, pd, guillem@paris
Date: Mon, 26 June 95 9:59:54 BST
Mailer: Elm [revision: 70.85]
Status: RO

Dear all,

                    Oh dear, I seem to have won our
little rugby world-cup competition.
you can all congratulate me when I get back :-)

          mike.
===============================================
Mike Edgington, Speech Technology
 Tel: +44 1473 640000
BT Labs, Ipswich, UK.
 E-mail: mde@dwarf.bt.co.uk
```

(b)

He held out his hand, and I saw in the light of the lamp that two of his knuckles were burst and bleeding.

"It's not an airy nothing, you see," said he, smiling. "On the contrary, it is solid enough for a man to break his hand over. Is Mrs Watson in?"

"She is away upon a visit."

"Indeed! You are alone?"

"Quite."

"Then it makes it the easier for me to propose that you should come away with me for a week to the Continent."

"Where?"

"Oh, anywhere. It's all the same to me."

There was something very strange in all this. It was not Holmes's nature to take an aimless holiday, and something about his pale, worn face told me that his nerves were at their highest tension. He saw the question in my eyes, and, putting his finger-tips together and his elbows upon his knees, he explained the situation.

"You have probably never heard of Professor Moriarty?" said he.

(c)

Fig. 6.2   Examples of three text extracts.

There are two basic, and opposing, approaches to the normalization problem.

In the first approach, a fairly simple normalized text format is defined, to which the input text must conform, and the normalization problem is pushed back on to the application calling the TTS. For example, the normalized form may allow only lower-case letters within words, spaces between words, very limited punctuation, and only full stops or question marks at the end of the sentence. At first sight this appears to be merely avoiding the problem, but the justification of this approach is that the application generating the text has the most comprehensive knowledge about the format of the text with which it is dealing. For example, the abbreviations used in pharmaceutical research are different from financial market information or directory enquiries.

The second approach is to allow unrestricted text into the system, and employ a set of rules to make decisions about the intended 'meaning' of the text. These rules may be switched on or off by the application. The major disadvantage here, is that the TTS cannot predict all possible text formatting issues, and may make wrong decisions.

In essence the two approaches outlined above define the role of a TTS system differently — how much implicit application knowledge must the TTS have to perform its job? In practical systems a compromise between the two extremes is usually taken, with a fairly intelligent normalization front end controlled by options set by the application, for example, an expanded abbreviation list. For most text there would be no need for an application to filter it, unless the format were very particular or unusual, e.g. e-mail. In this way the onus of text formatting is shared between the application and the TTS system.

6.2.2   Text segmentation

The most basic function of the text analysis process is to segment the input text into manageable units, and these units must also have some place within the linguistic framework(s) used in the rest of the TTS system. It is usual to split the text into sentences, and each sentence into individual words. Higher level structures, such as paragraphs, and possibly even larger units may also need to be identified. Below the sentence level, it is necessary to split the text into words, and possibly 'word groups'. (Word group is used to mean a group of words that are created to replace a non-letter sequence in the original text, e.g. the words 'seven hundred and four' for the input text '704'.) This issue of intra-sentence segmentation is closely tied to text normalization, as described below.

6.2.2.1   Sentences

For English at least, the most invariant substantial unit of written text is the sentence. Sentences are easy to segment most of the time, since they normally start with a capitalized word and end with a full stop or other termination character (question mark, exclamation mark or ellipsis). This simple approach produces reasonable results, but does not account for more complex cases where the presence of a termination character does not signal a sentence boundary. In the examples below, a full stop may be part of a number, *a*, date, *b*, acronym, *c,* or indicate an abbreviation, *d*. Most of these cases (*a,b*) can be resolved by considering the context immediately following the full stop; if it is a space, or certain other punctuation, *e*, then the sentence is terminated.

a   Please pay $43.23 every month.

b   Arrange the meeting for 23.08.95 in room 6.

c   Kilby was an M.I.5 informant.

d   How about Mon. 23rd?

e   He said, "I love you."

Acronyms, *f,* and especially abbreviations can still cause considerable problems, since they may also be potential words, as in the rather extreme example, *g*.

f   Colin did leave the N.A.S.A. building on time!

g   He tried to walk on the Sun. Howard died.

Without understanding the context it is impossible for even a human to tell whether the male subject of *g* was recovering from an accident on the Sunday that Howard passed away, or if Howard's foolhardy attempt at solar perambulation proved fatal! This example is particularly tricky due to the secondary cue of an initial capital letter on the word following the potential boundary. Decisions about sentence boundaries in such cases will be arbitrary unless the system has some knowledge of the text's meaning.

The final case to consider is where a sentence boundary occurs but there is no terminating character, as is often the case in electronic media such as e-mail, and some informal texts. In general the presence of a blank line or end of file implicitly marks a sentence boundary, as would systematic change in visual format, e.g. the change of font in the newspaper extract in Fig. 6.2(a).

6.2.2.2   Paragraphs and sections

For passages longer than four or five sentences, text is often split into larger logical groupings, typically paragraphs and sections, and this logical structure should be made available to the rest of the TTS system. Information about paragraph structure is necessary to model long domain intonation [6] and inter-paragraph pausing, which is longer than inter-sentence pausing. Paragraph boundary information is also useful in trying to resolve pronoun cross-references, which occur infrequently across paragraph boundaries.

Paragraph breaks are generally indicated by a blank line, or by some form of initial indentation, immediately following a sentence boundary. There are circumstances where these features do not indicate a paragraph break, such as some list formats.

6.2.3   Text normalization

Text normalization is sometimes referred to as text formatting, but this somewhat undervalues the nature of the task. It is in essence a process of 'de-formatting' the text, or unwrapping the strict token sequence from the visual presentation style, and encoding the useful parts of the 'style' in a defined and explicit way [7]. It should be remembered that much text is primarily intended for visual consumption, and aspects ranging from the typography through to the vocabulary are chosen to aid comprehension. For example, both italic and bold fonts are used for emphasis. By the time the text is in electronic form, in particular ASCII, much of this visual presentation has already been lost, although an increasing number of texts are created specifically for an electronic medium.

This stage also requires a limited amount of text generation, e.g. when converting a digit string into orthographic words. This can cause some significant problems, because there may not be enough information available at this stage to decide exactly what the generated text should be. For example, in many Romance languages, a number must agree in gender with its 'object', and it may not be clear what the object is without a full syntactic parse.

6.2.3.1   The process

The input text will have many of the following features:

- embedded control characters;

- symbols;

- indentation;

- white space;

- capitalization;

- numerals;

- punctuation;

- abbreviations.

The normalized form is typically a sequence of explicitly separated words, consisting just of lower case letters, with punctuation associated with some words, and enough other information to correctly 'read' the original text. An example is shown in Fig. 6.3.

In Fig. 6.3 the normalized words appear towards the bottom of the structure, with the preceding and following punctuation immediately above and below respectively. Words which had an initial capital in the input text are marked 'CAP', numerals with 'NUM', and abbreviations with 'ABBR'.

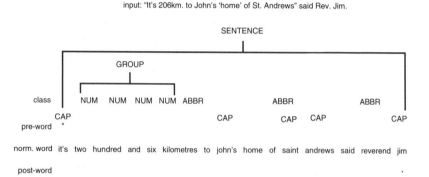

Fig. 6.3   Example sentence and its normalized representation.

Assuming that segmentation has been applied to sentence level, the most straightforward approach is to scan through the sentence using white space to separate out words. Some characters and strings will be converted during this process, dependent on their context and on which optional conversions have been set by the application. The following sections describe a typical set of conversions for an English TTS system.

6.2.3.2   Control characters and white space

Every control character, including the 'newline' character is converted to a space. This assumes that enough structural information, i.e. sentence and paragraph determination, has already been found in the segmentation stage. For example, newline characters are used to determine sentence boundaries.

6.2.3.3   Capitals

Capitalization normally occurs in five situations:

- the first letter of the first word in a sentence;

- proper nouns (all nouns have an initial capital in German);

- acronyms or abbreviations;

- initials, e.g. J. M. Smith;

- emphasized text, including section headings.

Abbreviations can be sensitive to upper-case letters, but otherwise capitalization cannot be relied upon as a precise indicator of a word, although it will bias later decisions, such as whether special dictionaries are searched. Therefore all words are converted to lower case, and capitalization information is retained.

Acronyms are identified as words where any non-initial letter is a capital, but if all letters are capitals, it could just be an emphatic word. Emphatic words tend to occur together, either as a phrase, or as a complete sentence (such as headings), so the form of surrounding words can bias the decision.

6.2.3.4   Numerals

Numerals occur frequently in text, and are used in a number of different ways, e.g as ordinals, dates, times, currencies, or in 'special' formats such as telephone numbers. The digit sequence is replaced by one or more words, according to the digit pattern.

Any simple digit strings, possibly including a full stop or comma, and matching none of the other patterns below, are treated as a default number and are converted to the corresponding words, e.g. '1245.67' becomes 'one thousand two hundred and forty five point six seven'. Ordinals are an integer followed by 'th', 'nd' or 'rd', e.g. '4th', '123rd'. The conversion is the same as a default number, but with the last word changed to its ordinal form, e.g. '123' becomes 'one hundred and twenty three', but '123rd' becomes 'one hundred and twenty third'.

6.2.3.5   Dates and times

Dates are very common in texts and occur in many formats, for example, '12-8-94', '12/8/94', '12.08.94', '12-Aug-94', and '12-aug-1994' should all be read as 'twelfth of august nineteen ninety four'. Most of these patterns should also be recognized as dates even without the year specified. The first three patterns have a different interpretation under the conventions of American English, where the standard date format is Month-Day-Year. This can clearly be automatically determined if the date were only meaningful using one convention, e.g. '13-10-94'. It is also useful to recognize a partly normalized date format, e.g. '12th December 1994', since the last digit string should be read as 'nineteen ninety four', rather than the default number 'one thousand nine hundred and ninety four'.

Times are defined by the patterns HH:MM or HH:MM:SS, where HH, MM and SS represent hours, minutes and seconds respectively. Range checking is performed on each of these values. If the 12-hour clock is used, the abbreviations 'am' or 'pm' may follow the time.

6.2.3.6   Currency

Currency formats start with a currency symbol, which may be a letter sequence, followed by a number, optionally followed by an amount abbreviation, e.g. '£12m', 'DM234bn', '$32'. If there is no amount abbreviation and the number has two digits following the point, then the interpretation is currency and sub-currency, e.g. '$12.34' becomes 'twelve dollars and thirty four cents', but '$12.3m' becomes 'twelve point three million dollars'.

6.2.3.7   Grouping

When a 'unit' in the original text is split into many normalized words, e.g. when expanding a number, it is important to remember that the words are grouped together. Any further structure of the grouping should also be stored, e.g. code and number format of telephone numbers, or parts of credit card numbers, etc. This information, along with the pattern type, can be used by the prosodic generation components later in the system.

6.2.3.8   Abbreviations

Abbreviations, which may end with a full stop, are normally expanded to their full value (as determined by a table supplied by the application), but there are some cases which require special treatment.

- Ambiguous abbreviations commonly occur in addresses, and can expand to more than one word depending on context. Common examples are St. which can expand to saint or street and Dr. which can expand to doctor or drive. A simple positional dependence can usually determine to which word it should be expanded, but there are still cases that are hard to distinguish; for example, consider the following pair: 'Mary St Mead' should become 'mary saint mead' but 'London St North' should be read as 'london street north'.

- Homographic abbreviations can also appear as words, e.g. 'sat' (saturday), 'in' (inch), 'is' (island), etc. This can cause a problem in sentence segmentation if the last word in the sentence is a homographic abbreviation (see example *g* above in section 6.2.2). The most reliable solution is to look for secondary cues to sentence termination, such as a blank line or initial capital on the following word. As there will be cases where the ambiguity cannot be resolved, a default expansion is used.

- Measurement abbreviations occur after numbers, and the expanded abbreviation must agree with the number, e.g. '1 cm' becomes 'one centimetre' but '7 cm' becomes 'seven centimetres'.

6.2.3.9   Punctuation

All punctuation is normally associated with the nearest word, and is stored as either pre-word or post-word punctuation. However, certain punctuation marks have a default binding which overrides this, e.g. open brackets (or upside-down question mark in Spanish) bind to the following word. It should be noted that it is perfectly possible for more than one punctuation mark to be associated with a word, especially when quotations occur, e.g. He said, "I saw John.", or multiple sentence terminators are used (Because... or Hello!!!). A single quote might not be punctuation, but part of the word itself, either internally in the case of contractions and possessives (didn't, John's), or externally in plural possessives (dogs' bowl) and heavy orthographic reductions ('tween or roamin').

6.2.3.10   Symbols

Various symbols are defined in ASCII which are not part of normal written text, such as curly brackets, hash, tilde, etc. These will be converted to the symbol name, which may depend on the application. Some symbols may already have been dealt with by earlier conversions, e.g. '$' in a currency format.

6.3   MORPHOLOGY AND PRONUNCIATION

The text segmentation and normalization task provides a stream of separate words from the original text. The morphology and pronunciation task analyses every word to determine its possible syntactic functions and pronunciation.

This process may produce alternative hypotheses for words or word sequences; it is the role of the tagger and parser (see section 5.4) to generate the correct unambiguous sequence.

6.3.1   Morphology

The morphological component is closely tied to the pronunciation component. It has three functions:

- grouping words into lexical phrases;

- marking the lexical syntactic information, i.e. part of speech for each word;

- generating the root-form, or stem morphology, of each word for subsequent pronunciation.

6.3.2   Lexical phrases

The word-tokenization used in the normalization component assumes that white space separates the text into distinct words. However, it is often more natural to treat certain word sequences as a single lexical entry, either because of a variant pronunciation, or more commonly, because the lexical phrase acts as a single unit with a single syntactic category. In most cases lexical phrases are idioms or phrasal verbs which are more naturally treated as single units. For example, the phrase 'up and over' can act as if it were a single adjective, as in the sentence:

It had an up and over door.

but it is important to also store the individual words, since this sequence of words may be used in another sense:

He climbed up and over the hill.

6.3.2.1   Lemmatization

For efficiency, and to avoid duplication, the lexicon is often lemmatized (reduced to headword entries). This achieves a significant reduction in space. For example, considering the word 'love'; in a full-form dictionary it would be necessary to store the words: love, loves, lover, lovers, loved, loving, lovingly, unloved, unloving, unlovingly, etc, whereas a lemmatized dictionary would just store the root-form, or headword, 'love' and an indication of the appropriate affixation rule sets for deriving all the other surface forms. This leaves the problem of deriving the affixation rules, but there is abundant literature on this subject, e.g. Chomsky and Halle [8], Gimson [9], Allen et al [10]. Rather than store a reference to every affix rule in an entry, rules can be grouped into overlapping rule sets, and a reference to allowed rule sets is stored in the entry. It should be remembered that, as this part of a TTS system is only used to analyse existing words, not to generate them, many of the complexities introduced to prevent over-generation of surface forms are unnecessary. Since this chapter considers practical systems (not just theoretically rigorous research systems), irregular surface forms can always be stored directly in the dictionary as a new headword.

6.3.2.2   Affixation and compounding

The morphological component can now use a lemmatized dictionary to search for any given word in the input text. This is a fairly easy task in theory, but there are some implementation issues to consider here. The lemmatized dictionary described above is effectively indexed on headword, not on surface form, but to analyse an incoming word, only the surface form is known. There are a number of possible solutions, which are described in the following paragraphs.

Since most affixation is actually suffix adding, the dictionary could be directly indexed on the fixed part of each headword, and the affixation applied, if necessary. There must either be some process to check for prefixes, or simply disallow prefixes, and add the extra words as new headwords. This solution is clearly not applicable for languages that make extensive use of prefixes.

An index could be created of all full form words, but this would defeat the object of lemmatizing the dictionary in the first place.

A recursive procedure can be used which tries any affix rule matching the current word, then attempts to find the resultant word as a headword. This can perform several recursions. If a match is found, it is necessary to check that the affix rule(s) used was valid for the matched headword. As an example, consider the case where there are two headwords, 'star' and 'stare', and two affix rules, '+ing' (which adds 'ing' to the root) and '−e+ing' (which removes the last 'e' from the stem then adds 'ing'). If the incoming word is 'staring', there are two possible analyses: 'star+ing' or 'stare-e+ing' (see Fig 6.4).

Clearly the first analysis is incorrect (adding 'ing' to 'star' would require consonant doubling, giving 'starring'). Therefore it is necessary to constrain the allowed affix rules, which in this case would mean only allowing the '−e+ing' rule to apply to the 'stare' entry.

A modification of the last approach uses a chained list to connect all entries which may use a particular rule. When a rule is activated, only entries on its chain are active, and can be searched. This way the forward search is constrained, rather than the backward constraint in the previous method.

The compounding process applies in a similar way, using the compounding rules appropriate for each entry.

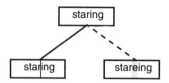

Fig. 6.4   Two analyses of the word 'staring'.

6.3.2.3   Lexical syntax

Every word in the lexicon will be marked with at least one syntactic category, or 'part of speech'. Many words have more than one syntactic category, for example 'man' and 'stall' can both act as verbs or nouns — so 'man the stall!' and 'stall the man!' are both reasonable sentences. The detail encoded in the syntactic category varies greatly between systems, but it will at least enable the system to distinguish between major parts of speech such as verbs, nouns, adjectives, adverbs, pronouns, etc. The entry may encode far more detailed syntactic characteristics of the word, such as a verb's transitivity, and this information can enable more sophisticated processing in later components. The main use of the syntactic category information is for the parser to build up the sentence structure, and if the parser includes a stochastic tagging component (see section 6.4), it is beneficial for the lexicon to include the relative frequency of each possible category for a word.

If the incoming word does not appear in the lexicon, even after a morphological analysis, it is still necessary for the system to have some syntactic information about the word. The result of the morphological analysis itself will often give a strong indication of the word's category, as many suffixes indicate a particular part of speech (e.g. words ending in 'iness' are very likely to be nouns). This characteristic leads to a robust method for determining the syntactic category of non-lexical words, since a table can be built giving the cumulative corpus category statistics for words ending in any given letter sequence.

6.3.3   Pronunciation

TTS systems traditionally used a set of letter-to-sound rules to produce pronunciations, and only resorted to a dictionary to hold exceptions to these rules. However, the current practice is to use a large on-line pronunciation dictionary, and revert to rule-based methods as a fall-back approach for non-lexicon words. There are three reasons for this change: increase in overall quality has made pronunciation errors more significant, the search time for a dictionary is often quicker than generation by rule, and, most significantly, a large dictionary is already needed to store other lexical information such as syntax, lexical semantics, and morphological information.

6.3.3.1   Representation of a pronunciation

Typically a pronunciation will be given to every word (or lexical phrase) fairly early on in the text analysis stage. Some later process will consider the effect of syntactic structure and of the contextual effects of surrounding words. This would imply that the pronunciation task should produce an 'archetypal' transcription for every word, often referred to as the 'citation form'. This is defined as the pronunciation given to a word when it is produced in isolation, rather than connected speech.

The issue of how best to represent phonemes, allophones[1], stress and syllabi-fication information for a pronunciation has been addressed many times within the speech community. The International Phonetic Association (IPA) publishes probably the most widely accepted symbol set, but it is not particularly suitable for computational systems, since it consists of a mixture of Roman, Greek and other miscellaneous symbols, not represented in a standard computer character set. Furthermore, the IPA alphabet is not 'theory-neutral' (it is motivated by a theory of word-phonology) so it uses a discrete cross-parametric segmental representation, where symbols are chosen mainly by their similarity to standard orthographic representations, not to any characteristics of the sounds themselves. The differing requirements of speech researchers has spawned a plethora of machine readable symbol sets [2, 11-13], which tend to have different coverage and scopes. Example pronunciations in this chapter are shown in SAMPA symbols for British English, shown in Table 6.1.

It is fairly clear from the discussion above, that there is no 'correct' way to represent even the agreed citation form of a word, since the selection of exactly what is represented is subject to a theoretical bias. Having made this caveat, there are, broadly speaking, two types of transcription:

---

[1] An allophone is a particular variant pronunciation of a phoneme, which does not change the meaning of a word, e.g. the final /p/ in 'stop' may be realized with or without a strong breath sound, but most listeners do not distinguish the pronunciations.

Table 6.1   SAMPA symbols for British English.

p	pear	t	tear	k	king
b	bear	d	dear	g	gear
f	fear	T	thing	s	sing
v	very	D	this	z	zing
S	sheer	l	leer	j	year
Z	treasure	w	wear	r	rear
m	men	n	near	N	wing
tS	cheer	=n	button	h	hear
dZ	jeer	=l	bottle		
@	ago	{	bat	E	bet
I	bit	Q	cod	U	good
V	bud	3	bird	A	bard
i	bead	O	bore	u	boot
@U	zero	aI	pie	aU	cow
E@	hair	eI	pay	I@	peer
OI	boy	U@	Ruhr		
"	Primary stress		$	Syllable boundary	
'	Secondary stress		#	Word boundary	

- phonemic, which attempts to detail the distinctive contrastive sound units used in the accent;

- phonetic, which details the actual sounds used in the production.

To a native speaker, who is at least subconsciously aware of how sound units are realized in their language, these two transcription types may seem artificially separated. However, it should be remembered that the phonemic description is based on an implicit analysis of the meaningful contrastive sounds in a particular language and accent, at a particular time in history, and only through this analysis can any claims be made about the actual production of speech based on the phonemic transcription.

For the purposes of synthesis, it is also necessary to mark lexical stress (or word stress) in the pronunciation. This gives rise to contrastive pronunciations of words which have essentially the same segmental description, e.g. compare the word 'increase' in the two phrases 'increase the pay' and 'pay the increase'. Lexical stress may be marked on a whole syllable, or in a non-syllabic analysis marked only on the nuclear element, which is invariably a vowel. There are many proposed systems of lexical stress [9] — using hierarchical stress, binary stress, ternary stress (primary, secondary and unstressed), or even four levels of stress. Some analyses further distinguish between the different realizations which stress may have in the speech signal, contrasting secondary stress which may take a pitch accent, and that which cannot.

6.3.3.2   Coverage of a lexicon

Since it is rarely feasible to generate an exhaustive list of all words in a language, one question that must be addressed in the design of a TTS system is how big the lexicon should be. There is no simple answer of course, but there are common trends in word distribution of which use can be made. Figure 6.5 shows the frequency of each of the 13 334 unique words in a newspaper corpus, C1, of 108 000 words [14]. It can be seen clearly that about half of all words only occur once in the corpus, and 90% occur less than 10 times. Many of these rare words are proper nouns, such as surnames and place names. The same data is presented as a graph of top-n words against corpus coverage in Fig. 6.6.

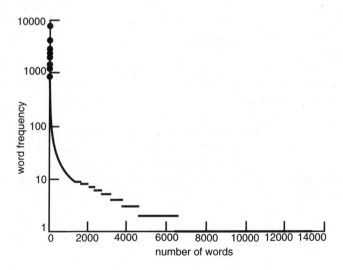

Fig. 6.5   Word frequency for corpus C1.

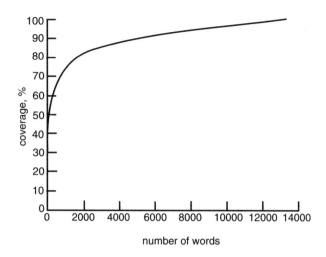

Fig. 6.6   Top-*n* word coverage for corpus C1.

Figure 6.6 shows that the top 1% of words make up about 50% of the corpus, the top 10% of words make up 75% and the top 40% make up 90% of the corpus. Clearly doubling the lexicon size from 1000 to 2000 words gives more coverage than doubling from 4000 to 8000 words. A very similar pattern can be seen in Fig. 6.7, which gives the word distribution for another similar sized newspaper corpus, C2, of 89 000 words, of which 11 377 are unique. Figure 6.7 also shows that the words from C1 only provide a partial coverage of C2 (87%), even though C1 is larger.

It is interesting to note that the first 1000 words from C1 or C2 provide approximately the same coverage of C2. In fact, the top 1000 words from C1 and C2 are almost identical, and will be similar in most texts; it is the low-frequency words which differ significantly between texts.

The observations above indicate that the lexicon should include the top few thousand words, but beyond about 5-10 000 the gain becomes quite small, and increasing the lexicon size will increase search times. The final size of the lexicon will also be affected by the accuracy and reliability of other fall-back pronunciation methods.

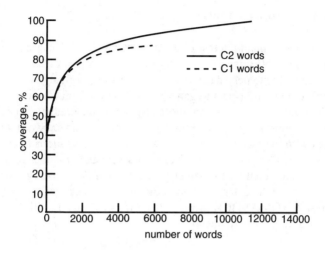

Fig. 6.7   Coverage of corpus C2 by top-*n* words from C1 and C2.

6.3.3.3   Fall-back methods

When a word is not found in the lexicon, it is necessary to generate its pronunciation by another method. The most common approach is to use letter-to-sound rules, which convert from the spelling of a word to the pronunciation. For some languages, such as English, this is a difficult process, as with the famous 'ough' words — through, though, rough and bough — where the 'ough' part is pronounced differently for each word.

The main emphasis in the past decade has been on increasing dictionary size, and the area of unknown word pronunciation has been relatively neglected. Klatt [2] provides a good description of non-lexical pronunciation methods for English, in which he describes the evolution and application of letter-to-sound rule techniques. Klatt also describes the early development of pronunciation-by-analogy techniques. These are based on the theory that when confronted with an unknown word, a human reader will pronounce it like any similar names they know. For example the name Plotsky would be pronounced as /" p l Q t $ s k i/ to rhyme with Trotsky /" t r Q t $ s k i/.

6.3.3.4   Proper names

There has been much interest in the pronunciation of proper names, mainly due to the many applications which require pronunciation of names and addresses from large customer databases [15]. Names often have a 'fossilized' pronunciation with respect to general words in a language since they are owned by individuals and are therefore more resistant to the gradual process of language change. Names are also continually moving around between languages and cultures with their owners, and may take several generations to assimilate into a new 'nativized' pronunciation. These features make pronunciation-by-analogy a good candidate method for names pronunciation [7]. Although there have been attempts to produce large name pronunciation dictionaries, in particular the European ONOMASTICA project [16], their coverage is still too small to make non-lexical methods redundant.

6.3.3.5   Spelling

The final fall-back method, if all other approaches have failed, is to spell the word. This seems a very natural thing to do, but it actually takes a while for a listener to realize that a spelling sequence is occurring if it comes in the middle of continuous speech, and is not 'signposted' in some way. If a difficult word appears when a human is reading, the speaker will normally attempt a pronunciation, then signal a switch to spelling mode with the cue 'which is spelt as...' or 'spelt...' , then spell the word in question.

The spelt sequence should be marked in such a way that later stages of the TTS system know that it is a spelt sequence, and came from a single word. This eases the job of the parser, which can treat the sequence as a single unit, and allows an appropriate prosody to be imposed on the spelt sequence.

Having generated the pronunciation(s) and possible syntactic categories for each word in the input text, and dealt with any unusual cases, the next task is to decide how the words relate to each other, i.e. determining the syntactic structure of the sentence.

6.4   SYNTACTIC ANALYSIS

The syntactic analysis task takes the sequence of words in a sentence and, from the lexical syntax, derives a structural analysis for the sentence. This is an important task for text-to-speech systems, in particular for prosodic generation (see Chapter 7), which can make use of the sentence structure to derive the sentence's prosodic structure. However, since language is so inherently ambiguous, knowing the structure of the sentence does not necessarily equate to knowing the meaning of the sentence. For example, consider the simple sentence:

If the display is not bright, change the batteries.

Ignoring context for the moment, 'display' and 'change' could be verbs or nouns, and 'bright' could be an adjective or adverb, giving eight possible part-of-speech (or tag) sequences for the sentence. However, an analysis of the possible meanings of the sentence based on the product of the meanings of each word [17], gives 6116 semantic interpretations of the sentence. If the correct part-of-speech sequence and syntactic structure is known, this number only reduces to 1764. Luckily, it is not necessary for a system to understand the sentence to make a reasonable attempt to read it; the system only needs a knowledge of the structural relationship between parts of the sentence. Of course, understanding the sentence would improve a system's reading ability.

Knowing the syntatic category of every word may also be important to determine pronunciation, for example:

He was *content* (adjective) with his lot.
The *content* (noun) was not exciting.

The *convict* (noun) escaped.
You *convict* (verb) him.

The first job of the parser component is to assign such syntactic categories to each word. The input to this task is a list of possible categories for the words — these categories coming from lexical sources, or if no lexical source exists, coming from morphological analyses of the structure of the word (see section 6.3.1).

6.4.1   Stochastic tagging

The most common method for tackling this task is the stochastic tagger. This uses a statistical model of language, which assigns words their most likely part of speech, given the probability of the words appearing in a certain context, and also the probability of the tags appearing on the words themselves. The model is derived from large corpora of correctly tagged text [18].

The probability that a particular sequence of words, W, are tagged with a particular tag sequence, T, is represented by the conditional probability $P(T|W)$. The conditional probability $P(T|W)$ can be expressed in Bayes theorem as:

$$P(T|W) = \frac{P(W|T)P(T)}{P(W)}$$

where the important parts are the probability that specific words are generated given specific tags $P(W|T)$, and the probability that the tags occur in that particular sequence $P(T)$. Note that, since W is the same for all hypothesized tag sequences, $P(W)$ can be disregarded. Obviously certain tag sequences are better than others, so the aim is to maximize $P(T|W)$.

The probability of each sequence can be rewritten as a product of the conditional probabilities of each word or tag, given all of the previous tags:

$$P(T|W)\,P(W) = \Pi\, p(t_0)p(t_i|t_{i-1}, t_{i-2},...)\; p(w_i|t_i,..., w_{i-1},...).$$

Now, the approximation can be made that each tag depends only on the immediately preceding tags (n-gram model), and that the word depends only on the tag. For a trigram model there would then be, for example:

$$P(T|W)\,P(W) = p(t_0)\,\Pi\, p(t_i|t_{i-1}, t_{i-2})\,p(w_i|t_i)$$

Once a tag is known to have been selected, no further information is gained from knowing the previous tags or words; so the above model is a Markov model.

With training data, n-gram sequence probabilities and the probability of each word being given a tag (lexical probabilities) can be estimated. However, the model must also cope with words that do not appear in the lexicon. For this, information from a morphological decomposition can be used to provide estimated probabilities based on suffixes and capitalization (see section 6.3.1).

Given these probabilities, the most likely tag sequence for a given word sequence can be found. One way would be to simply compute all possible tag sequences and find the maximum, but this would require an exhaustive search (although given that some words only have certain tags, the search space is reduced somewhat). Fortunately, standard algorithms, such as Viterbi [19] allow us to find the most likely tag sequence efficiently.

Given a sequence of tags for a particular language, although there are long distance dependencies between words far away in the written sentence, the probability of a tag occurring can be adequately determined by viewing a tag sequence history of two words. Such bi-gram taggers attain a useful level of accuracy, of about 95-96% on unseen text [18] (i.e. text on which they have not been trained). It is interesting to note that if the most likely tag is chosen for a particular word (the tag occurring most often for that word in a tagged corpus), accuracy is often about 92-93%.

Using higher level *n*-gram taggers does not seem to increase this figure enough to justify the investment in obtaining suitably large training corpora, nor the computational increase involved in their implementation. It is unlikely that current pure stochastic taggers can be improved on, since at the word sequence level, certain syntactic and semantic ambiguities arise which cannot be resolved without recourse to other levels of analysis. Fortunately, for pronunciation purposes, the number of words having multiple pronunciations is quite small — between 1 and 2% of words in a typical lexicon. In practice, there is less than a 0.3% chance of picking the wrong pronunciation for a word.

The robustness and speed of the stochastic taggers mean they are generally favoured over rule-based taggers in current natural language processing systems, and also provide an appropriate solution for commercial text-to-speech systems.

Rule-based taggers generally do not exist as such, but tend to form part of the higher level syntactic parsing. The purpose of this task is to assign a structure to the sentence which reflects the syntactic role of every word in it. By constructing a syntactic structure which considers the role of each word in the sentence, decisions can be made on individual word syntactic categories. Many rule-based parsers, however, use stochastic taggers as front ends, and parse with the tags given.

6.4.2   Parsing

The main approach favoured by syntactic parsers is the rule-based approach, where grammar rules, which may be recursive, describe valid sequences of symbols. These symbols correspond to word tags, or groups of words representing phrases, clauses or even whole sentences. There are many linguistic theories and corresponding rule-based grammars, but the fact that no one grammar formalism has come to the fore is an indictment of the state of the art with regard to rule-based parsing.

Statistical parsers exist which attempt to assign structures above the word level, by training on corpora of full syntactic structures [20]. The general lack of large syntactically analysed corpora have so far limited this approach, but perhaps a more fundamental problem is that the infinite theoretical structures possible in language limit the statistical approach — some generalizations must be made. There are several innovative approaches which incorporate probabilities into the application of grammar rules themselves [21], and these claim a 60% accuracy. Rule-based syntactic structure parsers can claim much higher accuracy than this, but they suffer from massive over-generation of incorrect structures, which limits their usefulness. Indeed stochastic parsers can also be viewed as suffering from over-generation, but, by their very nature, they impose a rank ordering, and often return only the most likely structure.

6.4.2.1   Ambiguity

The reason for this massive over-generation boils down to two main sources — lexical and structural ambiguity. Lexical ambiguity, where words have more than one sense, was demonstrated at the start of this section. Structural ambiguity is where there is more than one possible analysis of the attachment of words in the sentence. The two are not mutually exclusive. There are three main classes of structural ambiguity.

- Prepositional phrase attachment — "The man saw the boy with a telescope."

  where the telescope may have been used to observe the boy (Fig. 6.8 (a)), or the boy was carrying a telescope (Fig. 6.8 (b)). When more prepositional phrases are added, the number of ambiguities increases, e.g. "The man saw the boy in the park with the telescope."

- Noun-noun co-ordination — "Pressure cooker balance weight adjustment screw."

  [[[pressure cooker] [balance weight]] [adjustment screw]]

  It is fairly easy for humans to identify boundaries between noun compounds, but this task is extremely difficult for a machine.

- Co-ordination — "The old men and women."

  where the analysis is either "old men" and "women", or old "men and women".

Semantic restrictions, or world and contextual knowledge may be used to filter clearly incorrect structures, but, as has been the case with AI systems facing the same problem, these methods have only been proven in systems of limited domain. Therefore, it is not uncommon for parsers to return many thousands of possible analyses for even small sentences, only one of which is the correct one! This of course has performance implications in the number of nodes generated in a parse tree. Efficient parsing techniques such as chart parsing [22], extended by packing and indexing of edges, have been successful in addressing the generation of multiple analyses for one sentence.

The over-generation of multiple structures is related to the granularity and recursiveness of the grammar rules. No complete computational grammar of English exists, despite years of research, and probably never will. Human language is constantly evolving and expanding, and the notion of grammaticality is a theoretic device that does not always relate to how people actually use language. One area of language evolution is idiomatic expressions, but by changing the grammar rules to accommodate these exceptions, rules are often introduced which cause more structures to be generated — over-generation. At the other extreme, using fewer rules leads to the risk of the parser failing to find an analy-

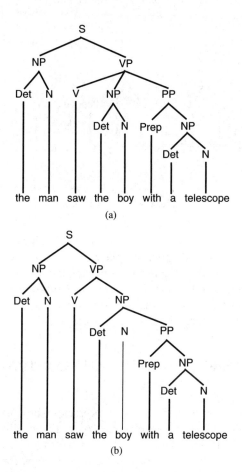

Fig. 6.8   Prepositional phrase structural ambiguity.

sis for what might be considered an acceptable sentence — under-generation.
Another possibility is to use fewer rules to recognize islands of partial linguistic
structures, and ignore the attachment problems, except when confident
attachment predictions can be made.

By combining the best of both worlds (stochastic word tagging and syntactic
structure parsing with a partial grammar), it is possible to derive a fast and efficient parser, albeit at the expense of a rich syntactic analysis. An important

aspect of this approach is that ill-formed input can be accommodated — an important feature for TTS. An example of such an analysis is shown in Fig. 6.9.

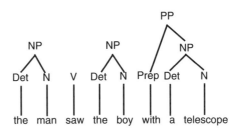

Fig. 6.9   Partial parsing.

This is a more sparse structure than in Fig. 6.8, but it provides a useful level of information for input to the performance parser. Current research would seem to be focused both on hybrid approaches selecting the best features of rule-based and stochastic parsers — for example, probabilistic parsing with grammar rules, with back-off if no parse can be found — and on incorporating novel techniques for treatment of idioms and collocational phrases which make up so much of normal language use.

6.5   POST-LEXICAL PROCESSES AND CONTINUOUS SPEECH EFFECTS

In the description in section 6.2.5, the processes of pronunciation and stress assignment work at the word level, i.e. they work on information which is described within the lexicon. However, there are a host of post-lexical processes that function across word boundaries. Lexical phonology, for example, divides the phonological analysis of language into two parts, a lexical component and a post-lexical component (sometimes referred to as phrasal phonology). This post-lexical phase may be further subdivided into processes which are phonologically motivated, and processes which are phonetically motivated. However, a clear boundary between the two subdivisions is difficult to derive.

The pronunciation task may be viewed as functioning like the lexical component described above, while the continuous-speech-effects component works at the level of the post-lexical component. Goldsmith [23] states that:

"The post-lexical phonology involves two major sorts of rule applications: (i) those operating crucially across word-boundaries or making

specific use of phrasal or syntactic structure, and (ii) those that fill in,
specify, or refer to non-distinctive features."

Goldsmith's rules of type (i) change a word's pronunciation due to its phonological and syntactic context. These changes are very common in normal speech,
and are necessary for natural-sounding speech synthesis. Rules of type (ii) are
applied primarily according to the style of speech (e.g. speech rate) and would be
used by a TTS system to model more subtle effects of fluid speech, requiring
careful application to avoid compromising intelligibility. Examples of both types
of rule in a TTS system are discussed below.

There is, in spoken English, a tendency for words to be pronounced differently depending on their position within the syntactic structure of a sentence. As
a result, most words are said to have a strong and a weak form. Some words,
however, are more prone to this form of reduction than others. These words, are
often called function or grammatical words, e.g. auxiliary verbs, prepositions,
pronouns and conjunctions, etc. For example, the determiner 'the' has weak
forms /D@/ as in 'the man' /D@ m{n/ when spoken before a word beginning
with a consonant and /Di/ as in 'the apple' /Di {p=l/ when spoken before a word
beginning with a vowel. The words 'a' and 'an' have the weak forms /@/ as in 'a
bag' /@ b{g/ when spoken before a word beginning with a consonant and /@n/
as in 'an apple' /@n {p=l/ when spoken before a word beginning with a vowel.
Notice how the surface pronunciation of the weak form is dependent upon the
underlying lexical pronunciation of the word and the phonemic context. Without
such reductions continuous speech sounds stilted and unnatural, while over-
application or inappropriate application of such post-lexical rules can reduce the
intelligibility of the synthetic speech.

The phonological rules described above are predominantly driven by the
grammatical structure of the utterance. There are a class of post-lexical processes, the application or degree of application of which is dictated by speech
style. The domain of application of such rules is across word boundaries. These
processes include elision, assimilation and r-insertion.

Elision is the process where, under certain circumstances, a phoneme may be
deleted or more accurately have zero realization. Elision can act both within
words and across words at word boundaries. Within-word elision affects the
weak syllables of a word, for example the word 'perhaps' may be realized as
[p^h"h{ps][1] where the unstressed vowel has been elided and the aspiration of the
initial plosive has taken up the whole of the middle portion of the syllable.
Elision often affects clusters of complex consonants such as 'looked back',
where /lUkdb{k/ may be pronounced /lUkb{k/.

---

[1] This transcription is phonetic: the p^h notation indicates an aspirated /p/, i.e. there is a strong breath
sound following the release of the /p/.

Assimilation affects consonants at word boundaries. For example, take the phrase 'that colour'; the citation form pronunciation of these words is /D{t/ and /kVl@/, yet, when produced as a phrase in casual speech, the process of assimilation changes the production of the consonants at the word boundaries so that the phrase is pronounced as /D{kVl@/. Elision and assimilation may act repeatedly on a phrase; for example, the phrase 'hand bag' may be pronounced /h{nd bag/ in over articulated speech, or as /h{nbag/ or even /h{mbag/ in casual speech. As with the production of weak forms, appropriate application of these processes increases the naturalness of synthetic speech.

R-insertion, like assimilation, is clearly a post-lexical process as it acts across word boundaries. The phonological rule for r-insertion can be simply stated as:

Zero $\rightarrow$ /r/ [@, 3, I@, E@, A, O] # V

This states that the phoneme /r/ can be inserted between a word ending in one of the vowels contained within the brackets and a word starting with any vowel, e.g. 'for Arthur' /fO r AT@/.

Recently a number of researchers have suggested that many post-lexical processes which traditionally have been described in terms of boundaries and syntax would be better represented under some form of prosodic hierarchy. Durand [24] cites the example of the domain of application of [r]-insertion and elision as a strong argument in favour of using a prosodic hierarchy.

6.6   CONCLUSIONS

The process of converting text into speech has been presented as a two-stage process, text and linguistic analysis, followed by prosody and speech generation. This chapter has described techniques used in current state-of-the-art commercial TTS systems to perform the analysis stage. In particular, the variability and variety of input text has been illustrated, and approaches were described to robustly extract and formalize the information encoded in the input text for later processing. Linguistic analysis could then proceed, starting with morphological analysis, which examines the structure of each word, aiding the pronunciation generation task. Syntactic analysis then derived a structural analysis for phrases, clauses and/or sentences. This must work robustly, providing the most appropriate analysis for subsequent prosodic generation. Syntactic information is also needed in order to model a number of continuous speech effects, which give the speech more fluent and natural characteristics.

The goal of this analysis is to extract all the information provided by the text deemed relevant for speech synthesis. Chapter 7 describes the process of generating an appropriate prosody for the synthetic speech, and discusses techniques used in speech concatenation and modification to create the final signal.

REFERENCES

1. Moulines E and Charpentier F: 'Pitch-synchronous waveform processing techniques for text-to-speech synthesis using diphones', Speech Comm, 9, No 5/6, pp 453-467 (December 1990).

2. Klatt D: 'Review of text-to-speech conversion for English', J Acoust Soc Am, 82, No 3, pp 737-793 (September 1987).

3. Lindstrom A and Ljungqvist M: 'Text processing within a speech synthesis system', Proc ICSLP '94, Yokohama, pp 1683-1686 (September 1994).

4. Jones B: 'Exploring the variety and use of punctuation', Proc 17th Annual Cognitive Science Conference, Pittsburgh (July 1995).

5. Carvalho P et al: 'E-mail to voice-mail conversion using a Portuguese text-to-speech system', Proc ICSLP '94, Yokohama, pp 1687-1691 (September 1994).

6. Silverman K: 'The structure and processing of fundamental frequency contours', PhD Thesis, University of Cambridge (1987).

7. Gaved M: 'Pronunciation and text normalisation in applied text-to-speech systems', Proc Eurospeech '93, Berlin, pp 897-900 (September 1993).

8. Chomsky N and Halle M: 'The sound pattern of English', MIT Press (1968).

9. Gimson A C: 'An introduction to the pronunciation of English', 4th Edition, Edward Arnold, London (1989).

10. Allen J, Hunnicutt W S and Klatt D: 'From text-to-speech: the MITalk system', Cambridge University Press (1987).

11. Seneff S and Zue V W: 'Transcription and alignment of the TIMIT database', DARPA TIMIT CD-ROM Documentation (1988).

12. Wells J: 'Computer-coded phonemic notation of individual languages of the European Community', J IPA, 19, pp 32-54 (1989).

13. Hieronymus J L: 'ASCII phonetic symbols for the world's languages: Worldbet', (unpublished).

14. Johansson S et al: 'The tagged LOB corpus: Users' manual', Norwegian Computing Centre for the Humanities, Bergen (1986).

15. Spiegal M and Macchi M: 'Development of the orator synthesizer for network applications', Proc AVIOS (1990).

16. Schmidt M and Jack M A: 'A multi-lingual pronunciation dictionary of proper names and place names: ONOMASTICA', Proc ELSNET Language Engineering Conv, pp 125-128 (1994).

17. Nirenburg S et al: 'Machine translation — a knowledge-based approach', Morgan Kaufmann, San Mateo, California (1992).

18. Charniak E: 'Statistical language learning', MIT Press (1993).

19. Viterbi A J: 'Error bounds for convolutional codes and an asymptotically optimal decoding algorithm', IEEE Trans Information Theory, 13, pp 260-269 (1967).

20. Bod R: 'Using an annotated corpus as a stochastic grammar', Proc EACL '93, Utrecht (1993).

21. Magerman D M and Marcus M: 'Pearl: a probabalistic chart parser', Proc EACL '91, Berlin (1991).

22. Kay M: 'Algorithm schemata and data structures in syntactic processing' in Grosz, Sparck-Jones and Lynn-Webber (Eds): 'Readings in natural language processing', Morgan Kaufmann, Los Altos, pp 35-70 (1986).

23. Goldsmith J A: 'Autosegmental and metrical phonology', Blackwell, Oxford (1990).

24. Durand J: 'Generative and non-linear phonology', Longman, London (1990).

7

OVERVIEW OF CURRENT TEXT-TO-SPEECH TECHNIQUES: PART II — PROSODY AND SPEECH GENERATION

M Edgington, A Lowry, P Jackson, A P Breen and S Minnis

7.1 INTRODUCTION

The conversion of text to speech (TTS) has been described in Chapter 6 as essentially a two-stage process (see Fig. 7.1). These are an analysis stage, which derives a linguistic structure from the input text, and a generation stage, which uses the linguistic structure to synthesize the speech, including the generation of intonation, rhythm, etc.

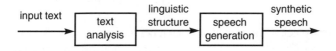

Fig. 7.1 Text-to-speech process.

The result of the text and linguistic analysis stage (see Chapter 6) is detailed linguistic information about the structure of the input text, in particular, syntactic, lexical, phonological and some semantic information. The following sections of this chapter describe how the linguistic structure is used to generate the synthetic prosody (section 7.2), and the final speech signal (section 7.3).

7.2   PROSODY GENERATION

The biggest single problem confronting TTS systems remains their poor generation of prosody for unrestricted text. The term 'prosody' covers a wide range of features characterizing the 'musical qualities' of speech, including phrasing, pitch, loudness, tempo and rhythm. These features are to some extent interrelated, but are often considered independently, placing conflicting demands on the ideal of a single prosodic structure. In practice, deriving all the relevant features within a unified framework is seldom attempted, mainly because the various relationships and dependencies are poorly understood.

Generation of synthetic prosody from text is limited by the capabilities of current text analysis algorithms. Natural prosody is a function of semantics (i.e. the meaning of the text), dialogue context, syntax, perception of rhythm, and human characteristics outside the realms of linguistic analysis (e.g. emotional state, age and state of health). For discourse-neutral prosody, syntactic and rhythmic analysis is probably sufficient, but long passages and spoken dialogue systems require semantic information for optimal performance. This discussion will concentrate on current methods for generation of discourse-neutral prosody, with the proviso that they should be capable of using semantic information appropriately.

The features considered here are prosodic phrasing, rhythm, intonation and segmental duration. Some progress has been made towards a unified framework for prosodic modelling, but there is still some way to go (this framework is based on metrical phonology, which is described in section 7.2.2). Prosodic phrasing and models of rhythmic and intonational phonology are represented in a single structure, although the initial phrasing can be developed independently. The durational model, although represented separately, depends heavily on information from this structure and has a strong metrical theme.

7.2.1   Prosodic phrasing

It is generally acknowledged that a sentence's prosodic tree is flatter than the syntactic tree. The classic example is the mismatch between the syntactic and prosodic structures for the embedded right-branching noun phrases (NP) in the sentence: 'This is the dog that chased the cat that killed the rat,' which are respectively:

[S [This is [NP the dog that chased [NP the cat that killed [NP the rat...
[This is the dog] [that chased the cat] [that killed the rat...

Prosodic phrases divide the sentence into smaller units for easier processing, and provide an aid to syntactic disambiguation.

Early work by Chomsky and Halle [1] ascribed the flattening of the prosodic structure to 'performance', and deemed this area a linguistically uninteresting problem. This view is now generally discredited, and there is agreement that every utterance, in addition to its syntactic structure, has a prosodic structure of some type, and that there is a mapping between the two (although this mapping is not well defined).

The empirical evidence for prosodic structures is from two sources — psycho-linguistics and instrumental analysis. On the instrumental level, the existence of prosodic breaks, which are discrete events associated with acoustic cues (such as duration lengthening, pause insertion and intonation markers), provides evidence for prosodic phrase structure. Furthermore, these phrases would seem to be arranged hierarchically with boundaries dominating, many phrases being more prominent in some way. Prosodic phrase structure can be represented by different phrase break indices, the location and relative size of these defining the prosodic structure. However, the acoustic cues are not always reliable, nor easily detected, and there is considerable debate as to how many 'levels' exist in a prosodic structure.

From a psycho-linguistic viewpoint, experiments indicate that intonation groups are planned in advance, since slips of the tongue tend to occur within these groups, and usually not at boundaries.

There are two main approaches to the prediction of prosodic structures — rule-based and stochastic.

7.2.1.1   Rule-based prediction

Many of the rule-based approaches stem from early work on performance structures based on experimental data, such as pausing and parsing values [2]. This work sought to account for the (then) disparity between linguistic phrase-structure theories and actual performance structures produced by humans, and focused on recreating the pause data of several analysed sentences from syntax (although they claimed that their method could easily account for other prosodic features). The central tenet of the work was that prosodic phrasing is a compromise between the need to respect both the linguistic structure and performance aspects of the sentence.

More recent efforts have extended the work on performance structure prediction to include prosodic phrasing. In this work, the basic rule-based approach is preserved, but other factors are introduced which are considered important for predictive purposes. For example, Bachenko and Fitzpatrick [3, 4] believe that syntax plays a lesser role in determining phrasing, and that certain prosodic performance constraints, such as length, override syntactic structure. They allow prosodic boundaries to cross syntactic boundaries under certain conditions, whereas early work was essentially inter-clausal. Other modifications include counting phonological words rather than actual words when determining node strengths [4]. (A phonological word effectively functions as one spoken item, as the internal word-word boundaries are resistant to pausing [2]. Typical examples are determiner-noun word groups, such as 'the + man'.) Further extensions incorporate punctuation into the predictive models [5], or assign more importance to specific features [6].

7.2.1.2   Stochastic methods

With the availability of large corpora, annotated with prosodic information such as location and salience of pauses, temporal information on durations, etc, the stochastic-based approach will come more to the fore. Recently methods for automatically predicting prosodic information using decision tree models have been described [7, 8]. Generally, decision trees are derived by associating a probability with each potential boundary site in the text, and relating various features with each boundary site (e.g. utterance and phrase duration, length of utterance — in syllables/words — positions relative to the start or end of the nearest boundary location, etc) [8]. The resulting decision tree provides, in effect, an algorithm for predicting prosodic boundaries and their saliences (i.e. relative importance) for new input texts.

It is interesting to note that evaluations of both rule-based and stochastic methods show these methods attaining similar results [7].

7.2.1.3   Predicting prosodic phrase boundaries

There are essentially two main components in both of these models — determining potential boundary sites, and assigning boundary salience. This defines a prosodic hierarchy, with strong boundaries typically being associated with pauses.

Boundary location is predicted in the rule-based method by grouping of words into phonological words, and then into phonological phrases, with reference to the syntactic structure for the sentence.

The stochastic approach determines potential boundaries by examining certain factors left to right, from the start of an utterance. An example is given in Fig. 7.2 [8]. Note that this decision tree was tailored towards recognition of prosodic boundaries, to provide clues as to the syntactic structure for speech recognition.

```
if seconds from word on left to end > 0.0495455
then phrase boundary (99.7%)
else  if [no of seconds from last boundary]/[no of seconds from
      last phrase] < 0.601564
      then not phrase boundary (94%)
      else  if node is noun, noun phrase
            then not phrase boundary (88.6%)
            else  if word to left is not accented
                  then not phrase boundary (81%)
                  else  if seconds from start to word on
                        left < 2.49455
                        then not phrase boundary (75.3%)
```

Fig. 7.2 Example decision tree for predicting prosodic boundary location.

Boundary salience is determined in the rule-based approach again with reference to the syntactic structure. A typical strategy is to balance material around the verbs in an utterance [3, 4]. However, this is complicated by the tendency to equalize prosodic phrase size, at least in English. This can be accommodated either as the prosodic structure is derived, or by adjusting the final structures [9].

The stochastic approach determines boundary salience by reference to training data. For example, after a certain duration, or after a number of syllables or words, had passed since the last boundary was predicted, there would be a high likelihood of a salient boundary at the next potential boundary site.

A typical performance structure, predicted using a rule-based approach is given in Fig. 7.3. The node values correspond to the number of phonological words they dominate, and translate directly into strong and weak boundaries, marked ‖ and | respectively.

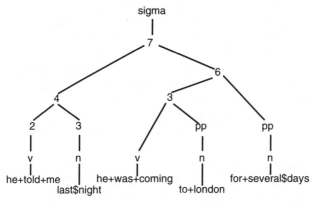

he told me last night ‖ he was coming to london | for several days

Fig. 7.3 Prosodic structure tree.

This performance structure may then be used as a basis for deriving metrical and rhythmic structures, intonation and duration.

7.2.2   Rhythm and metrical structure

One of the criticisms often levelled at synthetic speech is that "...the rhythm doesn't sound right...". Since intonation and duration are the most significant correlates of perceived rhythm, their generation models should be linked to a common framework of rhythmic analysis. Metrical phonology [10] provides such a framework, and a suitable metrical structure can be derived from a combination of performance and syntactic structures.

The fundamental concept of metrical phonology [10] is that stress is represented in terms of the relative prominence of paired constituents. Only two stress levels are considered — s, meaning 'stronger than', and w, for 'weaker than'. These pairs may be at any prosodic level (syllable, foot, word, phrase, etc) giving a hierarchical binary tree structure with w s or s w labelling at each branching node. This is known as a metrical tree.

7.2.2.1   Stress and the metrical tree

The major advantage claimed for metrical trees is that they encode relative prominences of embedded constituents directly, without the need for cyclic application of stress rules. A problem arises, however, when dealing with more complex structures than a single pair. Consider the tree for the compound 'coffee table book', shown in Fig. 7.4 (ignore the grid for now).

Using the w s relationship, 'book' is stronger than 'coffee table', and 'coffee' is stronger than 'table', but nothing can be said about the relative prominences of 'coffee' and 'book' since they are not sister constituents. The definition of a metrical head solves this problem. The metrical head is defined as being the stonger of a pair of sister constituents — hence 'coffee' is the metrical head of 'coffee table'. In this way, the w s relationship between 'coffee table' and 'book' is equivalent to a w s relationship between 'coffee' and 'book'. This allows relative prominences to be determined for any pair of strong constituents at the bottom level of the tree.

The metrical head of any sub-tree, i.e. the constituent dominated only by strong nodes, is called the designated terminal element (DTE). The DTE of a sentence tree has special significance for assignment of phrase focus (nuclear accent) in terms of intonation.

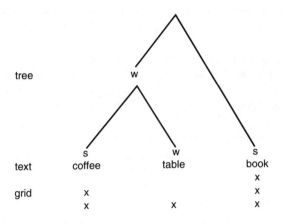

Fig. 7.4   Example metrical tree and grid.

7.2.2.2   Rhythm and the metrical grid

In many cases, relative prominences do not appear to be preserved under embedding in English. The classic example is 'thirteen men'. In isolation, 'thirteen' has a *w s* syllabic stress pattern, but appears to undergo a reversal to *s w* when embedded in 'thirteen men' (iambic reversal or stress shift). This can be viewed as a function of the rhythm of speech, and led to the definition of a linear structure for rhythm to complement the hierarchical representation of stress. This is the 'metrical grid', which shows relative prominences directly, as opposed to the tree which requires tracing of metrical heads to get equivalent information.

The grid is derived from the tree in three steps:

- assign a mark at level 1 on the grid to each terminal node;

- assign a mark on level 2 to each terminal node marked *s*;

- use the metrical head definition to assign marks at higher levels until all relative prominences are correctly represented in the grid (e.g. the DTE should have the highest prominence).

The last of the above steps is formally stated in the Relative Prominence Projection Rule (RPPR):

'In any constituent where the strong/weak relationship is defined, the DTE of its strong sub-constituent is metrically stronger than the DTE of its weak sub-constituent.'

This produces a minimal grid, representing the relative prominences with no redundant information. Referring back to Fig. 7.4, the RPPR ensures that 'book' has a greater prominence than 'coffee'. Accents are assigned to 'coffee' and 'book', with the latter being focal, as it is the DTE of the tree. It is possible to extend the grid without affecting relative prominences by adding levels to certain constituents and re-applying the RPRR. This flexibility allows manipulation of the grid to satisfy rhythmic constraints, but has inevitably led to problems in interpreting the role of trees and grids in predicting stress and rhythm in speech. Three interpretations have been proposed. Tree-only phonology [10] contends that all essential stress and rhythm information is present in the tree. The converse holds for grid-only phonology [11], whereas tree-and-grid phonology [12] maintains that both structures are necessary with trees and grids representing stress and rhythm respectively. The dual approach has been developed mainly by Hayes [12], and seems to produce a more compact phonology with fewer *ad hoc* rules and constraints than either of the single-structure approaches.

The concept of eurhythmy is central to the metrical structure, a detailed discussion of which can be found in Hayes [12]. Basically, this assumes that human perception finds certain rhythmic patterns more pleasing than others, and this influences the rhythmic structure of spoken language. This phenomenon is most evident in highly structured language such as poetry. Various rules can be incorporated to improve eurhythmy, such as the Rhythmic Adjustment Rule (RAR), which generates alternating prominence patterns by transforming *w w s* sequences into *s w s* sequences under certain constraints. This is useful for noun phrases of the form 'adjective, adjective..., noun' which would have a default right-branching structure with the noun as the only strong element.

7.2.2.3   Building a metrical structure

Metrical analysis can be tied to a performance structure, as discussed in section 7.2.1. For example, given the locations and boundary strengths of phonological words and phrases, and possibly higher phrase levels (major, minor, intermediate, depending on the prosodic model), a binary tree can be constructed which represents the phrase structure. Where relative boundary strengths are not defined (within phonological words, and sometimes phrases), rules must be developed to expand the tree fully to word or even syllable level. The default for phrasal constituents is a right-branching structure with *w s* labelling at each branching node. In practice, the labelling for branching nodes is determined by a set of syntactic rules to give an initial tree and grid, which are then adjusted according to eurhythmic rules.

Having built a metrical structure, it can be used to guide the generation of an intonation pattern for the synthetic utterance.

7.2.3   Intonation

Ideally, any discussion of synthetic intonation for TTS should be preceded by a definition of the role of intonation in natural speech. Unfortunately, this is a complex task in itself. For example, the simple sentence:

John drove to London yesterday

could be the answer to at least four distinct questions depending on the intonation. Determining which intonation pattern is appropriate in a given dialogue context is a sophisticated text analysis problem. The intonation algorithms in most practical TTS systems are based on syntactic analysis at sentence level, giving discourse-neutral intonation. Some systems perform rudimentary semantic analysis for given/new information, contrastive stress and other linguistic factors spanning more than one sentence. However, the development of successful semantic analysis algorithms for TTS is still some way off. Consequently, current synthesizers for unrestricted TTS applications concentrate on optimizing automatic discourse-neutral performance, with options to modify the intonation by direct annotation for dialogue applications.

With this in mind, two major requirements of an intonation component for TTS can be defined:

- pitch movements on accented words and at prosodic boundaries should closely mimic those of natural speech;

- accents should be distributed such that a natural rhythm is imparted.

The first of these has received the larger share of research effort to date, so that many systems have the potential to produce good intonation, but fail to do so, mainly due to simplistic accenting algorithms. This discussion will concentrate on the development of an intonational component which aims to satisfy both of the requirements above.

7.2.3.1   Basics of intonational modelling

The problem of generating synthetic intonation can be separated into two parts:

- the phonological component;

- the phonetic (or acoustic) component.

Intonational phonology provides a means of abstractly describing an intonation contour, whereas the phonetic component converts that abstract representation into a physically meaningful signal, e.g. a stream of frequency values (the pitch or F0 contour). Acoustically, the F0 contour appears continuous, and it is tempting to characterize it in terms of 'shapes' which may span whole phrases. However, the modern trend in intonational phonology is to specify a sequence of discrete tones, which are then joined by some interpolation process to form a continuous contour. These tones are usually of two types — pitch accents associated with prominent syllables, and boundary tones associated with varying degrees of prosodic boundaries (these may be derived from a performance structure as described previously). A widely used tonal sequence model was developed by Pierrehumbert [13]. A detailed understanding of tonal sequence models is not required in order to appreciate the essence of the intonational approach under discussion, but the salient points of the Pierrehumbert model are mentioned briefly for convenience.

A set of pitch accents is defined, based on two phonological levels — high (H) and low (L). The basic accents are labelled H* and L*, and can be combined to form bitonal accents with two-tone levels, e.g. H+L*, H*+L, etc (the * indicates the dominant tone, i.e. the one aligned with the accented syllable). Each accent is given a prominence on an abstract scale, which is eventually mapped to an F0 level within a specified F0 range. The model is essentially linear, in that each accent is specified independently, and the phonology does not cater for intonational dependencies at phrase level. One exception to this independence is intonational downstep, which is triggered by certain sequences of bitonals, where the prominence of an accent is a scaled version of the prominence of the preceding accent. As with phrase-level effects, there is no independent phonological control of downstep. These phonological constraints will be examined more closely later.

If a model can produce reasonable approximations to natural pitch contours, then the next step can be considered. This is to determine suitable locations, types, prominences and F0 ranges for the accents and boundary tones by automatic analysis of the text. This task can be broken into three stages. Firstly, pitch accents are assigned at word level, based on the metrical structure of the sentence. Secondly, a phonological model encompassing both accent downstep and phrase-level effects is imposed, which determines the type, prominence and F0 range for each accent. Boundary tones are not discussed in detail, as they are determined by the prosodic phrase structure. Finally, a phonetic model generates an F0 contour which can be imposed on the synthetic speech. The discussion will concentrate on the development of a phonological model which incorporates prosodic phrasing, rhythmic phonology and intonational phonology within a single metrical framework. The phonetic model used is that developed by Silverman [14] for Pierrehumbert's phonology, although other models are worthy of investigation (e.g. Taylor's rise-fall-connection model [15]).

7.2.3.2   Pitch accent assignment

For the discourse-neutral task, it is assumed that all the information necessary for natural accent placement can be derived by a combination of syntactic and rhythmic analyses of the text. Indeed, Hirschberg [16] asserts that good performance can be obtained on longer texts without the need for semantic, pragmatic and discourse information, by training classification trees using currently available text analysis techniques. Monaghan [17] uses a combination of accent over-generation followed by deletion rules to improve rhythm, whereas Dirksen [18] incorporates some aspects of metrical phonology to predict rhythmic structure on a more formal basis. All of these methods attempt to impart better rhythmic structure to synthetic speech, a goal completely ignored by the simple content/function word method used in many TTS systems. In that method, each content, or open class word (basically each noun, verb, adjective and adverb) is accented, and anything else is not.

Within the metrical framework, accents are assigned to words with grid levels of 2 or more (i.e. metrically strong words), as implied in the example of Fig. 7.4. Although this is presented in the context of discourse-neutral intonation, the metrical structure can easily incorporate *a priori* accent placements (e.g. from stored patterns, text annotation, or full semantic analysis when available), and distribute the remaining accents to optimize rhythm.

Having built a tree as normal from the performance structure, it would first be labelled to ensure that each accented node is strong. In particular, if phrase focus is pre-assigned, the focal word should be made the DTE. Additional accents would then be derived by applying eurhythmic rules and constructing the grid, ensuring that the best rhythmic structure is obtained with varying degrees of *a priori* information.

7.2.3.3   A phonological model of intonation

Gradient variability — phonological models generally have a finite set of tonal levels. Unfortunately, examination of natural F0 contours suggests that F0 targets can take a continuous range of values unpredictable from the phonology alone. This has been termed gradient variability by researchers, and is considered by Ladd [19] to be: '...the most serious empirical weakness of a great many quantitatively explicit models of F0.'

In the phonetic realization of F0 contours, Silverman's implementation [14] of the Pierrehumbert model uses two gradient parameters, prominence and range, which allow local and global F0 variability respectively. Although these make provision for gradient variability within phonologically equivalent contours, there is no reliable method of assigning meaningful values to them. In synthesis, a fixed alternating pattern of prominence patterns is generally applied, so that gradient variability in natural F0 contours is treated as essentially unpredictable.

On the other hand, Ladd [19, 20] has proposed another source of variability in F0 based on a single hierarchical structure to control accent downstep, register shift at phrase boundaries, and other effects previously ascribed to purely gradient variability.

The structure is a binary tree for control of pitch register (a register tree), and can potentially produce a much richer variety of F0 contours. The attraction of this approach is that register trees bear a striking resemblance in both structure and philosophy to metrical trees, and can be derived in a principled manner from the metrical structure described in section 7.2.2.

Register trees — since downstep is a relational phenomenon, it only makes sense when applied to pairs of accents or phrases — hence the binary tree structure. Analogous to metrical trees, each binary pair can be only labelled as H-L or L-H. In this case, however, the labelling is not symmetric, as H-L denotes a register downstep, whereas L-H represents no change as opposed to upstep. The highest terminal element (HTE) of a tree or subtree is defined as the terminal node dominated exclusively by H nodes.

To reinforce the analogy, Ladd linearized the tree such that a degree of downstep could be directly associated with each accent. This structure will be referred to as the register grid. The register grid is constructed by applying the Relative Height Projection Rule (RHPR):

- RHPR: in any constituent where the H L relation is defined, the HTE of the L subconstituent is

  — one register step lower than the HTE of the H subconstituent when the L subconstituent is on the right;

  — at the same level as the HTE of the H subconstituent when the L subconstituent is on the left.

With an initial register defined, application of the RHPR defines the register movement throughout the sentence in terms of register steps. The shift between accents within tone groups (true downstep) may differ from that between tone groups (for example, complete register reset may occur between major intonational phrases). This gives considerable flexibility in modelling F0 contours for speech synthesis.

Implementation of a register tree model — a register tree can be derived by collapsing the metrical tree so that the domain of each terminal node contains one downstepping at major boundaries. The register level associated with each accent applies over the domain of the corresponding terminal node. Since the metrical tree is derived from a prosodically based performance structure, register domains correspond to prosodic domains at various levels, ensuring that register shift only occurs at prosodic boundaries. In this way, elements of performance structure, rhythmic phonology and intonational phonology are combined in one cohesive structure, built to satisfy the requirements of all three.

Having determined a register for each accent, an appropriate level (intonational prominence) must be assigned within that register. The simplest method is to assign fixed prominences to each accent type, e.g. the focal or nuclear accent could receive a higher prominence than the others. A more sophisticated approach would be to derive intonational prominences from the grid prominences, thus modelling another aspect of gradient variability.

Assigning accent types — a varied set of accents is useful when characterizing F0 contours in natural speech, but predicting which accents are most appropriate for synthetic speech is a more difficult problem. Many models limit themselves to one accent (high) with interpolation, and some make use of a low accent, but obtaining consistent results with a larger accent set is not practical at present. In combining the Pierrehumbert accents with the register tree model, the following accent assignment has been found to work well:

- assign an H* to the first and last accents in a minor phrase;

- assign H+L* to intervening accents in downstepping minor phrases;

- assign a nuclear H* to the DTE in each major phrase.

This gives a natural downstepping contour within minor phrases, and the final H* signals the end of a tone unit with a definite upward pitch movement.

An example — Fig. 7.5 shows the complete intonational structure generated for the sentence:

years ago a young bride took a dowry to her new husband

with accents from a simple content word method for comparison.

Boundary tones are not explicitly shown in the accent sequence to avoid congestion in the diagram. Minor tone boundaries are indicated by the 'l' character in the text string, and are of the fall-rise type L-H%, with a final fall L-L%. The tree is labelled for both metrical and intonational register relationships (*w s* and H L) at each node, illustrating the unified structure. Collapsing of the metrical structure gives the register tree, the terminal nodes of which are denoted by horizontal bars. The bars also show the domain of the register associated with each terminal node. The metrical and register grids are built as previously described. This example illustrates several important features of the model:

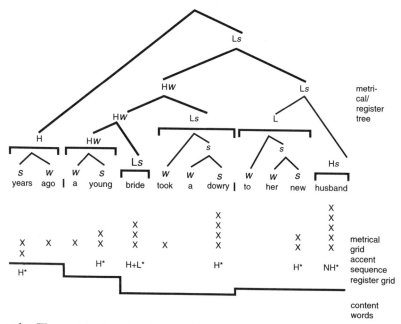

node. The metrical and register grids are built as previously described. This example illustrates several important features of the model:

Fig. 7.5   Example intonational structure.

- compared to the content word method, a more sparse accent distribution has been generated to satisfy rhythmic constraints;

- nested register shift between tone groups is evident — the last tone group is downstepped relative to the second, and both are downstepped relative to the first;

- because register shift is defined between HTEs, accent downstep within the second tone group results in an upward register shift at the boundary with the last group;

- although metrical and register trees share the same structure, they can have independent labelling strategies, e.g. *w s* does not imply H L.

The final stage of prosodic modelling is to predict segmental durations. By using information from the prosodic structure described above, the theme of a unified approach is preserved in the durational model.

7.2.4   Duration prediction

Natural speech clearly has a duration. It exists as an acoustic signal for a finite length of time. The rate at which it is produced may vary from the very slow to the very fast. When a speaker wishes to emphasize a given word or phrase, changing the duration of the speech signal is one of the techniques they employ. Natural speech has a particular rhythm to which we as humans appear to be very sensitive. Clearly, synthetic speech, if it is to sound natural, must model the duration of an utterance.

The first challenge is to decide on an appropriate unit (quantum) of speech. That is the smallest descriptor of speech for which the term 'duration' can sensibly be applied. Clearly a word spoken in isolation has a duration and is a well-understood linguistic unit, but it cannot realistically be used as a quantum in duration analysis. Words have structure, they may be said to contain syllables and depending on the phonological analysis applied, phonemes or possibly tiers of features. In linear phonology, the phoneme and its associated acoustic realization, the phone, act as convenient abstract quanta for durations. However, if the segmental approach imposed by linear phonology is rejected, then some other abstract unit of duration must be postulated which associates with bundles of features. The majority of work conducted on durations in speech have adopted the phone as the quantum for duration modification.

These units of speech are assumed in many statistical analyses of the speech signal to possess intrinsic or inherent durations, which vary due to a set of independent modification factors and hence have a characteristic probability distribution [21, 22]. The Gamma function, which has the property of describing random variables bounded at one end, has been shown to provide an acceptable model for the distribution of phone durations [23]. However, for computational expediency, many researchers use normal or log normal distributions to describe the behaviour of phone durations. The choice of distribution used by researchers depends upon the nature of the predictive model, i.e. whether the model assumes the factors affecting duration are additive or multiplicative. Such methods are exemplified in the classic model developed by Klatt [24] for the MITalk system. Techniques such as CART trees [25] and HMMs [26] represent alternatives to the additive-multiplicative models, and require fewer assumptions about the processes involved in duration changes. However, as a consequence of their simplicity such methods require large amounts of annotated data and inevitably suffer from the problem of sparse exemplar data. Comprehensive discussions on these topics may be found in van Santen and Olive [27] and Kornai [28].

All the techniques described above assume that the quantal unit of speech is being acted upon by a set of external factors which contribute to the observed duration. These factors include:

- phonetic identity;

- phonetic context;

- position within a syllable, word, phrase and sentence;

- speaking style;

- speaking rate;

- lexical stress;

- semantic prominence.

The power of probabilistic methods, such as HMMs, over predominantly deterministic methods, such as multiplicative models, is that much of the variance observed in the duration of segments can be accommodated within a sufficiently complex model, without the need for a clear understanding of how or what factors are affecting the quantal unit. CART tree-based methods represent a half-way house between probabilistic models, which have very few assumptions, and multiplicative models, which are very prescriptive. CART trees attempt to 'learn' the relative significance of a given set of factors.

7.2.4.1 Syllable-based duration models

A number of recent papers have suggested that the syllable rather than the phoneme is a more appropriate and theoretically robust structure for the prediction of durations. These papers suggest that the durations of segments within a syllable can be simply predicted once the syllable duration has been derived [29, 30]. These methods can be viewed as separating the phonetic identity and contextual effects from other prosodic effects.

The aim of phonological theory is not the prediction of segmental durations for synthesis; in fact, many phonological theories would deny the existence of a phoneme, let alone have anything constructive to say about any associated acoustic duration. However, theories of phonology can be used to provide a skeletal structure on to which pragmatic prediction models can be fixed, and, if the theories are appropriately motivated, then the more sophisticated the skeletal structure, the simpler the prediction model needs to be. As a concrete example, if the structures of rhythm and prominence proposed by a number of phonological theories are correct, then, once constructed, they provide sufficient constraints to ensure that any production model will sound 'reasonable', provided it adheres to these constraints.

This assumption underpins a particular model under development at BT Laboratories (BTL) [31], which is based on predicting the duration of a syllable; it is suggested that the syllable represents a better unit for mapping to the acoustic signal than the complex and ill-defined mapping between the phoneme and the phone.

In this duration model the quantal unit of time is not associated directly with any phonetic element or context. It is an abstract quantity which resides on a 'skeletal' or 'timing tier' [32]. The quantal unit is constant and is not affected by the internal structure of the syllable. The prominence or salience of a syllable is, however, affected by the metrical structure of the utterance in which it resides. The stress-timed tendency of English clearly suggests that syllable duration is affected by metrical structure. Given that this method assumes a single quantal unit which is constant for all syllables, internal syllable structure must have an effect and the degree to which it affects syllable duration is dictated by the metrical structure of the utterance. One explanation of this is that there is a clash between the desire to conform to the constant beat on the one hand, and the need to satisfy phonetic and perception constraints on the other. Phonetic constraints include the phonetic identity of the phone while perception constraints consider the cues required to perceive the rhythmic structure in the language.

The phoneme durations predicted by this method still exhibit known problems; in particular, the method of deriving phonemes from the predicted syllable duration is currently over-simplified. As a consequence, phrase final syllables exhibit an asymmetry in the duration of phones in the onset and rime [33]. Any approach such as this succeeds or fails on the accuracy and completeness of the phonological information supplied by the other components of the text-to-speech system. For example, a lack of information concerning the degree of prominence of words in a phrase leads to inappropriately short durations being assigned.

Finally, one constraint seldom mentioned is the interaction between the duration prediction method and the speech synthesis technique. This effect is particularly significant for unit inventory-based synthesis techniques. The duration of a phoneme segment is only one of a number of factors leading to the perception of rhythm in synthetic speech. Factors such as the degree of vowel reduction, assimilation, elision and the intrinsic duration of unit elements can all adversely interact with a predicted duration.

At this stage in the text-to-speech conversion process, a description of the linguistic realization of the original text has been constructed. The last remaining stage is to convert this symbolic representation into the final speech signal.

7.3   SPEECH GENERATION

The process of generating synthetic speech from a linguistic structure involves a mapping from an abstract symbolic representation of language to a parametric continuous representation. This mapping represents a demarcation point between the processes involved in describing language and the processes involved in the production of speech. The parametric information is used to drive some form of speech production model.

Speech production methods can be broadly divided into two categories — those which predominantly model the speech signal and those which predominantly encode aspects of the speech signal. There are two prominent methods of modelling speech — 'articulatory synthesizers' and 'formant synthesizers'. Both generate a synthetic speech signal purely from parametric information driving an abstract model of production. Articulatory synthesizers attempt to produce synthetic speech by modelling the characteristics of the vocal tract and the speech articulators, while formant synthesizers attempt to produce synthetic speech by modelling characteristics of the acoustic signal. Alternatively, examples of the speech signal may be encoded by some process. Encoding schemes vary from the LPC-based methods such as impulse driven LPC, RELP and CELP to simple PCM coders (see Chapter 3).

By far the most successful method of modelling the speech signal is the formant synthesizer. This class of model has been shown repeatedly to produce good quality copies of natural speech, provided appropriate parameters for driving the model are obtained. In theory, such models appear capable of reproducing most sounds used in speech. A number of commercially available systems are based on this technology — for example, DECtalk. The problem with such models has always been in the specification of appropriate control parameters. Currently, most systems which use such models, while highly intelligible, still sound robotic. The 'holy grail' for such systems, is a generalized mapping from some symbolic representation of language to a set of parameters which produces highly natural-sounding synthetic speech. The ability of such models to copy speech accurately has encouraged some researchers to use these models as complex encoders. There are a number of advantages and some disadvantages with such an approach. The advantages are that much higher quality synthetic speech can be produced with comparatively low storage requirements, while retaining the ability to modify the encoded speech signal. The main disadvantage is that producing high-quality copy synthesis is an arduous and time-consuming process, making the production of a large database of speech sounds difficult.

With the advent of cheap computer memory and more computationally efficient methods of manipulating and encoding the speech signal, such as PSOLA [34], the reasons for using a complex model are less clear. In reality, the only significant advantage in using a complex formant model as an encoder lies in its potential for modifying the encoded signal.

The remainder of this discussion will concentrate on concatenative speech synthesis, in which units selected from a stored-speech database are joined together, and artificial prosody is imposed upon the resulting speech sample. However, the unit selection techniques described could equally be used with other methods of speech generation.

7.3.1   Selecting speech units

The phoneme is still the most widely used symbolic representation of sound in text-to-speech systems. Depending on the phonological analysis applied, there are up to 44 phonemes in southern British English, and this set represents the minimum number of symbols required to uniquely describe any word spoken in that dialect. However, simply storing one example phone for each phoneme will not produce good quality speech synthesis. Processes, such as co-articulation, cause the production of one phone to be highly influenced by its neighbours. As an example consider the words 'spoon' and 'spin', which may be phonemically transcribed as /s p u n/[1] and /s p I n/. There are a number of co-articulatory processes acting within these words. The vowels /u/ and /I/ are being affected by the preceding /p/ phone, making the onset to the vowels slightly aspirated, and by the following nasal /n/ making the vowels slightly nasalized. While the vowels, in turn, are affecting the preceding /s/ and /p/ phones and the following /n/ phone. In the word 'spoon', the lips are rounded in preparation for the production of the back rounded vowel /u/, while in the word 'spin' the lips are retracted in preparation for the production of the front unrounded vowel /I/. Such co-articulation effects highlight the need to store phones in context.

The unit inventory methods of speech synthesis generate synthetic speech from a finite set of example sounds. Typically a unit is based around some combination of phones. The fundamental limitation of such systems is the number of example sounds that can be stored. The first and still one of the most common units of speech used in speech synthesis is the diphone. The diphone represents a computationally expedient compromise between the most immediate effects of co-articulation and the desire to reduce the number of units stored. Diphones make two fundamental assumptions about the speech signal, both of which are known to be wrong but are used as pragmatic approximations (see Fig. 7.6). The first assumption is that phones may be considered to be composed of an onset component, a steady-state middle component and an offset component. The second assumption is that the effects of co-articulation can be captured within the transition from one phone to another. The limitations of these assumptions can be seen by reference to the brief description of co-articulation given above. A diphone is defined as containing the transition from the steady-state portion of one phone to the steady-state portion of its immediate neighbour. A better approximation to the underlying speech signal can be obtained by defining a unit that contains the transitions from steady-state portions of neighbouring phones into and out of a phone. Such units are called triphones.

---

[1] Transcribed using the SAM-PA machine-readable symbol set (see Chapter 6).

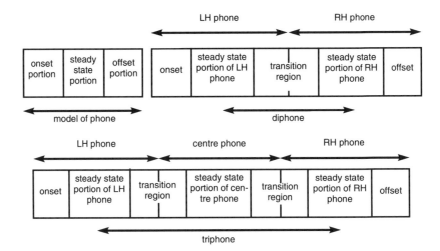

Fig. 7.6   Underlying model of *N*-phone units.

For the southern British English dialect described above which has 44 phonemes and silence, 1848 diphones and 79 422 triphones are required to cover every combination of phonemes. The triphone figure is a large overestimate, as many phoneme combinations are simply impossible in this dialect. A more representative estimate can be obtained by restricting the centre phoneme to one of the 22 vowels present in the dialect. This approximation results in a total of 38 892 triphones — still more than an order larger than the simple diphone set.

Neither the diphone or the triphone are linguistically motivated units, as no account of the underlying structure of language is considered. This has led researchers to examine the syllable as an alternative unit. The proposition is that retaining the syllable structure improves the naturalness of the synthetic speech. However, a database which contained all legal syllables of a dialect and the syllable transition glue would be substantially larger than either a diphone or triphone inventory. A compromise solution proposed by a few researchers has been to develop a unit called the demi-syllable. Such units exhibit some of the properties of the syllable but are much reduced in number [35].

The search to find a set of units which represents an acceptable compromise between the desire to store as few units as possible while maintaining high quality synthetic speech, has led to the introduction of the non-uniform unit or *N*-phone unit [36] systems. The method of non-uniform unit selection involves searching a database of annotated speech for the most appropriate unit sequence. The eventual sequence may contain phones, diphones, triphones or larger units, depending on the minimum cost solution available in the database. *N*-phone unit systems contain a pre-selected set of units where the size of the unit reflects the acoustic complexity of the cluster of phones to be stored.

The previous discussion has assumed that the phoneme is the most appropriate label for the process of unit selection; there are, however, alternatives. Recently, researchers have been investigating other representations of speech data to supplement the simple phonemic description, but the idea that the phoneme label is insufficient is not new. Systems based on formant models have been applying phonetic rewrite rules for many years [37], and more recently non-linear phonology has been applied to the specification of parameters for formant synthesis [38]. It is only with the growing awareness of the significance of prosody that unit-based systems have started to re-examine the sufficiency of the phoneme. Campbell and Black [39] have proposed supplementing the phonemic label with prosodic information, e.g. the type of intonation contour associated with a phone.

A phone carries with it a set of intrinsic characteristics. These characteristics are determined at the time of the recording, where the unit is subject to all the linguistic and paralinguistic forces normally present in natural speech. The significance of the recording has been known for some time. Researchers are constantly striving to define a controlled recording environment which produces units with the best compromise of intrinsic characteristics. One commonly used technique is to embed the unit within a carrier phrase, for example, 'say bat again', 'say bit again', etc. Using this technique database developers can more readily control the degree of emphasis on a word. Poorly controlled word emphasis can lead to overly articulated speech synthesis.

Increasing the amount of information on any one phone obviously adds substantially to the complexity of the unit selection process and blurs still further the minimum requirements of a unit inventory system. The danger is that with an increase in information about the environment of a given phone there will be an associated increase in the number of phones to be stored.

In the end, however complex or sophisticated the unit selection process, such systems represent a compromise between the specification of linguistic information and the size of the unit inventory. A question still remains as to whether some form of compromise can be found that will satisfy all the requirements of a unit inventory. If not, then signal manipulation will prove necessary to supplement missing environments.

7.3.2   Concatenation of speech units

In a concatenative text-to-speech system, once a set of speech units have been selected from a stored speech database, they must be joined to form a complete utterance. Since the units are selected from throughout the database, techniques are required to ensure that the resulting synthetic speech is free from sample-to-sample discontinuities and discontinuities in the pitch period, which would both result in unnatural sounding speech, and cause difficulties in the prosody modification stage (see section 7.3.3).

The simplest method of joining two units is to merge them together, by 'ramping' from one unit to the next, on a sample-by-sample basis. This simple joining process works reasonably well when one or both of the joining sections of speech are unvoiced. However, for joins in which both sides are voiced, it is necessary to employ a more sophisticated technique. One method of improving on the basic merging technique is to find the overlap giving the maximum correlation between the two signals, over the range of a single pitch period, and then carrying out the merge.

There are a number of problems with this method. Firstly, it is necessary to know the pitch period of one or both of the joining speech sections in order to decide on the range for carrying out the cross-correlation. Although it is reasonably easy to visually estimate this from a speech waveform, software methods which attempt to do this are prone to being unreliable and computationally complex. Secondly, the effectiveness of the method is dependent on how well matched the two signals are in terms of pitch period. If the pitch of the two speech sections is significantly different, the match achieved by the correlation process will not be very good.

The overlap used for the merge must be limited to a single period in order to minimize this pitch mis-match effect. This can result in relatively sudden changes in phoneme timbre in the speech after the joining process.

A solution to some of these problems is to synthesize segments from each side of the join in a manner which is pitch synchronous with the opposite side of the join, and then to carry out the merging process with an overlap which spans several pitch periods, as shown in Fig. 7.7.

This pitch synchronous approach solves the problem of limited overlap and also that of pitch period mismatch. However, it still requires that the pitch periodicity of each of the voiced sections is identified. In order to avoid having to carry out this computationally intensive task during the synthesis process, some systems make use of techniques which analyse the pitch characteristics of the entire speech database. This is carried out as an off-line process, which is only required once for each recorded speech database. For all recognizable voiced sections within the speech database a marker is created for a particular point in the pitch period. Using this information, it is possible to carry out the previously described pitch synchronous joining process in a computationally efficient manner.

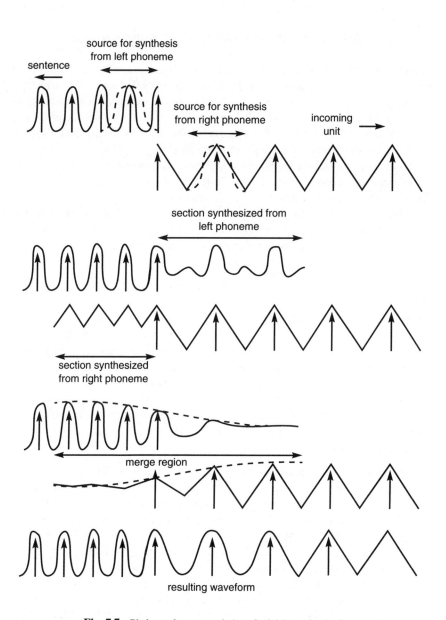

Fig. 7.7 Pitch synchronous technique for joining voiced units.

7.3.3   Pitch time and 'quality' modification

The concatenation technique described above will not generally result in speech with natural sounding prosody. This is because the units that are used to construct the synthetic speech are selected from throughout the stored speech database, on the basis of phoneme content and context. Prosodic characteristics are not a consideration in the unit selection process, so the resulting concatenated speech will have an apparently arbitrary prosodic pattern.

In order to produce a more acceptable sounding speech, it is necessary to apply synthetic prosody in the form of pitch, loudness and segmental duration. Each of these can potentially be controlled throughout the speech synthesis process, but in practice loudness is found to be the least significant in terms of perceived prosody [40]. The rest of this section will concentrate on techniques for modifying the pitch and phoneme duration of synthesized speech.

7.3.3.1   Time-domain pitch-synchronous processing techniques

Techniques of this type make further use of the pitch-synchronous overlap-add method used for joining units, but act on the complete speech signal. The definitive paper on such techniques was written some time ago [34]. The inputs to this process are the speech signal (and its pitchmark information), the boundaries of the phonemes within the speech along with their required duration, and the required pitch contour. The pitch contour consists of a series of frequency values, corresponding to the required pitch at constant intervals throughout the speech. The duration and pitch modification tasks may be carried out in two separate steps, or as a single operation.

- **Duration modification** — the speech waveform is decomposed into a sequence of short-term overlapping signals, which are obtained by multiplying the signal by a series of raised cosine windows centred on the pitchmarks. A new set of pitchmarks are calculated. The number of pitch periods in the new speech signal is modified to give the required duration. The spacing of the pitchmarks is derived from the original sample. The short-term signals are overlap added, centred on the new set of pitchmarks, to produce a new speech signal. Unvoiced sections of speech, which do not contain pitchmarks, first have artificial pitchmarks added, at constant spacing, before being processed in the same way. In the resulting speech, pitch periods from within the original speech signal are either duplicated (when increasing the phoneme duration) or omitted (when decreasing the phoneme duration), according to a mapping function.

- **Pitch modification** — the speech waveform is decomposed into a sequence of short-term signals in a similar way to the duration modification stage. However, the new set of pitchmarks are generated according to the required pitch, as defined in the pitch contour input to the prosody modification process. The resultant speech signal is then formed using the overlap-add process previously described. An example of both pitch and duration modification of a speech signal is shown in Fig. 7.8.

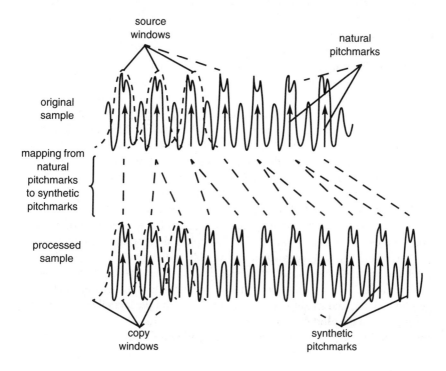

Fig. 7.8   Example of prosody modification using a pitch synchronous technique (duration and pitch increased).

7.3.3.2   Linear prediction techniques

Although time-domain overlap-add techniques are computationally efficient, they are known to introduce audible spectral distortion. Weighted averaging of time-shifted speech signals (or correlated signals in general) results in phase distortion, and has been shown to broaden formant bandwidths, particularly at higher frequencies [34]. If, however, the signal to be modified is uncorrelated and spectrally flat, phase distortion is less relevant as the phase is unstructured to start with, and there are no spectral peaks in the envelope. In speech signal processing, linear prediction (LP) is a widely used technique for separating the signal into a spectral envelope component and a spectrally flat residual or excitation component. By applying pitch and duration modification algorithms to the residual and then recombining with the envelope, distortion around spectral peaks can be reduced.

In the source/filter model of speech production, the LP residual and envelope signals can be roughly interpreted as the source and vocal tract components respectively. In practice, these components are not completely independent, and pitch modification should ideally be coupled to vocal tract modification to model the process more accurately. This interaction tends to play a more dominant role in higher pitched female and child voices, which suffer more degradation from time-domain methods than low-pitched male voices. Techniques based on vocal tract/source separation may therefore prove more robust over a wide range of voices, provided the vocal tract modelling is sufficiently good.

A particular method developed at BT is based upon pitch-synchronous re-sampling of the residual [41]. Smooth resynthesis is ensured by interpolation of filter parameters at each sampling instant between frame centres during analysis and synthesis to avoid transients. The pitch period is modified by resampling a portion of the residual, corresponding roughly to the glottal open phase, so that pitch modification is achieved without recourse to overlap-add. In this way, high frequencies injected at glottal closure are retained, as is the overall shape of the residual signal, and modifications are limited to the perceptually less important open phase of the glottal cycle. To further maintain smooth signal characteristics, the resampling factor is not constant over the modified section, but varies according to a sinusoidal function.

7.4   CONCLUSIONS

The conversion of text to speech has been presented as a two-stage process of analysis and generation. This chapter has described some techniques used in the prosody and speech generation stage of current state-of-the-art commercial TTS systems.

It is a widely held view that the greatest weakness of current TTS systems is in generating natural sounding prosody, and it is the authors' contention that improving the overall rhythmic quality of synthetic speech is the key to solving this problem. Metrical analysis was described as providing a common structural framework for the generation of synthetic prosody, in particular for accenting, so that rhythmic characteristics are maintained.

Current high-quality TTS systems all employ speech concatenation-based methods for generating the speech signal. This chapter has described a framework for the selection of appropriate speech units from a pre-recorded database, and an algorithm for smoothly joining them. It is then necessary to impose synthetic prosody on to the speech, and a popular time-domain method was described. However, there are a number of limitations to such methods, particularly the introduction of phase distortion and the inability to control the spectral shape of the resultant signal. A number of methods have been developed to avoid some of these pitfalls by separately modelling the spectral envelope of the speech, allowing finer control of the spectral shape in the final synthetic speech.

The perceived quality of synthetic speech has shown a significant improvement in the last decade, to the point where it is considered acceptable for many applications, such as telephone-based information provision. While there are many problems still to solve in TTS research in analysis, prosody and speech generation, the next ten years may well see a shift away from straightforward text-to-speech conversion towards more 'intelligent' speech synthesis systems. These systems will be designed to interact with a dialogue model, and will use text as an intermediate representation between an underlying machine 'intention', and the synthetic spoken output. This approach will require a deeper understanding of issues such as controlling of speech style, and the communicative function of prosody. It will then be possible to build machines that 'communicate with' the user, rather than just 'talk to' them.

REFERENCES

1.   Chomsky N and Halle M: 'The sound pattern of English', MIT Press (1991).

2.   Gee J P and Grosjean F: 'Performance structures: a psycholinguistic and linguistic appraisal', Cognitive Psychology, 15, pp 411-458 (1983).

3.   Bachenko J and Fitzpatrick E: 'A computational grammar of discourse-neutral prosodic phrasing in English', Computational Linguistics, 16, No 3, pp 155-170 (1990).

4.   Bachenko J, Fitzpatrick E and Wright C E: 'The contribution of parsing to prosodic phrasing in an experimental text-to-speech system', Proc 24th Annual Meeting of the Association for Computational Linguistics, pp 145-153 (1986).

5.   Minnis S: 'The parsody system: automatic prediction of prosodic boundaries from text', Proc COLING '94, Kyoto (1994).

6.   Wightman C W, Veilleux N M and Ostendorf M: 'Use of prosody in syntactic disambiguation: an analysis-by-synthesis approach', Proc of the DARPA Speech and Natural Language Workshop, pp 384-389 (February 1991).

7.   Ostendorf M, Wightman C W and Veilleux N M: 'Parse scoring with prosodic information: an analysis/synthesis approach, Computer Speech and Language, 7, pp 193-210 (1993).

8.   Wang M and Hirschberg J: 'Predicting intonational boundaries automatically from text: the ATIS domain', Proc of the DARPA Speech and Natural Language Workshop, pp 378-383 (February 1991).

9.   Nespor M and Vogel I: 'Prosodic structure above the world' in Cutler and Ladd (Eds): 'Prosody: Models and Measurement', pp 123-140, Springer-Verlag (1983).

10. Hogg R and McCully C B: 'Metrical phonology — a coursebook', Cambridge University Press (1987).

11. Selkirk E O: 'Phonology and syntax: the relation between sound and structure', MIT Press (1984).

12. Hayes B: 'The phonology of ryhthm in English', Linguistic Inquiry, 15, No 1, pp 33-74 (1984).

13. Pierrehumbert J: 'The phonetics and phonology of English Intonation', PhD Thesis, MIT Press (1980).

14. Silverman K: 'The structure and processing of fundamental frequency contours', PhD Thesis, University of Cambridge (1987).

15. Taylor P: 'The rise/fall/connection model of intonation', Speech Communication, 15, pp 169-186 (1994).

16. Hirschberg J: 'Pitch accent in context: predicting intonation prominence from text', Artificial Intelligence, 63, pp 305-340 (1993).

17. Monaghan A I C: 'Rhythm and stress-shift in speech synthesis', Computer Speech and Language, 4, pp 71-78 (1990).

18. Dirksen A: 'Accenting and de-accenting: a declarative approach', Proc COLING-92, pp 865-869 (1992).

19. Ladd D R: 'Metrical representation of pitch register', in Kingston J and Beckman M (Eds): 'Between the grammar and the physics of speech: chapters in laboratory phonology', Cambridge University Press, pp 35-58 (1990).

20. Ladd D R: 'In defense of a metrical theory of downstep', in van der Hulst and Snider (Eds): 'The phonology of tone: the representation of tonal register', pp 109-132 (1993).

21. Breen A P: 'A comparison of statistical and rule based methods of determining segmental durations', Proc ICSLP '92, Banff, pp 1199-1202 (1992).

22. Breen A P and Easton M: 'A method for estimating segmental durations', Proc Institute of Acoustics, pp 343-350 (1994).

23. Crystal T H and House A: 'Segmental durations in connected speech signals: current results', J Acoustical Soc Am, 83, pp 1553-1573 (1988).

24. Klatt D H: 'Synthesis by rule of segmental durations in English sentences', in Lindblom and Ohman (Eds): 'Frontiers of speech communications research, Academic Press, pp 287-300 (1979).

25. van Santen J: 'Using statistics in text-to-speech system construction', Proc Second ESCA/IEEE Workshop on Speech Synthesis, pp 240-243 (1994).

26. Sharman R A: 'Concatenative speech synthesis using subphoneme segments', Proc of the Institute of Acoustics, pp 367-374 (1994).

27. van Santen J and Olive J: 'The analysis of contextual effects on segmental duration', Computer Speech and Language, 4, pp 359-390 (1990).

28. Kornai A: 'Formal phonology', Garland publishing Inc (1995).

29. Campbell W N and Isard S D: 'Segment durations in a syllable frame', Journal of Phonetics: special issue on speech synthesis, 19, pp 37-47 (1991).

30. Campbell W N: 'Syllable-based segmental duration', in Baily and Benoit (Eds): 'Talking machines, theories, models and designs', North-Holland, pp 211-224 (1992).

31. Breen A P: 'A simple method for predicting the duration of syllables', Proc Eurospeech '95, Madrid, pp 595-598 (1995).

32. Goldsmith J: 'Autosegmental and metrical phonology', Blackwell (1993).

33. Wightman C et al: 'Segmental durations in the vicinity of prosodic phrase boundaries', J Acoustical Soc Am, 91, pp 1707-1717 (1992).

34. Moulines E and Charpentier F: 'Pitch synchronous waveform processing techniques for text-to-speech synthesis using diphones', Speech Communications, 9, pp 453-467 (1990).

35. Jurgens C and Wunderlich M: 'A comparison of different speech units for the German TTS system TUBSY', Proc Eurospeech '95, Madrid, pp 1105-1108 (September 1995).

36. Sagiszaka Y et al: 'ATR — u-talk speech synthesis system', Proc of ICSLP 92, Banff, pp 483-486 (1992).

37. Alan J, Hunnicutt W and Klatt D: 'From text to speech: the MITalk system', Cambridge University Press (1987).

38. Local J and Ogden R: 'A model of timing for non-segmental phonological structure', Proc Second ESCA/IEEE Workshop on Speech Synthesis, pp 236-239 (1994).

39. Campbell N and Black A: 'Optimising selection of speech units from speech databases for concatenative synthesis', Eurospeech '95, Madrid, pp 581-584 (1995).

40. Howell P: 'Cue trading in the production and perception of vowel stress', J Acoustical Soc Am, 94, No 4, pp 2063-2073 (October 1993).

41. European Patent No 94301953.9.

8

BEYOND INTELLIGIBILITY — THE PERFORMANCE OF TEXT-TO-SPEECH SYNTHESIZERS

R D Johnston

8.1   INTRODUCTION

Although text-to-speech synthesis offers the most flexible way of generating spoken messages, it has yet to find widespread use within telephony systems. There are several reasons for this, but the main one is that the perceived quality of synthetic speech has always lagged behind that of recorded natural speech. Only in very specialized niche applications has the advantage of flexibility been sufficient to outweigh the reduced performance.

However, text to speech (TTS) remains the only viable solution for all those applications where unconstrained text must be turned into speech, and the demand is increasing as more text becomes available on-line. The current growth of spoken e-mail provides a particularly important example (see Chapter 5) and there has been an increase in the number and variety of techniques now used (see Chapters 6 and 7). Given this interest there is a need to evaluate the performance of TTS systems, both to compare alternative methods and to determine how systems complement one another.

This chapter begins by reviewing the traditional methods used for telephony transmission performance and describes how they have been adapted for evaluating the performance of synthetic speech. It focuses on subjective methods and explains why tests based upon opinion scores have come to supersede other types of test for 'toll-quality' speech. It describes the reference system which has been developed as a standard for TTS evaluation and shows how it has been validated as a reliable technique. Finally the results of tests on a range of contemporary text-to-speech systems are presented.

8.2   SUBJECTIVE TESTING METHODS USED IN TELEPHONY

Methods based upon subjective experiments using opinion scales and reference systems are now so well established that they are routinely used for planning and improving the speech quality of telephone transmission systems [1].

The most general type of test is the 'conversation' test, in which pairs of subjects converse over various transmission links. Conversation tests take into account sidetone, overall loss, delay and echo as well as the quality of received speech, and so include all the factors which are likely to impact upon the perceived quality of a normal telephone conversation. Chapter 14 discusses some of the related perceptual issues, and Chapter 4 shows how such tests are used for transmission performance evaluation.

However, not all speech is conversational. Recorded announcement services have been available for over 50 years (the Speaking Clock was introduced in 1936) and although glass discs with analogue signals have been replaced by plastic discs with digital signals, many of the latest recorded information services still reproduce speech by concatenating prerecorded speech fragments.

In such cases listening tests provide an alternative to conversation tests, and over the years a number of procedures and recommendations [2] have been developed. It is these which have formed the basis for the methods presented here.

8.3   WHY SHOULD TRANSMISSION METHODS BE USED FOR EVALUATING SYNTHETIC SPEECH?

As the methods used for telephony speech quality evaluation have traditionally been so focused on coded speech, it may be argued that, as synthetic speech is not the same as coded speech, different methods should be used to evaluate it.

For some market segments (e.g. talking toys) this may well be the case, but for the vast majority of telephony-based applications, text-to-speech systems will inevitably have to compete 'head to head' with alternative technologies. These will include 'live' human speech, pre-recorded messages and concatenated speech. Such an approach is already successfully deployed in BT's current direc-tory enquiries systems where a concatenated digit string is appended to a pre-recorded message. Some call-centres already intersperse 'live' speech with pre-recorded and synthetic speech, and with current advances, which allow the syn-thesizer to have the same 'voice' as the recorded speaker, these hybrid approaches will become increasingly common.

For telephony applications, therefore, the assessment method must be capable of comparing and contrasting all types of computer-mediated speech. This requirement has been a major driver for the method developed here.

8.4   BACKGROUND TO SPEECH-QUALITY EVALUATION

There are two main groups of people for whom telephony-based speech-quality evaluation is useful:

- the developers of the technology;

- the users of the technology.

Their requirements are different. By and large the developers want to know if a change to an algorithm or technique has provided a real improvement. Ideally the test should be diagnostic — identifying what it is about the change that has brought about the desired effect.

Users, on the other hand, normally want to compare different commercial offerings of the same technology — often from different suppliers. Sometimes they may even want to differentiate between entirely different technologies and, ultimately, want standards which can be applied to as broad a range of candidate technologies as possible and encourage the growth of a commodity market.

Over the years a wide range of methods have been developed to serve the needs of both groups. All of these test methods tend to fall into one of a number of broad categories, each of which is sensitive to speech qualities ranging from the barely intelligible to high fidelity.

Some of the major 'quality bands' are shown in Fig. 8.1. The application areas served by the particular range are shown at the bottom of the figure and the double arrowhead lines highlight the ranges over which different test method-ologies provide discriminative responses. The categories are defined in the fol-lowing sections.

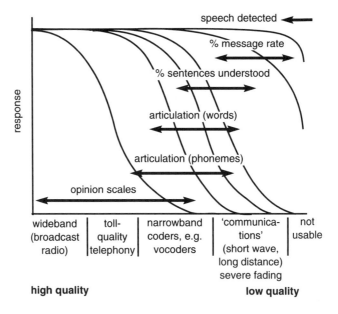

Fig. 8.1 Relationships between quality bands, test methods and typical applications.

8.4.1 Tests based upon detection of thresholds

Apart from total absence, the lowest quality of speech is that which is just detectable. At this level the speech cannot be understood — but a listener will be able to identify it as speech. Such speech need not just be very quiet (audible), it may be buried in noise or grossly distorted. This point is shown at the top right of Fig. 8.1. In principle it may be determined by asking a number of listeners to listen to the signals in question and register when they can detect speech. As the quality improves, other thresholds such as those determined by usability, intelligibility and comfort are reached. However, once any of these thresholds has been passed, the test becomes insensitive to further change. For example, once a system has passed the usability threshold, it is no longer useful to base quality on a usability index. The onset of a threshold effect can be quite sharp — as indicated by the 'knee' in the detectability characteristic — but may be blurred if different types of test material are employed. It may also vary considerably between individual listeners.

The range of sensitivity may sometimes be extended by exploiting masking phenomena where subjects are required to detect signals in the presence of added noise or other structured signals. In general, their role is limited to those cases where similar impairments are likely to appear in real applications. As they do not provide a smooth gradation of quality indices, they are not really suitable for overall quality evaluation [1].

8.4.2   Methods based upon measuring the performance of specified tasks

Just above the threshold of audibility, lies a range where speech is detectable, but the quality is such that there is information loss and frequent repetition or rephrasing of messages is required for meaning to be understood. 'Message rate' metrics such as transaction time, speed of response or job rate are sometimes considered appropriate here.

Apart from the fact that such tests span a range which borders upon being usable, there is also a question as to what is actually being measured, as task responses are confounded by factors such as the subjects' intelligence, stamina, motivation and dexterity. These are 'nuisance effects' in statistical terminology. It is possible to eliminate their perturbing effects by careful statistical design, but only at the cost of a large increase in the amount of work required.

One method occasionally used to try to extend such tests to a higher range of qualities involves the use of a simultaneous secondary task which increases the amount of work (cognitive load) required of the subjects. The secondary task is chosen to be simple and directly measurable, e.g. the subject may be required to repeatedly type the alphabet into a keyboard while simultaneously responding to separate spoken instructions from the synthesizer under test. The quality of the speech synthesizer is then assumed to bear an inverse relationship to the rate at which the secondary task is performed. Apart from the uncertainties highlighted above, this method makes the additional, rather dubious, assumption that subjects are limited by a 'finite cognitive capacity.'

There are some reports of such methods being used for text-to-speech evaluation [3], but most modern synthesizers have a quality above that measurable by such tests.

8.4.3   Articulation tests

In message rate tests subjects hear complete messages and so are often able to use context to 'fill in' those parts which are misheard. By shortening the messages to tokens consisting of single words, syllables or even phonemes, they can no longer do this.

Tests based on such tokens are called 'Articulation Tests', and like message rate tests rely upon there being information loss, the extent of which can be directly measured by asking listeners to label what they hear. Probably more than any other, this is the type of test which has been most widely used among text-to-speech researchers [4]. Typical of articulation tests are:

- intelligibility tests;

- rhyming/analogy tests.

The sounds used may be familiar, distinct words or syllables, in which case one range of qualities is covered. More often they are combinations of phonemes into nonsense syllables known as logatoms — this technique extends slightly upwards the range over which discriminating responses can be obtained. Usually the quality of the received speech is scored according to how well the subjects identify the correct token as part of a 'closed set' of candidates, although in rhyme tests they need only indicate a close match. Such tests have been used in synthesizer development where it is necessary to determine which algorithmic variant produces sounds closest to a 'target' sound [5].

Although the range of quality is higher than that spanned by message rate tests, by definition articulation tests are only valid where the speech quality is so bad that intelligibility is compromised. As such their role has been confined to evaluating very low bit rate vocoded signals, narrow-channel systems occasionally used in military or police applications, or very poor 'public-address' systems.

Once a speech signal has breached the 'intelligibility threshold' articulation tests lose their ability to discriminate. As most of today's text-to-speech systems are above this level, albeit still demanding effort on behalf the listener, methods based upon articulation scores can no longer be considered suitable for text-to-speech evaluation.

8.4.4   Tests based upon equivalence

Tests based upon equivalence form a very powerful and wide-ranging set of tests which can be used for almost every quality band. One of the simplest is based upon pair comparison, where subjects are required to decide which of two signals is preferred. Such tests are widely used when a new or alternative design (e.g. of microphone) or process (e.g. companding law) is being considered as a replacement for an established system.

The subjects' preference may be open (Which do you prefer?), or may be focused on a particular attribute — loudness, distortion, pleasantness, etc. In one variant the subject may adjust one source (the reference) until both are judged to be the same. Impairments being compared need not be identical — subjects may be asked to select which of two entirely different distortions is more pleasant or more comfortable. Such methods have been widely used for setting standards in telecommunications, for example, to compare the quality of a digital coding scheme to that of frequency division multiplexing.

In a single pair comparison, such tests only determine the ranking of systems, but tests based upon equivalence become especially powerful and general when they are used in association with a reference system. By finding the equivalent setting of a reference system it becomes possible to rate an unknown system against a fixed and well-defined standard.

Equivalence tests and preference tests (see section 8.4.5) are closely linked, the difference being the extent of generality. In equivalence tests, subjects are required to judge that two systems are similar 'head to head'. In opinion-based tests, comparisons between more than two systems can be made.

8.4.5   Tests based upon opinions

All of the above methods rely upon subjective responses, but, in most cases, the role of the subject is reduced to that of a 'detector'. However, the fact that humans remain sensitive to very small amounts of distortion even when quality is already high suggests that they could also be used directly as measuring instruments. The only problem lies in devising methods to exploit this sensitivity. The solution is simple — ask subjects directly for their opinion about the perceived quality. At first sight such an approach appears 'non-scientific' and surrounded with difficulties. It may be argued that every listener will have a different set of values and expectations, likely to cause their responses to fluctuate wildly and make analysis impossible. Moreover, their opinions will be influenced by all sorts of other factors — the 'accent' of the voice, their understanding of what is being spoken, even how comfortable they feel about giving their opinions.

However, it is precisely because people's opinions are so sensitive, not just to the signal being heard, but also to norms and expectations, that opinion tests form the basis of all modern speech-quality assessment methods.

Provided that the experiment is based upon a good statistical experimental design and the appropriate analysis methods are used [6], then not only can significant results be obtained from a small number of representative subjects, but the results will be a good predictor of how the total population will react. As a consequence, not only does the method directly address the quality dimensions of interest, but it also directly aligns the results with customer expectations.

By judicious choice of the opinion scale used, tests can be designed to cover any desired range of qualities, and, for telephony-quality speech, two opinion scales now tend to predominate and are established as ITU standards:

- the quality scale:

  — excellent;

  — good;

  — fair;

  — poor;

  — bad;

  which is a particularly general scale used for both conversation and listening type tests;

- the listening effort scale (**effort required to understand the meaning of sentences**):

  — complete relaxation possible — no effort required;

  — attention necessary — no appreciable effort required;

  — moderate effort required;

  — considerable effort required;

  — no meaning understood with any feasible effort.

There is a distinction to be drawn between the responses obtained from each. The listening effort scale applies to speech, whereas the quality scale may be applied to the communication link.

One other common scale is the loudness scale, which, although not a quality scale, may be used for determining the optimum level or range of loudness for comfortable listening. How the same method is used for evaluating very high quality speech is described in an ITU Recommendation [7].

Such methods can be standardized when coupling them to a separate reference system. Then, not only can results obtained at different times or places be more directly and reliably compared, but they may be used for all languages. Section 8.5 describes these in more detail.

8.4.6   Test based upon acceptability

This section would not be complete without mentioning 'acceptability'. All of the above tests are designed to allow comparisons to be drawn between different technologies, and the assumption is made that if one technique is perceived to be of higher quality than the others, then that will also be the most acceptable for any application. That acceptability is relative can be graphically illustrated by the plight of King Richard III:

'A horse, a horse: my Kingdom for a horse.'

It seems unlikely that in more normal times King Richard III would have considered his kingdom to be an acceptable exchange for a horse — yet clearly this had become the case.

This example serves to illustrate that any notion of acceptability is determined by the application, the alternatives on offer, user motivation (including cost), and how pressing the need. It is not an intrinsic property of a technology — nor a horse — and cannot be determined by direct measurement.

The only way in which acceptability can be 'fixed' is by comparison with an arbitrary standard — usually defined in operational terms. For example, in telephony, the level of acceptability was once defined as 'the quality at which 5% of subscribers on limiting connections (lines with more than 30 dB of loss) complained.'

Once established, the minimum specification of a system which meets this standard can be defined. No such operational standard has yet been set for text-to-speech systems — but, as will be shown later, the use of a reference system opens the way for such a standard to be defined and specified.

8.5   REFERENCE SYSTEMS

The first reference system designed for rating telephone speech quality was based upon the notion of the 'one-metre airpath' — see Fig. 8.2.

'Standard' speech was defined as being that which was received at the ear of a listener one metre from the lips of a speaker. This provided a 'gold' standard which had the merit of being simple, was linked to other physical standards, and was independent of language, talkers, listeners and the technologies to be tested.

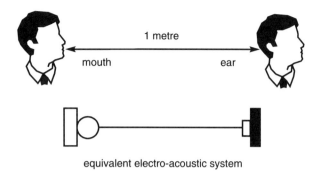

equivalent electro-acoustic system

Fig. 8.2   Electro-acoustic equivalent of one-metre air path.

However, the standard only defined a single quality. In order that it could be used to rate other speech qualities, some means had to be found whereby it could be controllably impaired. The simplest way to do this would have been to progressively increase the air-gap in standard steps — hence increasing the loss. Speech could then be rated as equivalent to that heard over 2 m, 3 m, 50 m, etc. However, this was not considered realistic and the standard eventually chosen was based upon an electro-acoustical equivalent of the one-metre air path. It was was known as SFERT and was first defined in 1928. A number of other models followed, e.g. ARAEN, SRAEN and NOSFER which embodied improved technical implementations [8] (see Fig. 8.3).

Today the standard is defined in terms of the measured electrical and acoustic properties of an intermediate reference system (IRS). The present IRS is defined by CCITT (Blue Book) Recommendation P.48 (Fig. 8.4).

8.5.1   The impairment principle

The impairment principle allows a real system to be rated against the standard — this is illustrated in Fig. 8.5.

The idea is that any 'real' telephone connection has a performance which is 'equivalent' to the reference system impaired by a certain amount. For example, a very long-distance connection might be found to be equivalent to the reference system in combination with a 20-dB loss in the transmission section. By an obvious extension of the same principle, a digital circuit might be found to be equivalent to the reference system with noise injected at 30 dB below the signal level.

Fig. 8.3 AREAN (1948) being used to rate a telephone connection. The specially trained speaker can be seen to be speaking into the handset being rated and into the AREAN 'ball and biscuit' microphone. The instrument on the right hand side of the picture is the meter of the Speech Voltmeter No 3 which allows the speaker to monitor, and keep constant, her speech level.

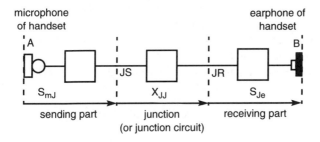

Fig. 8.4 Diagram of IRS.

The determination of such equivalents can only be obtained subjectively. To undertake such tests it is necessary to determine the reference impairment, the speech material, and the criteria upon which subjects base their decisions.

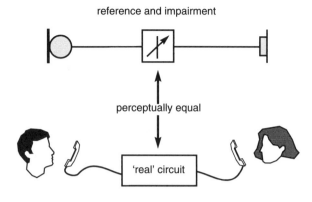

idea of impairment principle and reference

Fig. 8.5   The impariment principle.

8.5.2   Choice of impairments

For early telephony systems the dominant problem in the 'real' network was loss, and a simple attenuator provided an appropriate method for providing a realistic and controllable impairment. For other networks, however, alternatives had to be found. Added noise was used for some frequency division multiplex circuits, and injected 'babble' (noise shaped with a speech spectrum) was used to provide a reference for crosstalk. Probably the best known and most widely used of telephony reference standards is the Modulated Noise Reference Unit, which provides an impairment suitable for rating many digital networks [9].

8.5.3   Choice of material

A choice must be made concerning the speech material used in tests. If the application is known, then the best practice is to use speech that is representative of that application. For telephony systems (and for most TTS applications), this is almost a completely open set, but clearly excludes the use of 'nonsense' material, foreign-language material, and phrases which contain specialist terms or jargon.

The traditional material employed at BT has been based on 'no mental effort' (NME) sentences. These are simple sentences between 5 and 10 words in length selected at random from contemporary novels. They are examined, and any which appear to be complex, ambiguous or contain grammatical constructs, with which some listeners may not be familiar, are removed. Typical of such sentences are: 'The sun shone brightly in a blue sky', and 'The table was covered with a blue cloth'. The meaning of the sentences is deliberately made as simple as possible to try to ensure that the opinions of the subjects are mainly influenced by the signal quality of the material heard — not by its semantic aberrations.[1]

8.6   THE BT REFERENCE SYSTEM

Although general guidelines for all of the above are provided by CCITT [10], specific choices still had to be made to provide a system which was suitable for TTS evaluation. Those choices were determined as follows.

8.6.1   Choice of impairment

For a rating method to be successful, the impaired reference signal must sound similar to the systems under test and to find such an impairment a number of candidates were investigated. First among these was the modulated noise reference unit [10], already used for rating digitally encoded speech in telephony systems. As a well-established standard it would have been ideal if this could have been used without modification. However, it was immediately apparent that the MNRU impairments did not bear any similarity to those produced by any synthesizer; nor did any of the other 'classical' impairment methods — loss, injected noise, babble — transform 'real' speech into a form which could conceivably be considered similar to synthetic speech.

A number of 'novel' candidates were considered next. These included various modulation schemes, e.g. ring modulators to give 'Dalek'-like properties, and several low bit rate speech-coding schemes. Only two produced speech which was plausibly 'synthesizer-like'. These were linear predictive coding [11] and what came to be known as time and frequency warping (TFW).

LPC provided speech which, not unsurprisingly given that a number of synthesizers are LPC based, sounded closest to some synthetic speech systems. However, there was no known way by which an LPC system could be made continuously and monotonically degradable with the required precision. The use of an impairment so closely aligned with one particular method of synthesis was also undesirable.

TFW speech, however, could be varied over a wide range of qualities, and the parameters could be set to simulate some of the grosser surface effects, such as the durational and spectral aberrations present in synthetic speech.

---

[1] We try not to draw the subject's conscious attention to the signal quality any more than is necessary; what we want them to think about, at least in listening effort tests, is how easy or difficult they found it to understand.

8.6.2   Description of TFW

The time and frequency warping method is nothing more than carefully controlled 'wow' and 'flutter' caused by speeding up and slowing down the speech signal. In the original model the speech signal was recorded at a constant sampling frequency and then played back at a variable rate as shown in Fig.8. 6.

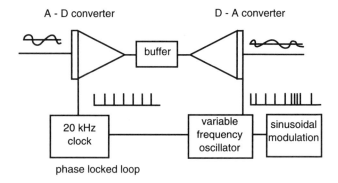

Fig. 8.6   Basis of the TFW impairment process.

The process has three variable parameters — modulation amplitude, modulation period and modulation waveform. After some initial trials a sinusoidal modulation waveform with a period of 150 ms was chosen so that the only variable parameter was the amplitude of the modulation about the original sampling rate. The specific algorithm is as follows.

7.6.2.1   Sampling of original speech signal

Let S_i be the ith sample taken at time T_i when sampled at a constant sampling rate F_s (Hz), where $i = 0, 1, 2,..., k$ and k is the total number of samples.

Hence $T_i = \dfrac{i}{F_s}$

7.6.2.2   TFW playback

Sample S_i is then replayed at time T'_i, where:

$$T'_i = \cfrac{1}{F_s + A_w \sin\left[\cfrac{2\pi i F_w}{F_s}\right]}$$

and A_w and F_w are the amplitude and frequency of the required warping $i = 0, 1, 2,..., k.$

8.6.3   Implementation and refinement

In the original design, special hardware was required to synchronize the clocks, but the present system is entirely software based. Speech sampled at a fixed rate of 16 kHz and 16 bits resolution is processed 'off-line'. The software resamples the processed file using appropriate interpolation to calculate the magnitude of the 'warped' samples.

The software TFW algorithm has three main parameters — the amplitude, the rate and the shape (sine, inverse sine, triangular, etc) of the deviation. After having conducted a number of pilot tests, it was decided to retain the original warping period of 150 ms and to sinusoidally swing the frequency to a maximum of 3.2 kHz either side of the 16 kHz, at which level of distortion the processed speech file is almost unintelligible to the naive listener.

8.6.4   Calibration and validation

Calibration and validation of the reference system was undertaken in a number of experiments to confirm that the system spanned the necessary ranges and to check for monotonicity. Figure 8.7 shows how the system performs using listening effort (LE) as a parameter and shows that the design is both sensitive to listening level and exercises the full range of the response scales.

8.7   APPLICATION EXPERIENCE

This software version TFW reference has now been used for several years to monitor and track performance of BT's own synthesizer (see Chapter 5), and to benchmark a number of commercially available systems with British English or American English output. So far five test series have been undertaken with three synthesizers (or variants of synthesizer) normally included in each, and in this section the results from several of these test series are presented and explained.

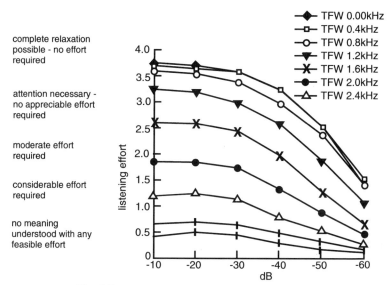

Fig. 8.7   Listening effort versus level (TFW as parameter).

8.7.1   Principal procedures adopted in the tests

In each test, subjects are drawn from a population of untrained naive listeners. The main factors (recorded reference speech, synthetic speech and choice of speech material) are arranged and balanced using standard statistical methods of experimental design and randomized orthogonal squares to minimize bias.

The tests are carried out using the procedures outlined in CCITT P.48 [12] in a sound-insulated room with 50 dBA Hoth noise.The material is presented over a range of fixed listening level loudnesses, from a loudness above that which is comfortable, to a level below that which would be encountered in modern telephone systems.

Each listener hears a series of samples of speech (synthetic and reference) in some predetermined random order. After each sample, the subject indicates his or her opinion of the effort required to understand the meaning of the speech. Subsequent treatment is based upon allocating numerical scores to these responses. A 'repeated measures analysis of variance' [6] is performed to identify which factors in the test are significant.

8.7.2   Main results of experiments

To illustrate the general trends, the results from a total of ten listening experiments are presented (see Table 8.1)

Table 8.1   Summary of experiments.

Experi-ment	Synthesizers	Opinion scale used	Speech material	Subjective results shown in
A	Laureate + A&B	Listening effort	NME Sentences	Fig. 8.8
B	Laureate + A&B	Listening effort	Paragraphs	Fig. 8.9
C	Laureate + A&B	Quality scale	NME Sentences	Fig. 8.10
D	Laureate + A&B	Quality scale	Paragraphs	Fig. 8.11
E	Laureate + C&D	Listening effort	NME Sentences	Fig. 8.12
F	Laureate + C&D	Listening effort	Paragraphs	Fig. 8.13
G	Laureate + C&D	Quality scale	NME Sentences	Fig. 8.14
H	Laureate + C&D	Quality scale	Paragraphs	Fig. 8.15
I	Laureate + E&F	Listening effort	NME Sentences	Fig. 8.16
J	Laureate + E&F	Listening effort	Paragraphs	Fig. 8.17

The first four experiments (A-D) were carried out on three synthesizers. One of these was Laureate — BT's in-house synthesizer — and the other two were commercially available systems running on PCs. The results are shown in graphical form in Figs. 8.8-8.11.

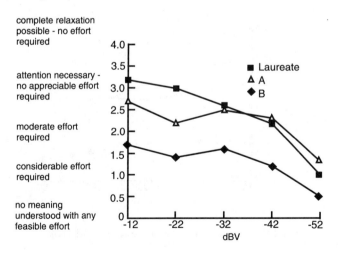

Fig. 8.8   Experiment A — average opinion of listening effort against level, NME sentences.

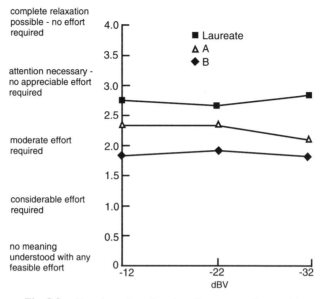

Fig. 8.9   Experiment B — listening effort, paragraph material.

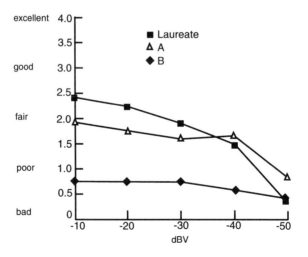

Fig. 8.10   Experiment C — quality scale, NME sentences.

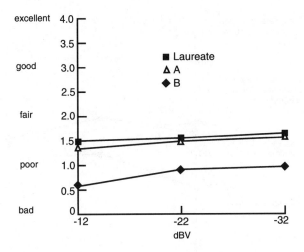

Fig. 8.11   Experiment D — quality scale, paragraphs.

The second set of experiments were the same as the first, except that two different commercial synthesizers were used, and results are shown in Figs. 8.12-8.15.

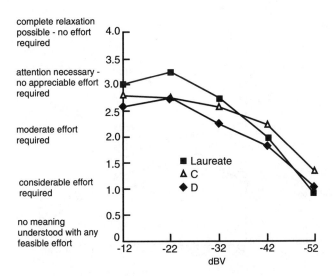

Fig. 8.12   Experiment E — listening effort, NME sentences.

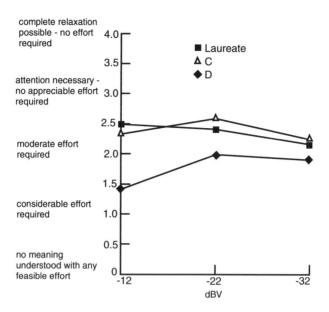

Fig. 8.13   Experiment F — listening effort, paragraphs.

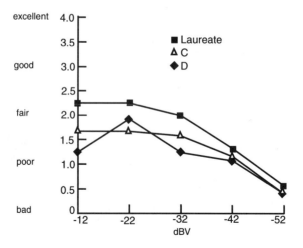

Fig. 8.14   Experiment G — quality scale, NME sentences.

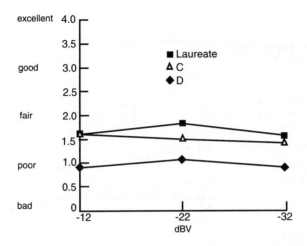

Fig. 8.15   Experiment H — quality scale, paragraphs.

The third set of experiments used another pair of commercially available synthesizers. Only sentence-based experiments were used in this instance. The results are shown in Figs. 8.16 and 8.17.

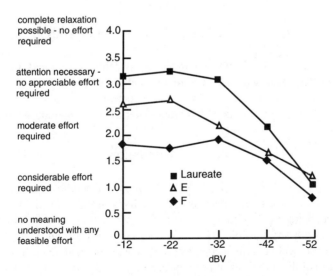

Fig. 8.16   Experiment I — listening effort, NME sentences.

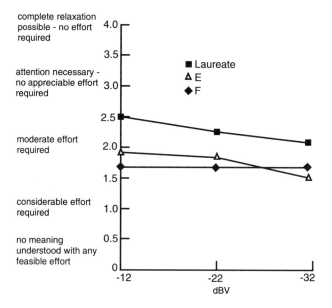

Fig. 8.17   Experiment J — listening effort, paragraphs.

8.7.3   Robustness of the reference system

Finally, in Fig. 8.18 all the 'reference equivalents' calculated for the one synthesizer (Laureate) which was common to all experiments is shown. There is a good correspondence for this in all the experiments with similar reference equivalents being obtained no matter what material or test method was used.

8.7.4   Discussion

The graphs of Figs 8.8-8.17 show the importance of listening level, as would be expected, and it is also clear that for most realistic listening levels the speech is regarded as being intelligible (i.e. it is scored above the category 'No meaning understood with any feasible effort') for every synthesizer when judged on the listening effort scale at reasonable levels. It is also apparent, even from a casual examination of the orderings of preferences, that subjects rank the synthesizers in a very similar way, no matter what material or opinion is used.

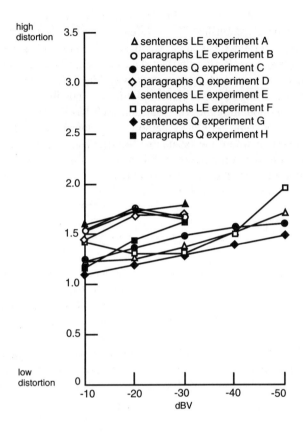

Fig. 8.18 Reference TFW equivalent for Laureate found in eight different experiments.

As would be expected, the use of a different opinion scale produces a different set of absolute scores in all cases. Also the opinion scores recorded on paragraph material are lower (i.e. subjects are more critical) than is the case with sentence material; in principle the use of a 'reference' distortion should normalize the effects. The results shown in Fig. 8.18 depict the extent to which these are normalized out by the reference system.

The listening effort scale appears to be more sensitive than the quality scale for this type of test, as measured by the range of categories used by subjects, and so is to be preferred for future tests. Also, despite the fact that the paragraph material is longer and elicited more critical responses it did not appear to provide better sensitivity than the original NME material.

8.7.5   Possible refinements and future developments

There are a number of areas where further investigation is required. In particular, it has not yet been established if the reference system is robust to different languages (a test in French is planned), and so far the system has only been used for 'male' voices — reflecting the fact that until recently there were very few 'female' voice text-to-speech synthesizers available. The reference system still lends itself to a number of refinements, e.g. it may be possible to find a distortion which is better at modelling the real distortions of TTS. The system is now being considered as a possible reference method for certain types of low-bit digital transmission system where the 'standard' MNRU distortion is inappropriate. It has also been considered for future multimedia assessment, particularly where it may be used to measure the perceived quality enhancement provided by having an associated picture of the speaker on a screen.

8.8   CONCLUSIONS

This chapter has described how a simple reference system based on time and frequency warping has been developed and how it has been used to provide a standard way for rating contemporary TTS systems. Results showing the extent to which the reference system is independent of test methodology have been presented, but clearly there are opportunities for refining the process further. The system has been successfully used to provide a technology-independent means for monitoring in-house technology and for evaluating systems available in the market-place. It has been proposed as an ITU standard for evaluating synthetic speech systems.

REFERENCES

1.   Richards D L: 'Telecommunications by Speech', Butterworths (1973).

2.   CCITT Blue Book (1988).

3.   Yaschin A et al: 'Performance of speech recognition devices — evaluating speech produced over the telephone network', ICASSP (1989).

4.   Speigel M et al: 'Using a monosyllabic test corpus to evaluate the intelligibility of synthesized and natural speech', Proc of the American Voice I/O Systems Conf, San Franciso, USA (October 1988).

5.   Voiers W D: 'Evaluating processed speech using the Diagnostic Rhyme Test', Speech Technology, 55, pp 30-39 (1983).

6.   Cochran W G and Cox G M: 'Experimental design', Wiley (1957).

7.   ITU: 'Methods for the subjective assessment of small impairments in audio systems including multichannel sound systems', Recommendation BS116 (November 1993).

8.   SFERT: Systeme Fondamental European de Reference pour la Transmission Telephonique.
      ARAEN: Appareil de Reference pour la Determination de l'Affaiblissement Equivalent pour la Nettete.
      SRAEN: Systeme de Reference pour la Determination de l'Affaiblissement Equivalent pour la Nettete.
      NOSFER: Nouveau Systeme Fondamental pour la Determination de l'Equivalent de Reference
      [See CCITT White Book, Vol V (1968)].

9.   CCITT Blue Book, Vol V: 'Modulated Noise Reference Unit, P. 81', (1988).

10.   CCITT Blue Book, Vol V: 'Methods used for assessing telephony transmission performance', (1988).

11.   Markel J D and Gray A H: 'Linear prediction of speech', Springer-Verlag (1976).

12.   CCITT Blue Book , Vol V: 'Sound rooms, P.48', (1988).

9

THE LISTENING TELEPHONE — AUTOMATING SPEECH RECOGNITION OVER THE PSTN

K J Power

9.1   INTRODUCTION

The impression often evoked by the term 'speech recognition' is of effortless spontaneous conversation with a computer. This ideal has long been inspired by examples from science fiction like Hal in '*2001: A Space Odyssey*', the ever-alert shipboard computer on the *Enterprise*, and, more recently, 'Holly' in *Red Dwarf*. These ideals embody the goals of today's speech researchers, but, while impressive advances have been made in recent years, achieving them is still some way off, requiring the facility not only to recognize speech but also to understand it. Nevertheless, with today's technology, if systems are designed carefully and the recognition domain is suitably restricted, tasks ranging from dictation of letters to catalogue shopping by telephone can be automated.

What makes automatic speech and speaker recognition (ASR)[1] such fascinating challenges is the variability seen (or rather heard) in the signals that must be classified. Firstly, speech signals exhibit enormous variability, even for a speaker repeating the same sound under identical conditions; then add to this different background noises, ranging from interfering speech to traffic noise, and compound the issue by passing the signal through a telephone handset of unknown type; finally, send it through the telephone network with its own noise and restricted bandwidth (300-3500 Hz) — the scale of the problem can then be appreciated.

---

[1] The abbreviation ASR refers to 'automatic speech and speaker recognition' throughout this chapter.

Even more variability arises if the speech originates from more than one speaker, because of differences in vocal tract, accent and speaking style. The level and type of background noise also have significant effects on the speech signal.

Some systems remove a lot of this variability by restricting themselves to working for a single speaker. Such **speaker-dependent** recognizers are trained on the speech from a single user and are usually only used over a single channel. Such systems require an enrolment session where users provide examples of their speech. A typical application of speaker-dependent recognition is PC-based dictation. The task of speaker recognition is, of course, necessarily speaker dependent. **Speaker-independent** recognizers, on the other hand, are designed to accept speech from any speaker and have the advantage that they do not require any enrolment from new users. They are therefore the only possibility for many telephony-based services where callers are unknown in advance.

There are many types of recognizer — for both speech and speaker recognition — each optimized for a different task. The mechanism for encapsulating task constraints (those limitations in the scope of the application domain that make recognition feasible) is described for several real systems and practical issues are outlined.

9.2   BASIC COMPONENTS

The basic principles underlying both speech and speaker recognizers are similar. They are illustrated in Fig. 9.1.

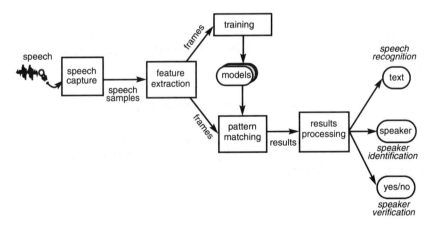

Fig. 9.1   Common components of speech and speaker recognizers.

9.2.1   Speech capture

The input to a recognizer is the physical speech signal. During speech capture the signal is sampled, digitally encoded and stored. In many telephony-based applications, detection of when to start and end speech capture is also necessary.

9.2.2   Feature extraction

Recognition is not performed directly on the sampled signal because it contains a great deal of redundant information. Instead, the signal is analysed to extract key features for either speech intelligibility or speaker characterization. This process differs slightly for speech and speaker recognition, but in either case the output is a sequence of speech **feature vectors** or **frames**.

9.2.3   Modelling speech sounds

Both speech and speaker recognizers attempt to identify patterns in the feature vectors corresponding to speech units such as words, phrases or phonemes. This is done using one or more models of each required speech unit, and a statistical model — known as the hidden Markov model (HMM) — is most widely used for recognition. Models must first be created in a separate training phase before being used for recognition.

9.2.4   Pattern matching

Speech signals are usually assumed to be a sequence of speech sounds. Pattern matching is the process of finding the best model sequence that matches the signal. This is usually done by a parser. The output is the recognition result — typically a set of hypothesized model sequences with accompanying scores.

9.2.5   Results processing

Further analysis is performed on the pattern-matching results to arrive at the recognition decision. The process is different for speech and speaker recognizers to produce the desired form of output as shown in Fig. 9.1.

9.3   SPEECH CAPTURE

Although the process of sampling the speech signal is relatively straightforward, accurate detection of when to stop speech capture is by no means easy. Similarly, in some systems a means of detecting when speech has started is needed. Both these scenarios are an integral part of the complete speech capture process often required for a live system.

9.3.1   End-of-speech detection

For many applications the recognition component does not run continuously. Generally, a recognition process once started must stop at some point, returning control to the application. One option is to stop recognition after a fixed time, but this introduces an undesirable delay if the user finishes speaking before the recognition window closes. Also, if the user delays before speaking, there is a risk of prematurely stopping speech capture.

A better strategy is to analyse the pattern-matching results as shown in Fig. 9.2. This requires models of non-speech sounds, i.e. background noise or silence, so that the entire speech signal can be better modelled. The overall results then reflect the results to be incorporated in the hypothesized model sequence. A sequence ending in one or more noise models indicates that the corresponding final portion of the signal is non-speech. Once post-speech noise — along with various other criteria — is detected, speech capture can be halted.

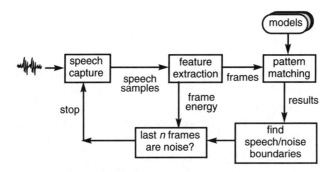

Fig. 9.2   End-of-speech detection.

9.3.2 Overriding a spoken prompt

In a spoken-dialogue application, users tend to become familiar with the system and it is therefore desirable that they can override prompts and so speed up their transactions. This is known as cut-through or 'barge-in'.

A major problem arises, unfortunately, because incoming speech becomes corrupted with echo from the outgoing prompt and this has detrimental effects on recognition accuracy. In a telephony application, for example, this is further exacerbated by both acoustic and electrical coupling of the incoming and outgoing signal inherent in the telephone connection, as illustrated in Fig. 9.3.

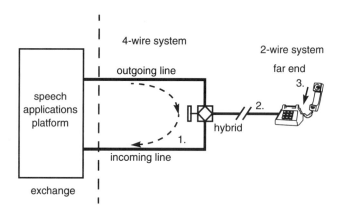

Fig. 9.3   Causes of echo in the PSTN (1. impedance mismatch in hybrid; 2. electro-acoustic coupling; 3. acoustic feedback in the handset).

One practical and effective solution is to detect an override as soon as possible and deactivate the prompt. If the system plays out a short (non-overridable) welcome announcement this may be used to estimate the echo level. Subsequent echo detection may then be performed sample by sample or on a per frame basis. Detection of echo is done on a per-frame basis for speed. A frame can be classified as speech, echo or noise. Energy-based features allow some discrimination — as users tend to speak louder as the background noise level increases (this is known as the Lombard effect [1]).

Even when the prompt is rapidly deactivated, some echo will still be mixed with the speech signal. This can reduce recognition accuracy by up to 10% for small vocabularies. Techniques similar to those used for noise robustness can be used to estimate the echo spectrum and (assuming signals are additive in the spectral domain) subtract it from the input spectrum and hence maintain recognition accuracy.

9.4 FEATURE EXTRACTION

During feature extraction, the input speech samples are transformed into a domain where information of importance for recognition is preserved and redundant information discarded. This signal-processing stage is usually quite computationally intensive.

To give some idea of the variability of speech signals, Fig. 9.4 shows time-domain speech waveforms for three speakers saying the word 'zero'. These waveforms are of similar duration but have no obvious common pattern.

Patterns in the speech signal become much more evident when the signal is transformed into the frequency domain. Figure 9.5 illustrates corresponding spectrograms — the vertical axis denotes frequency and the horizontal axis denotes time. The power of the frequency components is proportionally represented by darkness. The resonant frequencies of the vocal tract (formants) appear clearly as the dark bands flowing across each spectrogram.

The typical process of frequency analysis of the sampled speech signal is illustrated in Fig. 9.6. The samples are first assembled and windowed into overlapping frames. The overlap period is usually chosen to be in the range 10 to 20 ms during which speech is assumed to be quasi-stationary. A fast Fourier transform (FFT) is used to calculate the power spectrum.

The power spectrum is then organized into frequency bands according to a series of Mel-scale filters as shown in Fig. 9.7. These filters are spaced linearly to 1 kHz and then logarithmically up to the maximum frequency, reflecting the sensitivity of the human ear to changes in frequency. In this way the power spectrum can be represented by about 20 Mel-scale filter outputs — considerably reducing the data rate.

The dynamic range of the power spectrum is quite large and hence the logarithm is taken of the Mel filter outputs. This accords with human perception of sound intensity which is thought to vary with the logarithm of intensity.

Finally, a discrete cosine transform (DCT) is performed on the logged Mel filter outputs. This is given as:

$$C(k) = \sum_{i=0}^{N-1} f(i)\cos\left(\frac{\pi(i+0.5)k}{N}\right) \quad k\in[0, M] \qquad \ldots (9.1)$$

where $C(k)$ is the kth DCT output and $f(i)$ is the ith of N log filter-bank outputs. The DCT output is known as the cepstrum (though this is not, in fact, a true cepstrum which is the Fourier transform of the log of the power spectrum [2]). Two important functions are served by this transform. Firstly, it acts as a data reduction process. The power spectrum envelope varies slowly over the frequency range and so M is usually much less than N in equation (8.1).

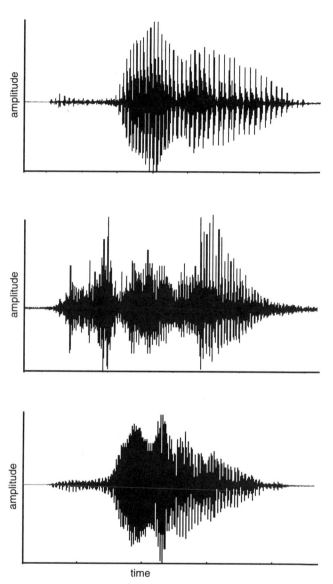

Fig. 9.4   Time-domain waveforms for three talkers saying the word 'zero'.

Secondly, the DCT outputs are relatively uncorrelated so that each output value can be assumed to be independent of every other value. This is ideal for most types of pattern matching [2], as the vector operations for training and recognition can be greatly simplified.

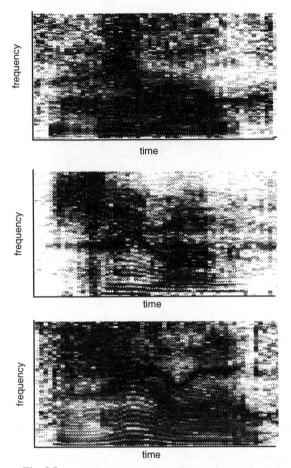

Fig. 9.5   Spectrograms for three talkers saying word 'zero'.

Each feature vector contains a subset of the cepstral coefficients. In addition, the time derivative of the cepstral coefficients computed over successive, non-overlapping, frames is often included. The differential logarithm of frame energies is also often included. The final feature vector may typically consist of:

$$\begin{bmatrix} C(0) \\ \dots \\ \\ C(M) \\ \Delta C(0) \\ \dots \\ \\ \Delta C(M) \\ \log \ \text{energy} \\ \Delta \log \ (\text{energy}) \end{bmatrix}$$

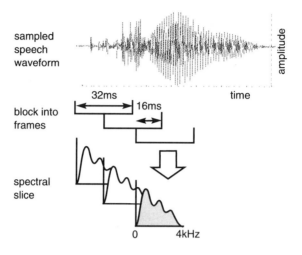

Fig. 9.6   Typical spectral analysis of the speech signal, showing 'overlapping' process.

9.5   MODELS — THE PATTERN-MATCHING ELEMENTS

For recognition, some form of model of the entities to be recognized is needed. The role of a model is to represent the common patterns of similar speech sounds while also allowing for variability.

This section first describes the application of statistical models known as hidden Markov models (HMM) for pattern matching. Next, the HMM training process is discussed, i.e. when the models 'learn' the sounds to be recognized. Then some of the limitations of HMMs are discussed together with details of an alternative model.

9.5.1   Hidden Markov models for recognition

An acoustic pattern, e.g. the spoken word 'fred', is represented by a sequence of feature vectors. Different acoustic realizations of the word 'fred' give rise to

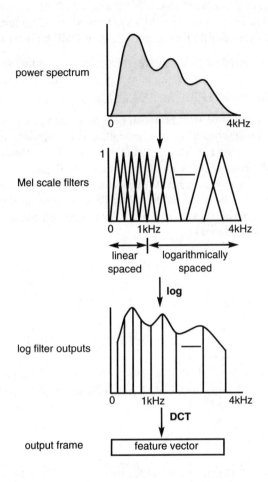

power spectrum

Mel scale filters

log filter outputs

output frame

Fig. 9.7   Transformation of the power spectrum.

different vector sequences. A good model must deal with the two following types
of variability presented by vector sequences.

- Temporal variability — different vector sequences corresponding to a sound
  will generally differ in duration. This is due both to differences in speaking

rate and to the inexact nature of human speech production. Even for vector sequences of the same duration the proportion of vectors for the /f/ sound, say, will vary as different parts of the word will be stressed differently.

- Acoustic variability — feature vectors corresponding to an /f/ sound will vary within a spoken example of the word 'fred' and even more so across different speakers and channels.

The great power of HMMs lies in how they can accommodate both types of variation. With HMMs, a signal is modelled as a **stochastic** process, i.e. can be characterized as a probabilistic function of a random variable.

An HMM consists of a number of states as illustrated in Fig. 9.8. In this example, four states are chosen to represent each of the four sounds of the word 'fred'. Transitions between model states have an associated probability (which can be zero). Transitions are usually chosen to reflect the temporal progression of speech sounds as shown.

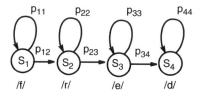

p_{ij} probability of going from state i to state j

S_i state i

Fig. 9.8   A hidden Markov model of the word 'fred'.

The simplest case arises when each speech frame can be uniquely assigned to one of the four states. Suppose there are six frames for an example of the word 'fred' producing the state sequence $S_1S_1S_2S_3S_3S_4$. For a Markov model the probability of a state sequence is simply the product of state transition probabilities, in this case $p_{11}p_{12}p_{23}p_{33}p_{34}$. A model output probability can be calculated for frame sequences of any length and with varying state durations.

In practice, the spectral properties of a sound are too variable to allow a unique mapping of frames to states. Instead, every frame is mapped to **all** states with an associated (non-zero) probability — the state-output probability. As each frame cannot be uniquely matched to one state, the model output probability is the sum of all state-sequence probabilities. The calculation of the state-sequence probability must also incorporate the state-output probability for each state in the sequence. The Markov model in this case is 'hidden' because the true state sequence traversed through the model is unknown.

The number of possible state sequences for a frame sequence is, in practice, unfeasibly large. Assume, for simplicity, that all valid state sequences start in state 1 and terminate in state 4. A word of this duration might span 30 frames of speech, giving rise to 9×4^{25} potential state sequences. A recursive algorithm is used to efficiently traverse all possible state sequences 'on the fly'. A recursive term $\alpha j(t)$ is computed for each state j, for each frame number t. This is illustrated in Fig. 9.9(a).

Alternatively, a state sequence can be thought of as a path through a state-time lattice. The term $\alpha j(t)$ is the total probability for all partial paths entering state j at time t as shown in Fig. 9.9(b). The model output probability for the six-frame sequence is the sum of the probability of each path through the lattice from state 1 at $t = 1$ to state 4 at $t = 6$. Note that the model cannot output a probability until a minimum of four frames has arrived. This can be seen from examining the transitions linking state 1 to state 4 on the left of Fig. 9.9(b). It is also clear from

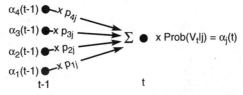

(a) Calculation of $\alpha j(t)$, the probability of being in state
     j at time t

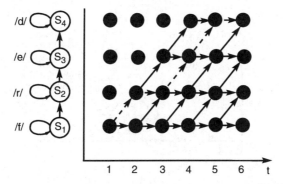

(b) Possible paths through the model with given
     transitions for six frames of input

Fig. 9.9   Viterbi decoding of best state sequence.

the lattice shown in the same figure that the first path into state 4 is propagated at time $t = 4$.

It is usual to replace the calculation of $\alpha j(t)$, shown in Fig. 9.9(a), by changing the summation operator to a 'maximum' operator. This is known as the Viterbi algorithm [3]. At each point in the lattice only the path with the highest probability is propagated forward. The model output probability thus represents the most likely state sequence rather than all possible sequences with the appropriate start and end states. The highlighted path illustrated in Fig. 9.9(b) corresponds to a most likely sequence of $S_1 S_2 S_2 S_3 S_4 S_4$.

9.5.2 State-output probability — Prob($V_t \mid j$)

The probability of feature vector V_t given state j ($Prob(V_t \mid j)$) in Fig. 9.9 is an indicator of how closely feature vector V_t matches the sound modelled by state j. A popular technique is to use a continuous probability density function (pdf). Such models are known as continuous density hidden Markov models (CDHMMs). The Gaussian distribution — shown for a one-dimensional feature vector in Fig. 9.10 — is frequently used. This models quantities which vary randomly about an expected or mean value. The popularity of the Gaussian pdf

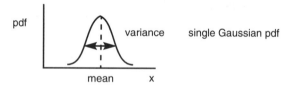

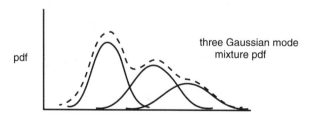

Fig. 9.10   Probability density functions of a one-dimensional feature vector.

arises because any pdf can be modelled by a weighted **mixture** of Gaussian pdfs. Each constituent pdf, or mode, of such a mixture can be thought of as representing a different variation (e.g. accent) of the sound within the state. A mixture with three Gaussian modes is also shown in Fig. 9.10.

As feature vectors usually have more than one component, a **multivariate** Gaussian mixture pdf is often used. Computation of the state output probability is greatly simplified if the feature vector components are uncorrelated. In this case each component is then described by a one-dimensional Gaussian mixture pdf and the probability of individual components are multiplied together to obtain the state-output probability for the feature vector.

9.5.3   Model training

The most difficult aspect of pattern recognition, generally, lies in training models for pattern matching. Broadly speaking, there are two forms of training — **supervised** and **unsupervised**. In supervised training, the pre-categorized examples of the item to be recognized are presented to the training process. Unsupervised training is data driven and consists of letting the training process derive categories for recognition from the training data. Humans tend to be very adept at unsupervised learning, i.e. noticing a new pattern in the environment (visual, auditory) and learning it for future recognition.

Training of HMMs is normally carried out in a supervised mode. As HMMs model the statistics of a signal, many examples are usually required for training. These are collected in speech databases which may be stored as files of speech data. In such instances, speech data must be annotated to be of use for supervised training. Annotation consists of specifying where each speech example starts and ends, and labelling the example according to how it is to be classified. Manual annotation of speech training data usually leads to best results as the expertise of a human listener at detecting word boundaries is captured by the process.

The Baum Welch algorithm [4, 5] is a powerful method used to determine model parameters such as state transition probabilities, means, and variances of Gaussian pdfs. This algorithm is iterative and requires an initial estimate of the model parameters. This initial estimate is generated during a separate phase.

9.5.4   Model initialization

This process starts with an arbitrary assignment of feature vectors to states for each training example (e.g. word, phoneme). The vectors assigned to a given state are pooled for all training examples. When all vectors from all examples have been thus assigned, the vectors in each pool are **clustered**. This is the process of searching for M clusters in N-dimensional feature vector space (for M Gaussian modes). In this way the mean of the vectors within a cluster becomes the initial estimate of a Gaussian mode.

The next phase of initialization uses this estimate of the parameters, but this time the assignment of vectors to states is not arbitrary. The Viterbi algorithm, discussed earlier, is used to decide the most likely state sequence through the model for a training example, and hence to which state pool each vector is to be assigned. After the vectors from all training examples are assigned, the state pools are again clustered. The process is iterated on the training data until a suitable convergence criterion is met.

9.5.5   Baum Welch training

The Baum Welch algorithm takes an initial model λ and estimates a new model λ' using all the training examples. During this process the model output probability $P(V|\lambda)$ is calculated for the vectors V of each training example. The algorithm guarantees that:

$$P(V|\lambda') \geq P(V|\lambda)$$

i.e. the model output probability for the training data set is at least as good as for the initial model. This algorithm is repeated using λ' as the initial model for the next iteration.

The principle behind Baum Welch training is fairly straightforward. In statistics, the mean, or expected, value of a variable x can be estimated by the summation:

$$\bar{x} = \sum_{\forall x} xP(x) \qquad\qquad ...(9.3)$$

where $P(x)$ is the probability of x for all values of x. The current estimate of the model is used to calculate $P_j(V_t)$ the probability of being in state j for feature vector V_t given an entire training example. Assuming a single Gaussian pdf per state (for example), its new mean vector is the accumulated product $V_t P_j(V_t)$ over all feature vectors V_t over all training examples. This is shown in Fig. 9.11.

For each iteration over the training data the probabilities in Fig. 9.11 will be different from the last iteration for a given training example. The probabilities more accurately reflect the 'best' assignment of a feature vector to a state.

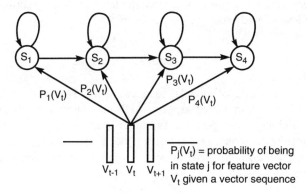

$\overline{P_j(V_t)}$ = probability of being in state j for feature vector V_t given a vector sequence

Fig. 9.11   Model training.

9.5.6   Limitations of HMMs

HMMs provide a very powerful mechanism for simultaneously modelling temporal and spectral variations of an acoustic signal. However, some assumptions are made in the application of HMMs to speech signals which seriously affect both robustness and recognition accuracy. Much research has been done in the last decade to overcome the theoretical limitations of HMMs for modelling speech.

The most dubious assumption is the independence assumption. This assumes that successive feature vectors matched to a model state are **independently** and **identically** distributed. This implies that the state output probability calculation makes no allowance for the influence of preceding vectors. The state probability calculation is indifferent to improbable events such as, say, a ten-vector syllable where each vector represents speech from a different speaker. HMM variants, such as the segmental HMM [6] and inter-frame dependent HMM [7], have been postulated to overcome this limitation.

However, the assumption that vectors mapping to a state are identically distributed is also a simplifying gross approximation to real speech signals. Essentially this models speech as a series of discontinuous jumps through a sequence of sounds. In reality the position of the vocal tract for one speech sound greatly influences how the following sound evolves.

Another weakness of HMMs is simply that sound durations are unrealistically modelled. This seems contradictory, as one of the strengths of HMMs is that they accommodate patterns which evolve through time allowing arbitrary duration signals to be matched to a model. However, the actual durations of the sounds themselves are inaccurately modelled by HMMs which impose an exponential model upon the duration. Consequently, implausibly long or short model-state durations are not penalized, which affects recognition performance. Strategies to incorporate better HMM durational models have been described in Levin-son [8] and Russell and Moore [9].

9.5.7   Neural networks

An alternative to statistical models, such as HMMs, is that of artificial neural networks (ANNs). As the name suggests ANNs are inspired by theories of how neurons in the brain are organized. The input to an ANN is typically a speech feature vector. Each output may represent a recognition outcome, e.g. a sound to be recognized. An example ANN known as a multilayer perceptron is shown in Fig. 9.12.

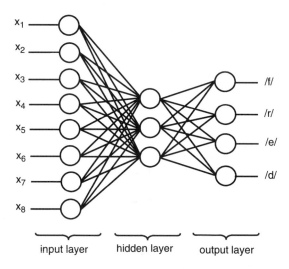

Fig. 9.12   Multilayer perceptron for recognition.

The inputs are the components of a feature vector. Each component connects to a node in the input layer. Each input node then connects to all the nodes in a **hidden** layer. Each output node represents a sound for the word 'fred' and is fed from all hidden nodes. The highest output node is usually taken to indicate the recognized sound.

ANNs distinguish different inputs by finding partitions between vectors for different patterns. This is the **discriminative** approach to pattern recognition. The number of nodes in the hidden layer limits the number of boundaries the perceptron can find. Adding more hidden layers allows more complicated boundary shapes to be accommodated.

A major drawback of ANNs is that a fixed size of input vector is required, making it difficult to deal with time-varying patterns. Several ANN architectures specifically deal with this problem, e.g. a time-delay neural network (TDNN) where the outputs feed back to the input layer [10].

A growing trend in pattern matching is to combine HMMs with ANNs in a hybrid system. This allows the temporal modelling capability of HMMs and the discriminative nature of ANNs to be exploited [11].

Alternatively, the standard HMM parameters can be estimated by discriminative training. This differs from Baum Welch training by optimizing some function of **all** the models to be trained. Examples of such training algorithms are maximum mutual information and minimum error-rate training [12]. Discriminatively trained models can be used for recognition without modification to the standard Viterbi algorithm.

9.6   SPEECH RECOGNITION

The previous section provided a brief overview of the 'low-level' pattern matching elements used to decode a speech waveform into a symbolic representation. This section first discusses the parser — a mechanism for searching possible model sequences to give a symbol transcription. The parser is central to tailoring speech recognizers to different applications.

Current speech recognition technology cannot transcribe spontaneous natural speech with anything like human performance — neither in terms of speed nor accuracy. However, with suitable constraints, recognition performance can be good enough to be usable in many situations. An example application is an automatic flight-booking system. In such a system the user can be prompted to speak an option, e.g. a destination airport or a departure time. The spoken input can then be verified by prompting the user to say either 'yes' or 'no'. Such a spoken dialogue constrains the recognition task and maximizes system integrity.

9.6.1   Parsing — the recognition engine

The previous section described how an arbitrary amount of speech can be matched to a model and a probability for that model calculated. Given a set of word models, recognition of a single spoken word could be carried out by matching the incoming speech to each model in turn, the model with the highest probability corresponding to the recognized word. More generally, the spoken input might be a word sequence. Some means of matching incoming speech feature vectors to the possible model sequences is needed to determine the most likely word sequence. This is the function of the parser.

As with state sequences within an HMM, the number of potential word sequences becomes intractable even for short utterances. An efficient parsing strategy is to extend the Viterbi algorithm to decode sequences of **models** rather than sequences of HMM states. Such a parser is described by a grammar network, as illustrated in Fig. 9.13, for a simple response to a prompt such as: 'Please state a departure airport'.

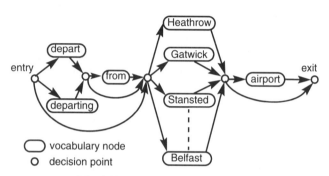

Fig. 9.13   A simple grammar network.

The network consists of nodes which connect to other nodes. Each node represents a model of a word. The parsing algorithm assigns frames to nodes where they are matched against the associated model. As frames get propagated through the model they eventually reach the end state, when a model score becomes available. Once this happens, the parser passes path information to connecting nodes. This contains at least the score and an identifier for the previous node. The frame time may be recorded as well to provide information for other processes.

If paths from more than one predecessor node are input to a node, the path with the best score is chosen. This parsing algorithm is completely independent of any type of pattern-matching element. The more general term 'score', rather than 'probability', is used to reflect this.

9.6.2   Viterbi search efficiency

The Viterbi search makes use of frame-synchronous processing, i.e. partial paths are evaluated at each frame. A very simple grammar to parse for 'yes' or 'yes please' is shown in Fig. 9.14.

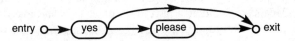

Fig. 9.14   A 'yes', 'yes please' grammar network.

If a minimum duration of two frames for both models is assumed, the parses shown in Fig. 9.15 are possible for a five-frame utterance.

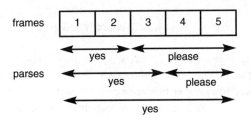

Fig. 9.15   Possible parses of five frames of input.

No output appears from the 'yes' node until two frames are processed, after which an output path emerges from the 'yes' node. At the third frame the path emerging from 'yes' at frame two is propagated into the 'please' node and hence into the entry state of the 'please' model as shown in Fig. 9.16.

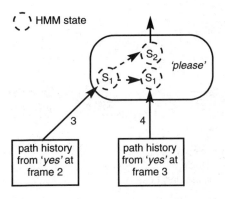

Fig. 9.16   Intra-node decision at frame four.

At frame four, however, the same model entry state is entered, but with a new path history. This state could also have been entered by the internal state transitions of the model. Only the better scoring path into state 1 at frame four is selected and further propagated. This decides between the first two parser outputs of Fig. 9.15 as early as frame four. An efficient implementation of this Viterbi search mechanism is described in Young et al [13].

Evaluating all partial paths at each time step affords great computational economy in searching the trajectories through the network for an utterance.

9.6.3   Isolated-word recognizers

The simplest type of recognizer is one where a single spoken word is expected. This is known as an isolated-word recognizer. If the words are each modelled by an HMM then recognition can be performed by matching the incoming speech against each model in turn. In practice, speech does not occur in isolation — it is embedded in background noise or silence. The noise arises from channel effects as well as ambient noise such as music, dogs barking, etc. A separate noise model is often used to match to non-speech regions of the signal. This is trained on non-speech portions of the training utterances.

A grammar network for an isolated-word recognizer is shown in Fig. 9.17. As can be seen there is a noise node both before and after the parallel vocabulary models. The same noise model is used for both these nodes, but a different instantiation of model calculations must occur, as the parser may pass different frame sequences to these nodes for mapping to different regions of the signal.

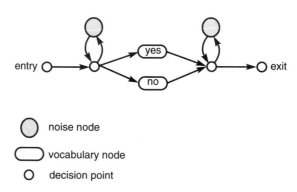

Fig. 9.17   Grammar for an isolated word recognizer.

9.6.4 Multiple parses

The network for the isolated-word recognizer of Fig. 9.17 can only propagate the best path to the exit node as the previous node selects its best entry path at each time instant. For many applications that use such a recognizer, it is desirable to know what the next best recognizer hypotheses are. To do this the exit node must propagate all input paths and all desired hypotheses must feed into the exit node. The same isolated-word network is shown in Fig. 9.18 for multiple candidate output.

As can be seen from Fig. 9.18, each vocabulary node must feed into a separate post-speech noise node so that all hypotheses are available at the output. This increases the computational cost as more noise-model calculations must be done. This network also features a separate silence, or noise-only, path. This enables the recognizer to detect if nothing was spoken.

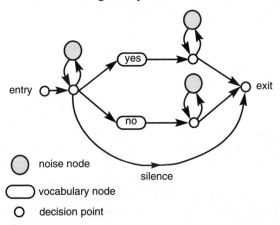

Fig. 9.18 Isolated-digit grammar with multiple candidate output.

This type of isolated-word recognizer is successfully deployed in the BT CallMinder™ service. The simplest vocabulary implemented is the 'yes, no' vocabulary for which the recognition accuracy is greater than 99% on live traffic.

Use of alternative recognizer hypotheses can improve overall system accuracy. For example, the top two paths can be checked for known confusable word pairs. If the top two contain a confusable pair (e.g. 'cat', 'bat'), it may be appropriate to prompt the user to confirm the top choice.

9.6.5 Connected-word recognizers

A useful component for many data retrieval systems is the ability to input a digit sequence. Recognition of connected words is considerably more difficult than isolated-word recognition. This is due to co-articulation — the influence of surrounding words over the sound of a word. A simple grammar network for a connected-digit recognizer is shown in Fig. 9.19.

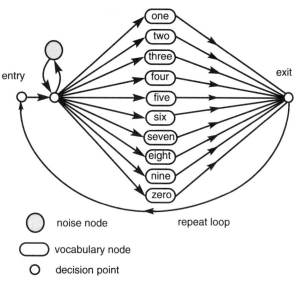

Fig. 9.19   Unconstrained connected-digit grammar network.

The example in Fig. 9.19 is of an unconstrained network with post-speech noise models omitted for simplicity. Any sequence of digit nodes with optional noise nodes between digits is allowed. The recognition accuracy for an isolated digit recognition task, as mentioned earlier, is typically above 95% for speaker-independent telephony recognition using CDHMMs. However, if the per-digit recognition is 95% for connected digits, the accuracy for ten-digit sequences is 0.95^{10} (~ 60%). Co-articulation effects result in this figure being lower in practice. The poor durational model of HMMs, mentioned earlier, has the effect that unconstrained recognizers tend to match more words to speech than were actually spoken.

Recognition accuracy can be increased if known constraints on the input are built into the recognition process. For example, if digit strings of length three are expected, then a network, such as shown in Fig. 9.20, can be used.

If all possible digit sequences are known, e.g. for account numbers, a huge tree network with all sequences can be built. Connected-digit accuracy for such a

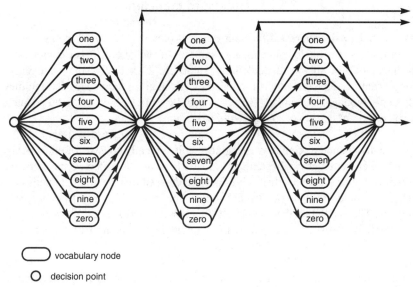

Fig. 9.20 Grammar for strings of up to three digits.

network of say 30 000 digit strings (each eight digits in length) is currently greater than 90%.

Again, obtaining the top N word sequences from recognizers, such as con-nected-digit or connected-alphanumeric recognizers, can be used to improve overall system integrity. Unless the network is a tree network, in which case the path to every leaf node is unique, special techniques [14] are needed to provide multiple candidate output.

9.6.6 Speaker-dependent recognizers

Speaker-dependent recognizers can be used for large vocabulary recognition, the task constraint being that they are optimized for only one user. A typical application is a dictation system. Usually a headset, with high-quality microphone (bandwidth of 0 to 20 kHz), is worn by the user. Some systems have advanced to allowing continuous speech (earlier systems requiring short pauses between words). Vocabularies of up to 60 000 words are possible. State-of-the-art dictation systems, such as those from Dragon, Kurzweil, IBM and Philips, allow the user to correct misrecognitions (on screen). This, combined with adapting a language model to the user's speech usage, provides very good 'net' recognition rates for large vocabularies.

9.6.7   Sub-word modelling

As vocabulary size increases for both isolated- or connected-word recognizers, it becomes less feasible to collect many examples of each word. This has motivated the technique of sub-word modelling. There are a relatively small number of constituent sound elements, or phonemes, in a given language. By building phoneme models a recognizer for any vocabulary can be built once the phonetic transcription of every vocabulary word is known.

To train phoneme models a continuous speech database is required. For speaker-independent telephony recognition this typically consists of phonetically rich spoken sentences collected across a wide range of different speakers and conditions. The sentences themselves may be devised to have good phonetic coverage of not only each phoneme but also the phonetic contexts in which it appears. The BT Subscriber database [15], collected across the UK over the public telephone network, is phonetically annotated. It contains five 'phonetically rich' sentences for each of 1000 talkers.

For recognition each vocabulary word is converted into a phoneme sequence using a phonetic lexicon. For words not in the lexicon, rules are applied to generate a sequence. Thus, new vocabularies can be generated 'on the fly' by what is known as rapid vocabulary generation (RVG). This offers huge advantages over the conventional approach of collecting a database for each new vocabulary and training new whole-word models. The price paid for the flexibility of RVG is that the achievable recognition accuracy is never as good as for whole-word modelling — for small, isolated-word vocabularies, recognition accuracy using RVG is typically 5% to 10% poorer. This is due to phoneme models being trained over many more phonetic contexts than specifically encountered in the target word.

More accurate modelling of a phoneme can result from considering the context in which it occurs. For example, instead of just one phone model for phoeme /a/, many bi-phone models can be created instead. A bi-phone is a model of a particular left or right context, e.g. /b/-/a/ and /c/-/a/ are both left bi-phones of phneme /a/. For a set of 50 phonemes there are 50 left contexts and 50 right contexts that can be modelled for each phoneme. A better sub-word model is the tri-phone. Tri-phone models account for **both** the left **and** right context of a phoneme e.g. /c/-/a/-/t/ is a tri-phone of /a/. There are 2500 (i.e. 50^2) possible tri-phones for **each** phoneme. Although not all of these contexts occur in practice, there are still a large number of models to train. Many of the contexts will have insufficient training data even in very large databases. The focus of much sub-word modelling research is to find a suitable compromise between model specificity and trainability. Techniques for this include decision-tree clustering [16], variable-mode CDHMMs [17] and phones in context [18].

It should be noted that both bi-phones and tri-phones are still trained on the data for the phoneme represented, e.g. the /a/ sound in the tri-phone example.

9.7   TOWARDS MORE NATURAL SPEECH INPUT

One of the greatest challenges of speech recognition is accurate recognition of natural speech. This section discusses two approaches to relaxing constraints of speaker-independent systems. The first of these is wordspotting where natural speech is searched for keywords. The second is a large-vocabulary system which uses a language model.

9.7.1   Wordspotting

Wordspotting is a form of search strategy on continuous speech input to detect any occurrences of a vocabulary keyword. Dialogue constraints are considerably relaxed while still extracting information from the signal. There are many applications for wordspotting. Voice control of consumer products, e.g. video recorders, televisions and PCs, can be achieved even with a small keyword vocabulary. Also, either live or recorded speech can be searched for particular topics. Even in many dialogue-controlled applications, the user's response to a prompt often contains extraneous speech such as 'ah, oh yes...' which a word-spotting recognizer is useful for filtering out.

The primary difficulty of wordspotting is accurate detection/elimination of non-keyword speech. The keyword vocabulary can be matched with specifically trained word models or with appropriate sub-word models. Generic models, trained up on a wide variety of non-keyword speech examples, are often used to match to non-keywords. These are known as sink models, as shown in Fig. 9.21. For simplicity the sink nodes in the diagram represent word models — more complex sink models can be built from either unconstrained phoneme loops or a large sub-network of phoneme sequences for common extraneous words.

This grammar allows any sequence of noise or sink models, the only con-straint being that keywords are assumed to be surrounded by either extraneous speech or noise.

An important feature for continuous-speech recognizers is the ability to out-put a hypothesized transcription as soon as it becomes available. This is known as partial traceback. To do this the parser traces back through all paths until they converge toward a common path. This common path can then be output as a stable partial hypothesis. Even for frame-synchronous parsing such stable partial hypotheses may lag behind the current frame by well over half a second.

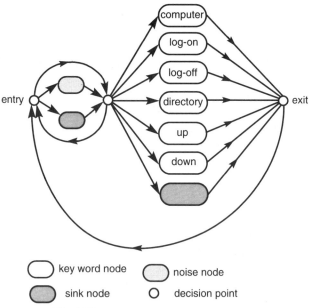

Fig. 9.21   Wordspotting grammar network.

Wordspotting is particularly prone to false triggering on non-keyword speech. Information from the partial path, such as keyword or sink model scores and durations, can be processed by a rejection strategy to validate both keyword and non-keyword hypotheses and minimize the tendency to false trigger.

9.7.2   Large-vocabulary recognition

Perhaps the most exciting application of sub-word modelling techniques is for large-vocabulary, speaker-independent, continuous-speech recognizers. Here, a language model is used as a search constraint, and usually the application is constrained to a particular domain, e.g. medicine. A system developed by Cambridge University Engineering Department [20] performs impressively on the US ARPA Wall Street Journal task at vocabulary sizes in excess of 20 000 words. The object of this is to recognize spoken sentences (from any speaker) taken from the Wall Street Journal.

Large-vocabulary, speaker-independent recognition requires accurate sub-word modelling — often using thousands of tri-phone CDHMMs. In addition, a language model is applied to the parser output. This is typically a statistical model, an example of which is the N-gram language model which captures the probability of sequences of N words from the training data.

The computational requirements and memory overhead are clearly enormous for this type of task. This can be reduced in a system with large numbers of context-dependent models by sharing model parameters both within and between models [19]. At the parsing level some form of network **pruning** must be used to improve search efficiency. With pruning, any nodes with output path scores below a threshold are deactivated and the corresponding paths removed. Careful tuning is needed to avoid the best path being pruned. Dynamic network creation, i.e. creating new nodes 'on the fly' and destroying redundant ones, also renders computation more manageable. Despite this, real-time performance is often unattainable even on platforms such as a Unix™ workstation.

An in-depth overview of language models and decoding strategies for large-vocabulary continuous-speech recognition is given in Ney [21].

9.8   RESULTS PROCESSING

The parser output results (i.e. word sequences with associated scores) are continually processed to detect the end of speech. Before returning the decoded speech to an application, a further processing stage known as a **rejection strategy** is often applied to moderate the recognition decision. Many factors can be combined to make the rejection decision as illustrated in Fig. 9.22.

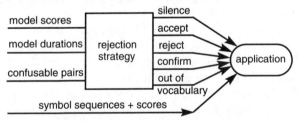

Fig. 9.22   Rejection of recognizer decisions.

Both the score (and score ratios) reflect the extent to which the data matched to the models. The score ratio helps to determine whether the best candidate is a clear winner or not. The rejection decision may be based on heuristic rules, found by trial and error. Alternatively, this stage can be a sophisticated pattern-classification stage in itself, using a Gaussian classifier, neural networks or fuzzy logic. The controlling application can then verify the recognizer decision by a dialogue depending on the rejection output.

One of the most difficult conditions to detect is when the speech contains an out-of-vocabulary (OOV) word [22]. OOV speech often satisfies all the input criteria to the rejection stage, including scoring highly on an incorrect model. Detection of OOV speech is vital for robust telephony-based recognizers as well as for wordspotting. The limitations of standard CDHMMs (see section 9.5.4) undoubtedly undermine robustness to OOV speech.

Ultimately, as large-vocabulary continuous-recognizer technology improves in both accuracy and efficiency, tasks such as wordspotting and OOV rejection will become obsolete.

9.9   SPEAKER RECOGNITION

Acoustic pattern-matching techniques can also be applied to recognizing the speaker rather than the speech content of a signal. Speaker recognition falls into two categories — speaker **verification** and speaker **identification**. Verification is the task of validating a speaker's identity claim. Identification, on the other hand, is deciding who the speaker is. Identification systems usually assume a fixed-size speaker population. Speaker-recognition systems can be text dependent where the vocabulary is pre-specified, or text independent where arbitrary speech is recognized.

- Speaker verification

    For text-dependent verification, one or more templates (or CDHMMs) are trained for each vocabulary item — usually also for each repetition.

    To verify an identity claim, the user is prompted for a vocabulary word or phrase and the input is then matched against the appropriate templates (or models) for the expected speaker. If the average score from the models for that speaker exceeds a threshold, the user is accepted. For text-independent verification, a vector quantized (VQ) codebook (see below) for the claimant is used. The performance metric for verification systems is usually the equal error rate (EER). This is the operating point at which imposter acceptance equals valid user rejection.

- Speaker identification

    Speaker identification is very similar to verification. During recognition, speech is matched to the templates (CDHMMs or VQ codebook) for the prompted vocabulary word (text dependent) or against all templates (text independent). The speaker associated with the 'closest score' is output as the identified person.

As would be expected, speaker recognition is different in many ways to speech recognition. The problem is to extract information from the speech which is unique to an individual rather than that specific to the intelligibility of the spoken input. Interestingly, a very similar front end is often used — the main difference being that a different subset of feature vector coefficients is selected.

Statistical models such as HMMs are difficult to train for speaker recognition tasks due to the paucity of available training data. A widely used technique (the precursor to HMMs for speech recognition) is dynamic time warping (DTW). With DTW the actual frame sequence for each enrolment utterance is stored in a template. During pattern matching, incoming frames are 'warped' in time to align with each template, and an accumulated frame distance used as a score [23].

CDHMMs can also be used as pattern matching elements for speaker recognition. This is an advantage if a system is to perform both speaker and speech recognition as the basic pattern matching software can be used throughout. Best performance is obtained if the number of states approximates the expected number of frames for the word. This forms a specific model of the unique sound progression for the speaker. The lack of training data makes it difficult to obtain a good estimate of the variances of the state probability distributions. For this reason a model resolution of only a single Gaussian mode per state is used. Also, only one variance — known as a grand variance — is estimated from the entire data and shared by each state's Gaussian pdf.

Using specific vocabulary templates (or models) for each speaker leads to the best performance for text-dependent systems. For text-independent systems, it is generally futile to model the training words explicitly — instead a VQ codebook is generated for each speaker [24]. The codebook is generated by pooling all speech frames from the training data for a speaker and clustering the frames to form the desired number of clusters. Pattern matching then consists of accumulating the distance of each speech frame to the nearest cluster centre. The advantage of VQ codebooks is efficiency, but at a cost of losing temporal information of speech production. An in-depth discussion of speaker recognition is given in Furui [25].

9.10   WHAT NEXT?

This chapter has presented an overview of the technology behind telephony-based ASR. State-of-the-art feature extraction and pattern-matching components were described first. Next, the use of parsers to guide a recognition process along suitable task constraints was explored, and finally some issues of results processing and speaker recognition were discussed.

So, where lies the future of speech and speaker recognition technology? The best speech recognition systems of today, although far short of human performance, demonstrate the power of the purely statistical acoustic and language model. On the other hand automatic speaker recognition possibly already outperforms humans. One vital direction is to improve the acoustic pattern matcher — humans learn new words with few repetitions and generalize to different speakers and conditions rather than being trained on many representative samples. There is plenty of scope for overcoming the theoretical obstacles of current techniques. Moreover, combining different classifiers — by running them in parallel and comparing their output — may improve recognition accuracy.

It seems likely that continued research into statistical techniques will lead to steady ASR performance gains. Moreover, as computing power becomes cheaper, the best systems of today will soon appear in real-time applications.

In the longer term, perhaps an alternative to statistical models will bridge the gap between human and machine performance. Such models might employ techniques inspired by artificial intelligence to discover rules based on the features to characterize different sounds.

If such a change of focus occurs in the next decade, real-time conversations with a machine may be a commonplace occurrence not long into the next century.

REFERENCES

1.   Lombard E: 'Le signe de l'elevation de la voix', Ann Maladies Oreille, Larynx, Nez, Pharynx, 37 (1911).

2.   Parsons T: 'Voice and speech processing', McGraw-Hill, p 175 (1987).

3.   Viterbi A J: 'Error bounds for convolutional codes and an asymptotically optimum decoding algorithm', IEEE Trans on Information Theory, IT-13 (April 1967).

4.   Cox S J: 'Hidden Markov models for automatic speech recognition: theory and application', BT Technol J, 6, No 2, pp 105-115 (April 1988).

5.   Huang X D, Ariki Y and Jack M A: 'Hidden Markov models for speech recognition', Edinburgh University Press (1990).

6.   Holmes W and Russell M: 'Experimental evaluation of segmental HMMs', ICASSP-95 (1995).

7.   Ming J and Smith F: 'Inter-frame dependent hidden Markov models for speech recognition', IEE Electron Lett, 30, pp 188-189 (1994).

8.   Levinson S: 'Continuously variable duration hidden Markov models for automatic speech recognition', Computer Speech and Language, 1, No 1 (March 1986).

9.   Russell M and Moore R: 'Explicit modelling of state occupancy in hidden Markov models for automatic speech recognition', ICASSP-85 (1985).

10. Robinson T, Hochberg M and Renals S: 'IPA: improved phone modelling with recurrent neural networks', ICASSP-94 (1994).

11. Morgan N and Bourland H: 'An introduction to the hybrid HMM/connectionist approach', IEEE Signal Processing Magazine, 12, No 3 (May 1995).

12. Reichl W and Ruske G: 'Discriminative training for continuous speech recognition', EUROSPEECH-95 (1995).

13. Young S et al: 'Token passing: a simple conceptual model for connected speech recognition systems', Cambridge University Engineering Department Report (1989).

14. Ringland S: 'Applications of grammar constraints to ASR using signature functions', in: 'Speech Recognition and Coding (New Advances and Trends)', NATO ASI Series F: Computer and System Sciences, 147 (1993).

15. Simons A D and Edwards K: 'Subscriber — a phonetically annotated telephony database', Institute of Acoustics, Speech and Hearing (1992).

16. Lee K F et al: 'Allophone clustering for continuous speech recognition', ICASSP-90 (1990).

17. Ollason D: 'Variable pool size tied parameter systems for context dependent sub-word unit speech recognition', Institute of Acoustics Speech and Hearing Conference (Autumn 1994).

18. Moore R et al: 'A comparison of phoneme decision tree (PDT) and context adaptive phone (CAP) based approaches to vocabulary independent speech recognition', ICASSP-94 (1994).

19. Farhat A and O'Shaughnessy D: 'A shared-distribution approach in a hidden Markov model-based continuous speech recognition system', EUROSPEECH-95 (1995).

20. Woodland P et al: 'Large-vocabulary continuous-speech recognition using HTK', ICASSP-94 (1994).

21. Ney H: 'Architecture and search strategies for large-vocabulary continuous-speech recognition', in: 'Speech Recognition and Coding (New Advances and Trends)', NATO ASI Series F: Computer and System Sciences, 147 (1993).

22. Sukkar R and Wilpon J: 'A two-pass classifier for utterance rejection in keyword spotting', ICASSP-93 (1993).

23. Furui S: 'Cepstral analysis technique for automatic speaker verification', IEEE Trans Speech and Signal Processing, ASSP-29, No 2, pp 254-272 (April 1981).

24. Matsui T and Furui S: 'A text independent speaker recognition method robust against utterance variations', ICASSP-91 (1991).

25. Furui S: 'An overview of speaker recognition technology', ESCA Workshop on Automatic Speaker Recognition, Identification and Verification, Martigny (1994).

10

ADVANCES IN TELEPHONY-BASED SPEECH RECOGNITION

M Pawlewski, B P Milner, S A Hovell, D G Ollason,
S P A Ringland, K J Power, S N Downey and J Bridges

10.1 INTRODUCTION

In recent years, people's expectations of speech and speaker recognition systems have grown rapidly (see Chapter 1), and the underlying core technologies have had to evolve accordingly. In the early 1980s speech recognizers were only expected to cope with limited, isolated-word vocabularies. However, as components of natural language systems, modern recognizers are now required to deal with continuously spoken sentences, while operating in adverse conditions, e.g. people telephoning from moving vehicles. As a consequence, today's speech recognizers must incorporate many new techniques, together with improvements over their predecessors.

At its simplest level, a speech recognizer consists of just two components, the front-end feature extractor and the pattern classifier, or parser. These fundamental components are described in Chapter 9. However, for the recognizer to operate in difficult conditions additional components are required. Figure 10.1 illustrates the typical constituents of a robust recognition system.

To compensate for environmental distortions, such as noise and non-ideal channels, the speech signal can be enhanced after the parameterization stage, or alternatively the speech models themselves can be modified to reflect the prevailing conditions using model adaptation techniques. To provide robustness to unexpected words, adjustments may be made to the modelling components and additional post-processing stages included after parsing.

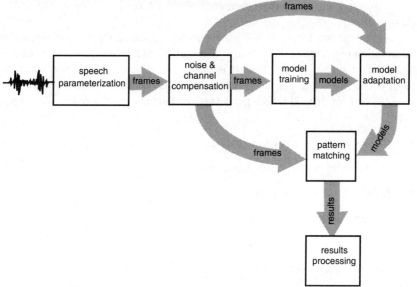

Fig. 10.1   Constituents of a robust speech recognition system.

The following sections of this chapter present some of the technical advances which differentiate today's recognizers from those of only a few years ago. The individual sections are written so that they may be read either independently or as a logical sequence.

10.2   ADVANCES IN FEATURE EXTRACTION

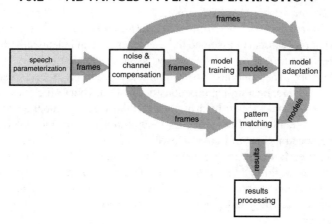

The first stage of an automatic speech recognition system is parameterization, or feature extraction. This is the process whereby the time-domain speech signal is transformed into a representation containing the key information required for classification.

In the time domain, speech is highly correlated and contains much redundant information (e.g. that of the identity of the talker) that is not required in speaker-independent speech recognition. It is important to select the feature extraction process carefully, in order to obtain the most useful type of speech feature. Parsons [1] lists six criteria which good features for pattern recognition should possess:

- wide variation from class to class;

- insensitivity to extraneous variables;

- stability over long periods of time;

- frequent occurrence;

- ease of measurement;

- non-correlation with other features.

However, in practice it is difficult to find features which meet all of these requirements, and a compromise is usually made.

10.2.1   Design of a high-performance front end

One problem is that, after feature extraction, successive feature vectors remain correlated, and a well-known deficiency of hidden Markov models (HMMs) is the lack of an efficient mechanism for the utilization of this correlation. The left-right HMM provides a temporal structure for modelling the time evolution of speech spectral characteristics from one state into the next, but within each state the observation vectors are assumed to be independent and identically distributed (IID). The IID assumption states that there is no correlation between successive speech vectors. This implies that within each state the speech vectors are associated with identical probability density functions (pdf) which have the same mean and covariance. This further implies that the spectral-time trajectory within each state is a randomly fluctuating curve with a stationary mean. However, in reality the spectral-time trajectory clearly has a definite direction as it moves from one speech event to the next. This violation of the IID assumption contributes to a limitation in the performance of HMMs. Including some temporal information into the speech feature contradicts the assumption that speech is a stationary-independent process, and can be used to improve recognition performance.

The conventional method of including temporal information is to augment the standard speech feature with first- and second-order time derivatives [2]. An alternative method for including temporal information is to use a matrix-based feature, namely the cepstral-time matrix [3], which implicitly includes speech dynamics. Several other features exist which can offer increased performance. These work by incorporating extra information into the speech feature, such as dynamic and cross-spectral information. This section describes several of these 'high-performance' features.

10.2.2   MFCCs and differential parameters

Currently the most popular feature set used in speech recognition is the Mel frequency cepstral coefficient (MFCC), augmented with higher-order time derivatives. Typically the feature vector comprises eight MFCCs plus eight first-derivative MFCCs and eight second-derivative MFCCs. The first derivative is computed from the instantaneous cepstrum using regression:

$$\partial c_t(n) = \frac{\sum_{k=-K}^{K} k c_{t+k}(n)}{\sum_{k=-K}^{K} k^2} \qquad \text{... (10.1)}$$

where $\partial c_t(n)$ is the first derivative of the nth MFCC at time frame t, and $c_{t+k}(n)$ is the nth coefficient of the $t+k$th cepstral vector.

In a similar manner, the second derivative, $\partial\partial c_t(n)$, is calculated using a weighted summation of the first derivative, typically where the window width $K=1$:

$$\partial\partial c_t(n) = \partial c_{t+1}(n) - \partial c_{t-1}(n) \qquad \text{... (10.2)}$$

The first derivative is often referred to as the velocity of the cepstra through time and the second derivative as the acceleration. Figure 10.2 illustrates a typical speech vector, comprising instantaneous, velocity and acceleration cepstra.

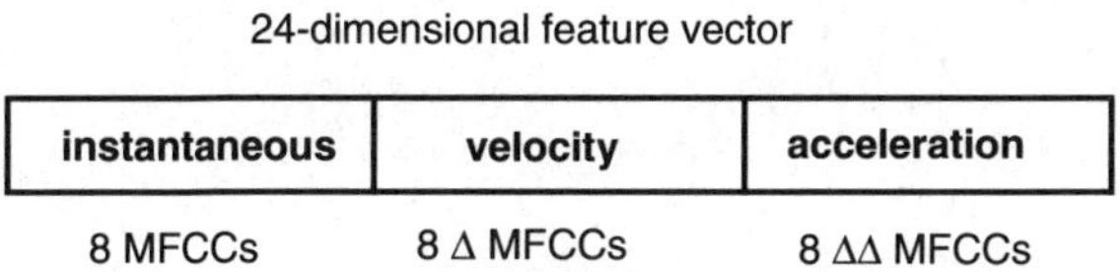

24-dimensional feature vector

instantaneous	velocity	acceleration
8 MFCCs	8 Δ MFCCs	8 ΔΔ MFCCs

Fig. 10.2   Cepstral feature vector.

10.2.3   Cepstral time matrices

An alternative method for including the temporal information of a speech signal is to use a cepstral time matrix. A cepstral time matrix, $c_t(m,n)$, is obtained either by applying a 2-D discrete cosine transform (DCT) to a spectral time matrix or by applying a 1-D DCT to a grouping of conventional MFCC speech vectors [3]. M log filter bank vectors are stacked together to form a spectral time matrix, $X_t(f,k)$, where t indicates the time frame, f the filter bank channel and k the time vector in the matrix. The spectral time matrix is then transformed into a cepstral time matrix using a 2-D DCT. Since a 2-D DCT can be divided into two 1-D DCTs, an alternative implementation of the cepstral time matrix is to apply a 1-D DCT along the time axis of a matrix consisting of M conventional MFCC vectors.

In the cepstral time matrix, the lower index coefficients along the axis, n, represents the spectral envelope, whereas the higher coefficients represent the pitch and excitation, as is the case for conventional MFCCs. Along the axis, m, the lower coefficients represent the longer time variation of the cepstral coefficients, and the higher coefficients the short time variation. The column, $m = 0$, represents the average or steady-state of the spectral time matrix. Figure 10.3 shows these regions.

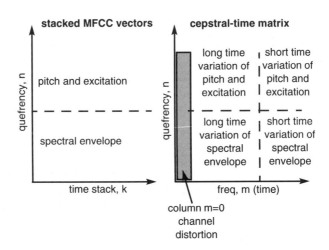

Fig. 10.3   Regions of the cepstral time matrix.

Only a sub-matrix of the cepstral-time matrix is useful for speech recognition, thus the matrix can be truncated down to $N' \times M'$. Additionally, if the steady-state column, $m = 0$, of the matrix is also discarded, any distortion of the speech signal by a non-ideal channel will be removed, making the cepstral time matrix a channel-robust feature.

Contained within the cepstral time matrix is information regarding the transitional dynamics of the speech, as well as the instantaneous values. The normal way to include speech dynamics is to augment the instantaneous cepstral vector with differential parameters. A comparison can be made between the first- and second-order cepstral derivatives and the columns of the cepstral time matrix. Equations (10.3) and (10.4) show that the cepstral derivatives are produced by weighted summations. As described, the cepstral time matrix can be obtained by applying a 1-D DCT along the time axis of a matrix containing M MFCC vectors. The first and second columns of the cepstral-time matrix, $c(n,1)$ and $c(n,2)$, are given as:

$$c(n, 1) = \frac{2}{M} \sum_{k=0}^{M-1} c_k(n)\cos\frac{(2k+1)\pi}{2M} \qquad \text{... (10.3)}$$

$$c(n, 2) = \frac{2}{M} \sum_{k=0}^{M-1} c_k(n)\cos\frac{(2k+1)2\pi}{2M} \qquad \text{... (10.4)}$$

where $c_k(n)$ is the nth coefficient of the kth MFCC vector in the matrix containing the stacked MFCCs. Thus it can be seen that the first-order derivative and the $c(n,1)$ column of the cepstral time matrix are both produced by a weighted summation of the static cepstral vectors, $c(n)$.

Substituting equation (10.4) into equation (10.3) shows that the second order cepstral derivative is generated in a similar manner to that of the $c(n,2)$ column of the cepstral time matrix. Figure 10.4 illustrates the similarity of the basis functions for producing the differential cepstra and the first and second columns of the cepstral time matrix.

The similarity of the basis functions of the DCT and the differential cepstra can also be extended to higher orders [3]. It is interesting to note that for higher-order cepstral derivatives, larger time series of cepstral vectors are required. However, with the cepstral time matrix, the number of cepstral vectors in the stack remains constant.

Additionally, speech transitional dynamics are produced implicitly within the cepstral time matrix, compared to the explicit representation achieved with a cepstral vector augmented by derivatives. Thus, models trained on cepstral time matrices have the advantage that inverse transforms can be applied which allow transforms back into the linear filter bank domain for such techniques as parallel model combination (PMC) [4].

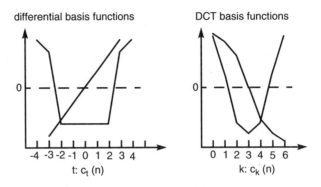

Fig. 10.4   Differential and DCT basis functions.

10.2.4   Experimental results

The experimental results shown in this section highlight the improvements which can be gained by including temporal information into the speech feature. To examine the recognition accuracy of the various feature types a telephony-based speaker-independent isolated-digit task was chosen. Table 10.1 shows experimental results for a number of different feature types.

Table 10.1   Comparison of various feature types.

Feature type	*Accuracy*
8 MFCC	77.3%
8 MFCC + 8 Δ MFCC	91.0%
9 × 4 CTM (a)	94.8%
9 × 3 CTM (b)	96.9%
8 × 4 CTM (c)	92.6%
8 × 3 CTM (d)	94.4%
9 × 2 CTM (e)	96.7%

Standard six-state, seven-mode, diagonal-covariance HMMs were used to model the speech.

The most obvious conclusion to draw is that including temporal information with standard MFCCs improves the performance considerably — in this case from 77.3 to 91.0%. Table 10.1 also shows performance for cepstral-time matrices which implicitly include temporal information. Figure 10.5 illustrates the regions of the cepstral-time matrix that have been used in each experiment.

The results show that best performance is obtained when a 9×3 cepstral-time matrix is used. This matrix effectively contains first-, second- and third-order time derivatives of the cepstrum, with no instantaneous cepstrum. It is interesting to observe that removing the third column of the matrix only reduces accuracy marginally — from 96.9 to 96.7% — but reduces computation by a third. Higher accuracy is achieved when the *zero*th column of the cepstral-time matrix is removed, i.e. compare 9×3 with 9×4 and 8×3 with 8×4. Contained within this column is the steady-state component of the MFCCs, which includes the cepstrum of the channel.

all spectral-time matrices 19x8

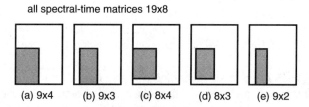

(a) 9x4   (b) 9x3   (c) 8x4   (d) 8x3   (e) 9x2

Fig. 10.5   Spectral time to cepstral time truncations.

10.2.5   Discussion

Many improvements can be made to conventional cepstral features to improve their performance. Perhaps the most important of these is to include some temporal information as this helps to overcome the limiting IID assumption of HMMs. A simple way of incorporating this information is to explicitly augment time derivatives of the cepstrum on to the instantaneous cepstrum. A mathematically more implicit method of including temporal information is to use cepstral-time matrices. These are shown to give considerably higher recognition accuracy compared to conventional cepstra augmented with higher-order derivatives.

10.3   THE PATTERN-MATCHING PROCESS

Once the incoming speech signal has been parameterized, the actual recognition process involves trying to match it to some word or set of words within a predefined vocabulary. Hidden Markov models representing words or sub-word units within the vocabulary can be combined as a **network**, and the problem then reduces to that of finding the path through that network that best matches the signal in question.

Modern tasks often involve very large vocabularies, comprising many thousands of words and high perplexities[1]. In order to reduce the computational requirements of these tasks to a level where real-time performance is feasible, a new generation of recognizers has been developed which implements a two-pronged speed-up strategy.

The first tactic employed is to use as many *a priori* knowledge sources as possible. Such sources typically include the implementation of statistical grammar, where the probability of a given word following a previous set of words is known, or a finite state grammar, where knowledge of syntax and semantics is taken into account.

The second tactic is to reduce the amount of network in use at any given time. One established route that helps achieve this is the use of a phoneme look-ahead or fast-preliminary network pass in order to reduce search space. A more recent technique is to move from a **static** network recognizer to a **dynamic** network recognizer [5]. It is this approach which is discussed in this section.

Static network recognizers load a predefined network at the commencement of recognition. This network describes fully all possible utterances that can be recognized. A dynamic network recognizer, however, creates a network during recognition.

---

[1] In this chapter perplexity is defined as the average number of nodes fanning out from a given node, i.e. the average network branching factor.

At any given time, the network can be extended by adding extra words or phrases. Similarly, those parts of the network not in use can be disassembled and need no longer be considered. This approach ensures that only the minimum amount of network is ever in use.

Unavoidably, a significant amount of processing is involved in the dynamic creation and destruction of the network, and this can result in slower recognition times for small tasks. When applied to long utterances and large vocabulary tasks, the cost of dynamic network creation becomes trivial compared with the memory savings made.

10.3.1   System architecture

Figure 10.6 presents a typical block diagram of a modular dynamic network recognizer. The architecture is designed around a central network, which provides a repository for all information about the system. As mentioned in the previous section, a network comprises a lattice of nodes, each of which has an associated HMM representing a word or sub-word unit. Nodes may also contain other information, such as a word name, phoneme name and language model penalty score. Other modules are able to operate upon the network using the information found therein in order to produce a series of word hypotheses with associated probabilities when attempting to recognize an utterance. The operation of each of these modules is, ideally, independent of the others, and a given recognizer need only contain a subset of the modules shown. The advantage of a modular design such as this is the ease with which subsequent additions to the system can be incorporated.

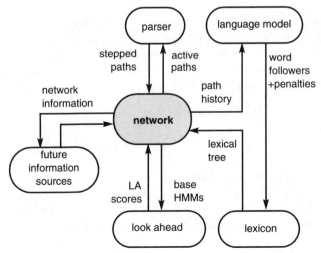

Fig. 10.6   System architecture.

10.3.2   The parser

The core module used in all recognizers, both static and dynamic, is the parser. Typically the parser performs a conventional time-synchronous Viterbi beam search on the network [6], constrained by a dynamic pruning algorithm. During recognition, this generates a set of partial-path hypotheses, each with an associated likelihood, or 'score', based on the acoustic match between the models in the partial path and the incomplete utterance. These partial paths are extended on a frame-by-frame basis, and those with scores that fall below an acceptable threshold are removed by the pruning algorithm.

10.3.3   The language model

As observed at the beginning of section 9.3, the language models used are usually either statistical or finite state. An n-gram statistical language model is one which contains information about the probability of any word occurring after a string of $n-1$ previous words. The problem with this approach is that there is a vast number of n-grams — for a k word vocabulary there are k^n n-grams. Even for the most widely used bigram (2-gram) language models, a 5000 word vocabulary implies a potential 25 million bigrams. As it is clear that even the largest training sets are unlikely to contain examples of all of these, a compromise frequently adopted is to use a back-off language model. An n-gram back-off language model copes with the absence of statistics for a given length n word string by backing off to a modified $(n-1)$-gram probability for that word string [7]. This process can be iterated as necessary, until simple unigram (1-gram) weights are used.

A finite-state grammar restricts the choice of words following a given string of previous words. The list of allowed words can be dictated by semantic or syntactic considerations. In a dynamic network recognizer, the string of previous words typically corresponds to some partial path through the network, and the list of following words encompasses all the possible ways this path might be extended.

10.3.4   The lexicon

To use the list of words generated by the language model, it is necessary to create a small fragment of the network representing those words. One way is to use a lexicon, which firstly develops phonemic representations of the list of following words, and then, from these representations, is able to generate a set of sub-word models representing the words. Finally, this set of sub-word models can be attached to the end of the partial path, ready for the next frame of data. Together, the language model and the lexicon provide the capability to extend a network dynamically during the recognition process.

An example of this is shown in Fig. 10.7, where (a) shows a partial path comprising the phoneme models making up the words 'the' and 'cat', and (b) gives an example of the set of allowed following words that might be returned by a call to a language model. In Fig. 10.7 (c), a network fragment derived from this set of words is shown. Note that in this example, the lexicon has created this fragment in the form of a tree. This enables different words with the same initial sequence of phonemes to share the same models, and thus reduces the size and perplexity of the network. Finally, in Fig. 10.7 (d) the dynamic network extension process is completed by attaching the new network fragment to the end of the partial path.

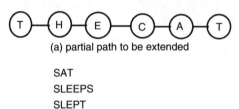

(a) partial path to be extended

SAT
SLEEPS
SLEPT

(b) following word list provided by language model

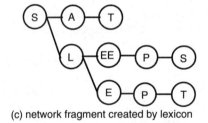

(c) network fragment created by lexicon

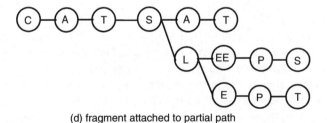

(d) fragment attached to partial path

Fig. 10.7   Dynamic network extension example.

10.3.5   Predictive strategies

The goal of any predictive strategy is to reduce the number of active states by performing a quick, rough, or fast match upon all current hypotheses to ascertain their likelihood. Only those which match reasonably well to the acoustic signal need be considered for the subsequent detailed match. Such strategies can be broadly divided into two categories — those that use a two-pass search algorithm, and those that use a phoneme look-ahead.

Two-pass search algorithms typically define the network with a fast first pass, and then do a detailed match constrained by that network. Because the first pass can produce rough estimates of path scores, a popular method is to reverse the direction of the second pass, and use an A* stack decoding algorithm to extend partial hypotheses in a non-time-synchronous fashion [8].

Phoneme look-ahead algorithms reduce the number of active states by estimating how well a particular phoneme matches the next few time frames. At a given time frame, only those paths about to be extended into relatively likely phonemes need be considered [9].

10.3.6   Path merging

The above technique results in a tree structure with numerous branches that differ only slightly. Clearly a considerable saving could be made if similar branches could be merged together. However, if path merging is implemented without due consideration, the quality of the top-N output of an N-best[1] recognizer can be degraded.

For the purposes of discussion, path merging can be divided into three approximate levels:

- instant path merging;

- merging paths with similar recent histories;

- no path merging.

In the case of instant path merging, all paths that reach a word boundary are propagated into a single set of following models. Only the best partial hypothesis up to the merge point is continued, and all other partial paths are stored for subsequent use in trace-back. Although the propagation of the top path is guaranteed, the final N-best output may be sub-optimal, as other path hypotheses can compete directly with (and hence be destroyed by) the winning hypothesis.

---

[1] One in which the N-best scoring results are output at the end of recognition..

The root of the problem observed above lies in the potential for alternate hypotheses to have different time segmentation at the merge point. Optimal N-best is assured if differing hypotheses are only merged when their segmentation has become aligned. Unfortunately, there is no hard and fast way of ascertaining this. An interim step towards establishing the point at which two partial hypotheses may become time aligned is to split the history of each partial hypothesis into **distant** and **recent** history. Recent path history is defined such that the segmentation of two partial hypotheses which share the same recent path history will have become time aligned by the end of that recent history.

From this definition, it is evident that the merging of only those paths with identical recent histories will ensure the optimal N-best output of a recognizer. The remaining difficulty is to establish exactly the length of the recent history. Schwartz and Austin [10] suggest using a 1-word recent history, with encouraging results, although the variation in word length may cause some errors.

In practice, this minimum length is often more than satisfied due to constraints imposed by the language model. In section 10.3.3, it was observed that the language model of choice is often an n-gram statistical one. When implementing path merging in conjunction with an n-gram language model, the recent history of hypotheses must also equal or exceed the span of the language model ($N-1$ words) before path merging between two hypotheses can take place. This ensures matching penalties on the resulting node tree shared by both hypotheses.

10.3.7   Integration into spoken dialogue systems

The output of a recognition pass is an acyclic graph of word and phrase hypotheses. This can be searched efficiently (using, for instance an A* algorithm) to extract a top-N candidate list, or can be passed in its entirety to a speech understanding unit as part of a spoken dialogue system.

When employed as part of such a system, the flexibility of a dynamic network recognizer comes to the fore. The need to switch between domains, or re-recognize a particular utterance using tighter constraints (to ascertain a place-name, for example) is achieved simply by employing — and, if necessary, creating — the appropriate language model, rather than having to load up a specific predefined network exclusively for one recognition pass.

The use of a dynamic network recognizer thus presents an efficient solution to dealing with large vocabularies or long utterances. Several strategies can be employed to minimize the size of network in existence at any given time. In addition, dynamic recognizers can be integrated easily and efficiently into natural-language dialogue systems.

10.4   MODEL TRAINING

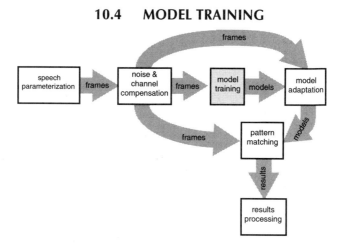

Recognition system vocabularies can either be based on whole-word or sub-word models. Whole-word models are derived from several training examples, covering all possible variations. Although this approach is fairly straightforward, and generally yields better results than sub-word models, it is limited by the fact that each vocabulary item requires a separate model and separate training data.

Vocabularies based on sub-word models use a different approach. Individual vocabulary items are formed from combinations of phoneme models. If it is assumed that the phonemes are context independent, then only 44 models, corresponding to the 44 phonemes in the standard British English phoneme alphabet, are required. In practice, this is not a very good assumption since phonemes can be highly dependent upon context.

Contextual effects may be accommodated by moving to a triphone-based approach. The transition from a context-independent to a context-dependent modelling system inevitably results in an increase in the number of models to be trained. Consequently the amount of training data afforded to each model can become grossly inadequate.

The problem of insufficient training data may be overcome by using tied parameter models. The following section considers a method of relating the structure of tied-parameter context-dependent systems to the amount of available training data and to the variability within it. Three separate tied-parameter systems are compared. The systems are shown to be more accurate than their uniformly structured counterparts.

The trade-off between model specificity and model trainability has been the subject of much research for a number of years [11-13]. One trend has been to develop context-dependent HMM sub-word unit systems which compensate for the lack of allophone[1] training data by tying the parameters of each group of allophones, associated with the same monophone[2], to a separate pool [14, 15]. However, systems which keep the number of shared model parameters constant across groups of tied allophones fail to address two important aspects of the training problem. Firstly, there is typically a large variance in the distribution of training data across the allophone groups, and secondly the variability of sounds to be modelled within each group is different.

The approach presented here aims to link the number of parameters shared by each group of models to the trainability of that model group and also to the variability within the group.

The parameter pool size is calculated via the 'rule of thumb' given by equation (10.5). Trainability is defined as the number of training tokens available for the allophone group and variability is defined as the number of different triphone contexts present within the group:

$$S_m = F \times \sqrt{T_m} \times V_m \qquad \qquad ... (10.5)$$

where:

S_m is the parameter pool size for allophone group m,
F is the constant which is used to control the total number of parameters,
T_m is the number of training tokens associated with allophone group m,
V_m is the number of triphone contexts present in group m (measure of variability within group).

This strategy has been applied to two tied-parameter systems. The task used to evaluate the performance of this technique was the speaker-independent recognition of examples of 697 different surnames collected over the UK telephone network (see section 10.4.4).

10.4.1   Context-independent seed models

Sets of context-independent models are required as seed models for the tied-parameter systems and also to provide baseline context-independent performance levels. The context-independent symbol set used for this work comprises 44 phoneme symbols and one noise symbol. Two main systems were created based on this symbol set.

Model Type 1 was a 3-state 12-mode system containing 1584 modes in total. Model Type 2 was a 3-state variable-mode system also with a total of 1584 modes. This system was created by applying equation (10.5) during the alloca-

---

[1] One of two or more forms of the same phoneme.

[2] Context-independent realization of a phoneme.

tion of modes to each state of each context-independent unit. In this case each context-independent unit is treated as an allophone group.

The performance of the two systems is presented in Table 10.2 in section 10.4.5.

10.4.2   Tied-parameter systems

The following sections describe the three types of tied-parameter system investigated.

10.4.2.1   Shared-mode system (SMS)

In this system the corresponding states of all the allophones, derived from the same monophone, share a common pool of modes. The application of equation (10.5) to this system results in a variation in the size of the shared-mode pool from one allophone group to the next. Figure 10.8 shows the arrangement for a single allophone group where the corresponding states in each allophone share a pool of three modes. Allophones within a group differ only in their mode weights.

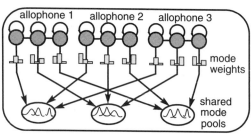

Fig. 10.8   Shared-mode system.

10.4.2.2   Shared-mode and clustered-state system (SM&CSS)

All the allophones derived from the same monophone share a common pool of modes and a common pool of states for corresponding states. Figure 10.9 shows the arrangement for a single allophone group where the number of corresponding states has been reduced from 3 to 2 by the state clustering process, and the states in each pool share a pool of three modes. Comparing Fig. 10.9 to Fig. 10.8 it can be seen that the two systems are essentially the same except that the SM&CSS has fewer sets of mode weights. State clustering applied to a shared mode system effectively results in the clustering and sharing of sets of mode weights while the modes themselves remain unaltered.

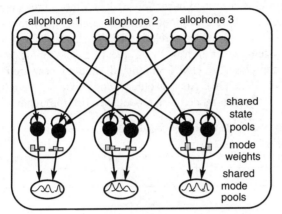

Fig. 10.9   Shared-mode and clustered-state system.

10.4.3   Triphone construction

The recognition network for this task contains 2210 different triphones. Of these, 23% occur three or fewer times in the training corpus and only 14% have 50 or more occurrences. For this reason it was decided that for each experiment two shared-parameter systems would be constructed, one for left biphones and one for right biphones. Following the re-estimation of both shared-parameter systems the triphones required for recognition are synthesized from the appropriate biphones.

10.4.4   Database

The training corpus used for these experiments consisted of 13 448 examples of isolated surnames collected over the UK telephone network from callers spread throughout the country. There were in total 8000 different surnames in the training corpus. The test data was a separate set comprising 494 examples of surnames collected in the same manner, and there were 697 of these in the recognition network. Some of the surnames had multiple pronunciation transcriptions resulting in a network size of 1000. The 95% confidence interval for this task is 4.3%, based on an assumed error rate of 58%. A standard cepstral-based signal parameterization was used.

10.4.5   Experiments

To compare each of the shared parameter systems, the total number of modes in each system was kept constant and equal to the number of modes in the context-independent systems described in section 10.4.1. Although the number of modes is kept constant the number of mode weights varies from system to system. All the recognition times quoted are relative to the standard fixed-mode context-independent system shown in row 1 of Table 10.2. The timing information was acquired by applying the UNIX command time to each run of the HTK recognition tool HVite [16]. All vocabulary models are 3-state left-to-right with no skips. A single-state 6-mode noise model was used throughout.

Table 10.2   Context-independent results.

Model type	Percentage accuracy	Recognition time
1	54.7	1.0
2	56.3	0.8

10.4.5.1   Context-independent experiments

For comparison with the context-dependent systems, Table 10.2 shows the results obtained by a 3-state 12-mode context-independent system (model type 1) and also those achieved by a 3-state variable-mode context-independent system (model type 2) containing the same total number of modes.

10.4.5.2   Shared-mode system experiments

Using the context-independent models as seed models, two separate context-dependent shared-mode model sets were created:

• set 1 — left and right biphone SMSs generated using the fixed-mode context-independent models as seed models,

• set 2 — left and right biphone SMSs generated using the variable-mode context-independent models as seed models.

The SMSs were then re-estimated and the triphones required for recognition were synthesized from the two biphone SMSs. The results for sets 1 and 2 are shown in Table 10.3.

Table 10.3   Context-dependent SMS results.

Set	Percentage accuracy	Recognition time
1	58.5	2.8
2	60.9	3.2

10.4.5.3   SM&CSS experiments

State clustering was applied to the left and right biphone SMSs generated as set 2 in the previous experiment and three separate context-dependent, shared-mode and clustered-state model sets were created:

- set 3 — the number of states in each system was reduced by 25%;

- set 4 — the number of states in each system was reduced by 50%;

- set 5 — the number of states in each system was reduced by 75%.

The SMSs and CSSs were then re-estimated and the triphones required for recognition were synthesized from the two biphone SMSs and CSSs. The results for sets 3, 4 and 5 are shown in Table 10.4.

Table 10.4   SMS and CSS results.

Set	Percentage accuracy	Recognition time
3	61.3	1.8
4	59.7	1.7
5	58.5	1.5

10.4.6   Discussion

There are some advantages to be gained in modelling contextual variation within sounds, especially when this technique is coupled with a modelling structure which reflects the amount of training data available and the variability within the sounds to be modelled.

This work also demonstrates that it is possible to produce accuracy improvements without the computational overhead normally incurred with context-dependent modelling.

Compared with fixed-mode context-independent modelling, the two shared-parameter systems produced differing performance in terms of both recognition accuracy and computational efficiency.

Shared-mode systems were found to give gains in terms of recognition accuracy at the cost of increased computation at recognition time. Recognition accuracy rises by 6.2% (from 54.7% to 60.9%) with a corresponding rise of 220% in recognition processing time.

SM&CSSs succeed in dramatically reducing the computational overhead of the shared-mode systems while maintaining the accuracy gains achieved. Recognition accuracy rises by 6.6% (from 54.7% to 61.3%) with a corresponding rise of 80% in recognition processing time.

Taking both recognition accuracy and processing time into consideration it would appear that the SM&CSS approach provides the best overall performance.

It was mentioned in the introduction that lack of training data and model inflexibility are two factors that determine how well the contextual variations within a phoneme can be modelled. This work has shown that variable pool size shared-parameter systems provide increased model flexibility, allowing the models to make more efficient use of the limited training data available.

10.5   USER-SELECTABLE VOCABULARY

In addition to system defined vocabularies there are situations where the end users choose their own vocabularies and hence where the vocabularies cannot be trained in advance. The most common application where this situation arises is in 'voice dialling' where the user just speaks the name of the person they wish to call and the system then places the call. Clearly it is impossible to predict in advance the choice of names that each user will select, so the system must be conditioned to recognize each user's particular selection of names. This type of recognition is termed 'user-selectable vocabulary'.

The traditional approach to user-selectable vocabulary recognition has been to use speaker-dependent whole-word modelling techniques. In an enrolment session the user is asked to say each of the words one or more times and from these examples a model or template is created for each word. Dynamic time warping (DTW) against these templates is the technique then typically used to perform the subsequent recognition.

An alternative approach which employs sub-word HMM techniques is presented here. The principle behind the new technique is to generate phonetic transcriptions of the user's words and then create a recognition network which allows only those phoneme sequences seen in the enrolment utterances. The sub-word technique is described below and the results of experiments comparing it with a DTW whole-word system are presented.

10.5.1   Theory

The principles underlying a sub-word HMM-based user-selectable vocabulary system are relatively simple. The first step is to generate one or more sub-word transcriptions for each of the words in the vocabulary. This can be done by presenting the utterances of the user to a sub-word HMM recognizer with a grammar that imposes little or no constraint on the possible sequences of sub-word units recognized. The transcriptions generated can then be used to define a recognition grammar which will allow only the words that the user has chosen.

10.5.1.1   Transcription generation

In the simplest case, the words that a user speaks may be transcribed using a sub-word HMM recognizer that imposes no grammar constraints whatsoever. In this case phonemes have been used as sub-word units and modelled with HMMs, trained in a speaker-independent manner. In practice, however, the accuracy of an unconstrained phoneme recognizer operating on telephony speech is currently not very high. Furthermore, unconstrained recognizers are prone to phoneme-insertion errors in 'silence' periods (which often are not very silent). Consequently, the approach adopted here was to impose a weak grammar upon the sub-word recognizer. Figure 9.10 shows the form of the grammar used. This grammar allows only a single connected sequence of phonemes surrounded by noise.

Due to the weakness of the grammar, the recognizer can generate phoneme sequences that do not occur in the English language. To improve the transcriptions generated it is possible to impose a stronger grammar that restricts the possible phoneme sequences. One possibility is to use a phoneme bigram grammar that constrains which phonemes may follow which others. The example in Fig. 10.11 shows the sort of transcriptions that the system generates for a few spoken digits.

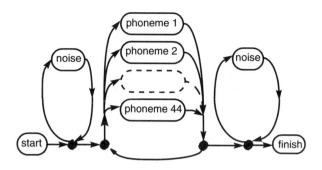

Fig. 10.10   Network used for generation of phonemic transcriptions.

Without bigrams	With bigrams
one = W AA MA	one = W AA M A
two = F B S UU A	two = T UU A
three = TH R EE	three = TH R EE
four = V V V F AW E	four = B V F AW
five = F AR AI A TH	five = F AR AI A TH
six = S AI K S	six = S I K S

Fig. 10.11   Examples of system-generated transcriptions.

Figure 10.11 shows that the transcriptions generated using the phoneme bigram restrictions look more plausible than those from the totally unconstrained network.

10.5.1.2   Recognition network generation

When one or more transcriptions have been generated for each of the words in the desired vocabulary, a recognition network may be generated. This is a straightforward process which may simply combine the individual transcriptions into parallel alternatives in the network or, more efficiently, into a tree network.

In a practical system it may be desirable to request several examples of each word from the user. These can then be checked to ensure that the transcriptions generated are at least reasonably similar to each other, thus providing a measure of robustness against bad enrolment utterances. They can be further checked to ensure that they are significantly different from those for other words, thus protecting the user from selecting a vocabulary which is too difficult for the recognizer. Figure 10.12 shows an example of a recognition network generated from automatically generated transcriptions of the words 'one', 'two' and 'three'.

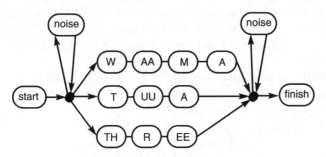

Fig. 10.12   Example of a network generated from transcriptions of utterances of the words 'one', 'two' and 'three'.

10.5.1.3   Advantages of the sub-word technique

The sub-word method has a number of advantages over the conventional whole word techniques.

It lends itself easily to mixing predefined words (which need no user training) with user-selected ones because the sub-word models are speaker independent. For example, a voice-dialling application developer could predefine the words 'call', 'help' and 'exit' and, if the user then enrolled with the names 'Mrs Jones' and 'doctor', a recognizer could be constructed that would allow the user to say 'call Mrs Jones', 'call doctor', 'help' or 'exit' even though the user had not previously given any training examples of the predefined words. This feature also gives the application developer the flexibility to add new control words after the service is up and running without requiring any further enrolment data from the users. This would be much more difficult with the whole-word approach as it would either require the mixing of speaker-dependent and speaker-independent models (which might have a detrimental effect on accuracy) or require the user to provide enrolment examples of all the predefined words.

The sub-word technique also requires over an order of magnitude less storage than the conventional technique to store the vocabulary of each user. Sub-word uses only about 10 bytes per word to store the transcription rather than a few hundred or thousand bytes to store a model or template.

10.5.2   Experiments

Experiments were carried out to compare the sub-word HMM technique with a more conventional whole word DTW system.

A 14-word test vocabulary was chosen, consisting of the digits 'one' to 'nine', 'oh', 'zero', 'nought' plus two control words, 'highline' and 'chequebook'. The data was taken from the BT Brent telephony database. For each of 100 talkers, between one and three repetitions of each word were used for enrolment, each repetition being taken from a separate phone call. The test set consisted of seven repetitions of each word, again each taken from a separate phone call, where the calls were placed over a six month period.

The results in Table 10.5 show that at best the conventional DTW system is 1-2% more accurate than the sub-word system, but much more significant is the effect of the number of training examples used.

Table 10.5   Comparison of sub-word HMM and HTW systems.

Recognizer	Number of training examples per word	Accuracy, %
sub-word HMM	1	88.8
sub-word HMM	3	94.1
whole word DTW	1	90.8
whole word DTW	3	95.3

A second set of experiments was performed to see how well the sub-word HMM system would fare with vocabularies that were part user-trained and part predefined. This meant replacing transcriptions generated from the user's enrolment utterances with 'idealized' pronunciations for a proportion of each user's vocabulary.

The results in Table 10.6 show that use of predefined words reduces the accuracy to about the same as that obtained when only a single training example is used. This observation lends support to conclusions reported elsewhere [19] that for speaker-independent recognition, performance can be improved if the lexicon of standard pronunciations is supplemented with automatically generated transcriptions of real utterances.

Table 10.6   Performance of sub-word HMM system when part of the vocabulary is predefined.

Vocabulary predefined, %	Number of training examples per word	Accuracy, %
0	1	88.8
0	3	94.1
50	1	88.3
50	3	89.3
100	0	88.2

10.6   OUT-OF-VOCABULARY REJECTION

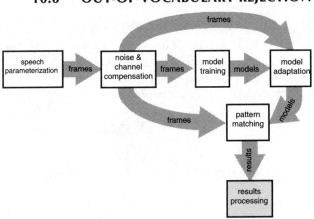

In automatic speech recognition, one of the most difficult conditions to reliably detect is when the input speech contains a word not in the recognizer vocabulary. The problem of out-of-vocabulary (OOV) rejection remains a challenging one, crucial to the robustness of commercial recognizers. An OOV condition occurs when an input speech signal cannot be correctly decoded into any of the possible recognizer outputs. This can happen in several ways. For example, the spoken input to a 'yes/no' (isolated word) recognizer may be an utterance of the word 'banana'. Alternatively, a spoken digit sequence may be input to a connected digit recognizer which is not in the expected set of digit sequences. The ideal solution is to attempt to recognize all possible spoken input. Clearly this method would be prohibitive in terms of the computation and storage requirements.

A more conventional approach is to process the model scores after the recognition stage. One would expect a poorer score for speech not corresponding to a model than for a valid match to the model. A prescribed threshold, applied to each model score, would enable discrimination between vocabulary and OOV elements of the input speech. Score post-processing turns out to be a surprisingly poor arbiter of OOV occurrences (see section 10.6.1).

The solution proposed here operates at the level of acoustic pattern matching. Generic models of speech, or sink models, are incorporated into the recognition process to detect extraneous speech.

The issues involved in training sink models and incorporating them into the recognition framework are discussed in section 9.6.2. This technique dramatically improves OOV performance but at greater computational cost than score post-processing.

10.6.1   Score post-processing

An example of a grammar network for an isolated-word recognizer is shown in Figure 10.13. This network includes a noise-only, or silence, path. Any input to this network can only be 'recognized' as one of vocabulary models V_1 to V_n, or silence. OOV occurrences can only be detected by processing information arriving at the output node, e.g. path scores or model trace-back data.

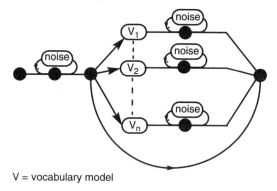

V = vocabulary model

Fig. 10.13   Grammar network for n parallel vocabulary models.

Score-based rejection is performed by the two following tests:

$$\text{reject if } S_{best\ path} < t \qquad \text{or} \qquad \frac{S_{best\ path}}{S_{next\ best\ path}} > t,$$

where $S_{best\ path}$, $S_{next\ best\ path}$ are the best and next best path scores respectively and t is a threshold. Varying t allows the rejection rate to be controlled.

One would expect the best path score to be poor for OOV speech input. This is tested by the first condition. The second condition examines the score ratio between the best and second best paths. This detects when the best path is not a clear winner which may suggest the data is not matching particularly well to any of the models.

The effect of applying score-based rejection techniques to an isolated-digit vocabulary is illustrated in Fig. 10.14. The vocabulary consists of 13 words: digits 'one' to 'nine', 'zero', 'nought', 'oh' and 'stop'. The test set consists of 1209 spoken digit examples. The OOV test set consists of 369 spoken town name examples. This set contains 143 different OOV words making a good OOV test set, as the OOV detection performance is not likely to be biased towards a particular word.

As can be seen from Fig. 10.14, rejecting any OOV words incurs a cost of rejecting correctly recognized words. To compare the different rejection techniques a comparison point of 5% vocabulary rejections — denoted by the vertical dashed line — is shown. Anything greater than this is usually unacceptable in a real system.

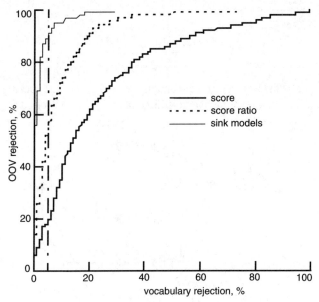

Fig. 10.14   OOV rejection performance curves.

At 5% vocabulary rejection, the score and score ratio reject about 20% and 55% of OOV words respectively. This poor performance may be attributed to the simplifying assumptions inherent in the modelling of speech signals with standard continuous-density hidden Markov models (CDHMMs). Furthermore, the need to train models over a wide variety of speakers and channels for telephony applications greatly increases the variability of valid speech signals matching to a particular model. The poor performance of the score-based classifiers suggests a large overlap between model scores for valid matches and model scores for OOV matches.

10.6.2   Sink models

As noted earlier it is not feasible to attempt to recognize all possible OOV inputs. Even if recognition accuracy permitted, the computational overhead would be prohibitive. A suitable compromise is to use sink models as generic models of OOV speech.

Sink models are used in parallel with vocabulary models in a network to provide alternative hypotheses at a decision node. The original network, with parallel sink models added, is shown in Fig. 10.15. Speech is classified at the exit node as follows:

$$\text{reject if} \quad \frac{S_{vocabulary}}{S_{sink}} > t$$

where $S_{vocabulary}$ is the best scoring vocabulary path, S_{sink} the best scoring sink path and t is the rejection threshold. The rejection rate is again controlled by varying t.

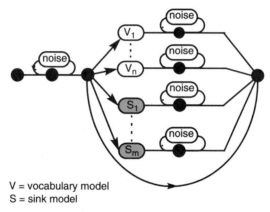

V = vocabulary model
S = sink model

Fig. 10.15   Grammar network including m parallel sink models in distinct parallel paths.

Sink models are not intended to model a specific sound. The approach to training is to train on many utterances (as for vocabulary models) and also on a wide variety of different speech sounds. The effect of this is to widen the state probability densities compared with the vocabulary models. This is shown in Fig. 10.16 for a single feature.

If the speech data (represented by feature x) consistently occupies a region where the pdf for state j of the sink model $P(x|S_j^{oov})$ is greater than for state j of a vocabulary model $P(x|S_j^{voc})$, the sink model will tend to score better relative to the vocabulary models.

The choice of sink model topology is a contentious issue. In principle, a single state model with a self-loop should suffice as any temporal structure is meaningless for such a generalized model.

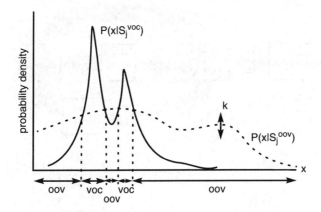

Fig. 10.16   Example pdfs for state j of a competing vocabulary and sink model.

Experimentation shows that multi-state sink models invariably give better results. This is possibly because the sink paths accurately segment regions of speech as effectively as the vocabulary paths.

The effectiveness of the sink path at segmenting speech can be fine-tuned by weighting the sink model. This takes effect in the HMM forward probability calculation $\alpha(S_j, t)$ of being in state j of the sink model at time t:

$$\alpha(S_j, t) = \text{Max}_i \{k.\alpha(S_i, t-1)\, a_{ij}\, P(\underline{X} \mid S_j)\} \qquad \dots (10.6)$$

where a_{ij} is the transition probability from state i to j; $P(\underline{X}|S_j)$ is the likelihood of observation $\underline{X}$ given state j and k is the model weighting factor. The weighting factor k effectively slides the sink model probability density up or down as shown in Fig. 10.16.

The OOV rejection results from the same isolated-digit and OOV test sets are also shown in Fig. 10.14. Two parallel sink models were used — one trained on 5558 words of sets A-B of BT's TADS surname database (4212 different words); the second on 1585 words of sets A-E of BT's LETRASET surname database (100 different words).

The performance gained from using both sink models is a dramatic improvement over score-based methods rising to ~90% OOV rejection at the 5% vocabulary rejection level.

Future improvements in modelling the acoustic signal are likely to complement the sink-model approach and lead to increased recognizer robustness.

10.7   SPEECH RECOGNITION IN ADVERSE CONDITIONS

Leading-edge speech recognition systems are capable of highly accurate performance for particular tasks within well-defined operational environments. However, to fully exploit the technology, practical systems must be capable of operating in the range of different environmental conditions encountered in everyday use. For example, a problem associated with the use of car phones and mobile equipment is that the background noise and communications channel response are likely to be highly volatile not only between calls, but potentially may change every few seconds of the call duration.

Although current designs tailored to a particular environment (known as 'matched' recognizers) will give good performance in the local ambient conditions, the performance of these dedicated systems falls off severely as background conditions move away from the training environment.

The problem is clearly illustrated in Fig. 10.17, which presents the results of an experiment outlining the performance of an HMM-based isolated digit recognizer trained in relatively noise-free conditions and tested with noisy utterances over a range of signal-to-noise ratios (SNRs) from 'clean' (>25 dB) to very noisy (~0 dB). As a comparison, a second experiment presents the matched (or 'optimum') recognizer performance at each point. For this experiment, sets of speech models were trained with data at each of the SNRs considered in the first experiment. The results of this experiment are also plotted in Fig. 10.17.

The experiments demonstrate that good performance can be obtained in very noisy environments, if the speech models and testing environment remain appropriately balanced. It is the aim in this section to present several different techniques for achieving this balance, each leading to a more robust recognition system design.

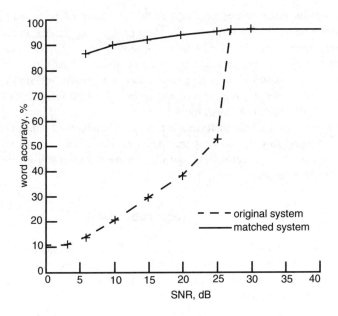

Fig. 10.17 Recognition performance in noisy conditions.

10.7.1 Dealing with adverse conditions

A range of different techniques for accomplishing noise-robust speech recognition has been established. These may be divided into two separate groups. The first approach applies some form of preprocessing (such as filtering) to the incoming speech to compensate for the interfering signal. The second attempts to modify the pattern-matching stage (for instance, by adapting the speech models to the new conditions).

The choice of compensation technique depends upon the range of adverse factors likely to be present in the operating environment of the recognition system. These fall into three basic categories — interfering background noise, communication channel effects (such as echo and distortion), and the altering of human speech production in challenging conditions. The last factor, known as the 'Lombard effect', is the most complicated of the three, as it is difficult to isolate the instantaneous changes in pitch, intensity and spectral shape which result from the 'stressed' speech production. Compensation methods for the Lombard effect involve a different approach from those of noise and distortion, being more related to methods for dealing with varying talker styles (quiet, loud, angry, etc). The interested reader is referred to Junqua and Angelade [20].

This section concentrates on addressing the issues of noise and channel effects. It is shown that methods of filtering the signal to remove the unwanted sounds are highly successful for constant, or slowly varying noise sources (e.g. line characteristics, or computer background noise), whereas for more complicated interfering noises (i.e. background speech, television or music) model adaptation techniques are required. These more advanced techniques have the overhead of higher computational requirements.

Communication-channel distortion is a convolutional effect, and as such cannot be dealt with using the above mentioned techniques. The second half of this section describes a 'channel-robust' speech parameterization and its application to a speaker-verification system.

10.7.2   Noise robustness

Acoustic noise is usually considered to be additive. This means that in any transform domain which is linearly related to the original time-domain waveform, the speech and noise signals may be considered as separate sources having no cross-correlation:

$$F(\text{Speech} + \text{Noise}) = F(\text{Speech}) + F(\text{Noise})$$

where $F(.)$ is any linear transform.

Removal of the interfering noise, in theory, should not alter the characteristics of the speech signal in the linear case. Unfortunately, the conventional MFCC representation of speech is not linearly related to the original waveform domain because of the log operation in the feature generation. Therefore noise robust recognizers operating in the MFCC or log spectral domains would typically invert the log transform before noise compensation is applied.

10.7.3   Stationary and non-stationary noise

An interfering noise source is termed 'stationary' if the statistics of the noise do not vary with time in the observation period. Examples of stationary noise include mains hum and fan noise from computer equipment. For slowly varying noise sources, it may be possible to update the noise parameters of the interfering signal, and hence track such a source. This type of noise may be termed 'quasi-stationary', and would represent, for example, the varying drone of the noise present in a car as its speed changes.

A typical example of 'non-stationary' noise would be a radio or television set playing in the background; it is not, in general, possible to update the noise parameters quickly enough to compensate for the changing nature of the noise.

10.7.4   Stationary-noise compensation techniques

One of the simplest techniques for dealing with stationary or quasi-stationary noise is that of **spectral subtraction** [21, 22]. This technique attempts to take account of the spectral shape of the interfering noise present in the signal, and operates in the linear power-spectrum domain. An estimate of the noise mean in each channel is subtracted from the incoming signal; if the result of this process is less than a predefined spectral floor, the spectral floor value is substituted for the subtracted value. The clean speech estimate may be thus defined:

$$\hat{S}_k = \text{Max}((S^2_k + N^2_k) - \alpha \overline{N^2_k}, \beta_k) \qquad \qquad \qquad ... (10.7)$$

where $S_k{}^2$ and $N_k{}^2$ are the power spectra of the speech and noise components in frequency channel k respectively, β_k is the spectral floor in that channel and $N_k{}^2$ is an estimate of the mean of the noise. Inclusion of the α factor in equation (10.7) allows the amount of noise energy subtracted from the signal to be varied.

To obtain an estimate of the noise mean, a method for discriminating speech from noise in a signal is required. One way to do this is to make use of a 'noise-tracking' system [23] which makes decisions on a frame-by-frame basis, typically using an energy-based measure to determine the information content of the frame.

Determining appropriate values for the spectral floor threshold is very important to the success of the subtraction technique — the majority of non-speech frames will be replaced by the floor values in each channel, hence the spectral floor provides a desired 'environment-independent' noise image. A set of spectral floor parameters may be determined by basing the threshold values on the dynamic range of the speech energy in each frequency channel, measured under ambient conditions. The floor values for each filterbank channel can then be set at a level below the majority of the energies associated with speech frames in each channel.

A drawback of this spectral floor is the distortion effect that the threshold has on the speech waveform; this is illustrated in Fig. 10.18 where signal energies are compared in a particular channel before and after subtraction processing. The spectral floor for the channel is set at 32.5 dB, as shown by the dotted line. Also shown in Fig. 10.18 is the output produced by setting the spectral floor to zero, indicating the potential for severely distorting the speech if the floor is not carefully chosen. The effects of the signal distortion can significantly alter the cepstral representation of the speech data, it is therefore advantageous to include subtraction data during model training if cepstral parameters are to be used for recognition. In this case, a subsequent processing stage must be applied to the post-subtraction speech data in order to transform the parameters to the cepstral domain.

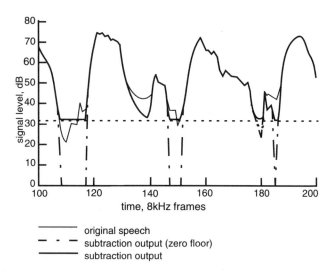

original speech
— - - — subtraction output (zero floor)
subtraction output

Fig. 10.18   Time-domain distortion due to spectral subtraction.

Employing the spectral subtraction technique into the isolated-digit experiment of section 10.7.1 improves recognition performance in noisy conditions greatly — a recognition accuracy of over 75% is obtained at SNRs as low as 15 dB, a figure some 50% higher than the 'uncompensated' system performance. Figure 10.23 in section 10.7.5 plots the recognition accuracy as a function of signal-to-noise ratio.

If, instead of subtracting the noise estimate from the incoming data frames, it is used as a mask to screen out low-energy detail from the signal, a second noise robust technique is obtained — **'noise masking'** in this manner was first proposed by Klatt [24].

The masking is generally implemented by using the noise estimate in each frequency channel to replace the channel input if its value drops below that level. Similarly, the means of the speech models (trained on unmasked data) are adjusted so that any state output mean less than the associated noise estimate is replaced by the noise mean for that channel:

INPUT: if $O_k < \mu_{nk}$: set $O_k = \mu_{nk}$
HMMs: if $\mu_{sik} < \mu_{nk}$: set $\mu_{sik} = \mu_{nk}$

where O_k is the observation in a particular channel, μ_{nk} is the estimate of the noise mean in that channel and μ_{sik} is the mean of the channel pdf associated with state i of a particular speech HMM.

The 'masked' speech models are desensitized to the low-energy detail present in quiet training environments, but which would be swamped in noisy conditions. The recognition environment is thus stabilized in a predictable manner, leading to increased noise robustness.

One inherent advantage of the masking process over that of spectral subtraction is that it is applied in the log-filterbank domain, thus requiring less computational overhead in data transformation, if recognition is being carried out in the cepstral domain. Figure 10.19 details the processing stages required.

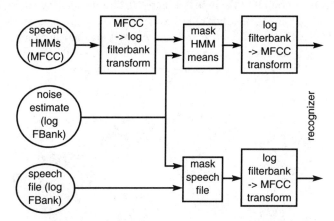

Fig. 10.19 Transform stages in the noise-masking system.

Implementing the masking technique on the benchmark isolated-digit experiment results in similar performance gains to those obtained with spectral subtraction. The results, shown in Fig. 10.23 in section 10.7.5, show an increase in useful operational range of the recognizer compared to the earlier method, providing word accuracies of 80% at 15 dB SNR, and over 70% as low as 10 dB.

The performance of both systems declines quickly as the signal-to-noise ratio falls below 10 dB. This occurs when useful information in the speech signal starts to be removed due to either subtraction or masking with increasingly large noise estimates. Subtraction data will become steadily more distorted, and result in a poorer match with the speech models.

The effect on the noise-masking technique is twofold. Firstly, as the noise estimate increases, more and more speech model means will be masked, and hence their discriminational ability will be reduced. Secondly, masking of the incoming utterances has the effect of reducing the variance of the non-speech regions of the data, and hence reducing the variance of the 'noise'. A consequence of this is that the recognition process will increasingly favour word models over noise models due to the tighter variances associated with speech models (variances are not altered by the masking technique).

10.7.5   Severe and non-stationary noise compensation

The loss of low-energy detail in noisy conditions can be clearly illustrated by comparing the log filterbank HMM 'profiles' of a speech model trained in both clean and noisy conditions. Figures 10.20 and 10.21 present such profiles for the isolated-digit model 'eight'; the plots show the log filterbank energy (z-axis) for each frequency coefficient (y-axis) as a function of model state (x-axis). Hence Figs. 10.20 and 10.21 represent the change in characteristic spectral shape of the model over the duration of the model. Figure 10.22 illustrates the adapted model.

Figure 10.20 represents the 'clean' version of the digit, and can be seen to have clearly defined contours in the low energy regions. In comparison, the 'matched' model of Fig. 10.21, trained in severe noise conditions (i.e. below 10 dB SNR), has lost large amounts of the low-energy detail which can characterize a particular word.

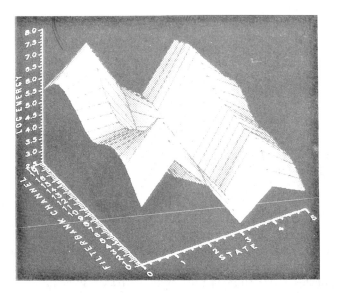

Fig. 10.20   Log filterbank profile for HMM 'eight' trained in clean conditions.

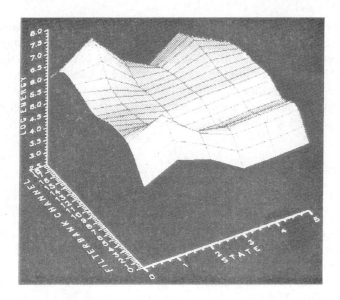

Fig. 10.21   Log filterbank profile for HMM 'eight' trained in 'matched' conditions.

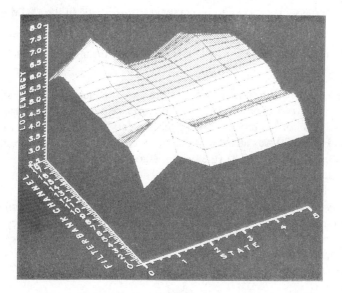

Fig. 10.22   Log filterbank profile for adapted HMM 'eight'.

Achieving the 'matched' system levels of performance in these conditions requires some method for simulating the effects of training the recognizer in these environments. Various methods for achieving this have been reported in the literature [6, 25]. The latter presents a technique known as '**Wiener Adaptation**', an implementation of Wiener filtering [26] in the context of an HMM.

The conventional Wiener filter equation operates in the spectral domain, where the filtered speech signal is given by:

$$\hat{S}(f) \;=\; Y(f)W(f) \;=\; Y(f)\frac{\mu_s(f)}{\mu_s(f) + \mu_N(f)} \qquad\qquad \dots (10.10)$$

where $\mu_s(f)$ and $\mu_N(f)$ represent the mean of the clean speech and interfering noise spectra respectively. The noisy speech signal is denoted by $Y(f)$.

It is shown in Milner and Vaseghi [25] that Wiener filtering of the noisy speech is equivalent to the adaptation of the mean cepstral vectors of the states of an HMM. The adaptation technique overcomes the two key drawbacks of the Wiener filter, namely that:

- the filter assumes speech to be stationary (this is true in the HMM context, as each state represents a quasi-stationary event);

- a clean speech estimate is available (the HMM means provide such 'clean' speech estimates).

Modifying the clean speech models using the technique effectively adapts the speech models to an approximation of the current environment, thus simulating 'matched' conditions as required. Using the adaptation method, the 'clean' model 'eight' of Fig. 10.20 was adapted to the adverse conditions used for training the 'matched' model of Fig. 10.21 (the same method for obtaining the noise estimate as in the earlier techniques is applicable). The 'adapted' model is illustrated in Fig. 10.22 for comparison with the 'matched' HMM.

Adding Wiener adaptation into the recognizer for the isolated-digit experiment delivers the best overall performance, achieving a word accuracy of over 80% at 10 dB SNR, a figure within 8% of the matched performance and an improvement of over 63% from the original uncompensated system. The results are again shown on Fig. 10.23.

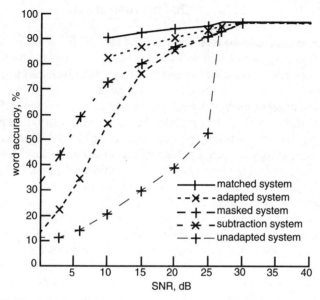

Fig. 10.23   Results of the isolated-digit experiments for various noise robust techniques.

10.7.6   Noise-robust results and conclusions

A comparison of the various methods for dealing with adverse conditions has been made. The relative performance of each method is compared using a baseline HMM-based isolated-digit recognizer.

Noisy conditions were simulated by adding noise data recorded in a car travelling at 60 mph to the digit utterances at various levels; this particular noise type has a quasi-stationary spectrum.

The results, shown in Fig. 10.23, clearly indicate the benefit of adding noise robustness to a recognition system if it is to operate successfully in a 'real-world' situation. The trade-off between recognition performance and computational complexity of the robust techniques is also captured by the graph.

The Wiener adaptation system consistently achieves the highest recognition accuracy over all the conditions tested here, and has been shown to compete well with the more complicated model adaptation techniques [6, 27] used for highly non-stationary noise.

10.7.7   Channel robustness

This section investigates the effect of non-ideal channel frequency responses on the accuracy of telephony-based speaker recognition systems. It is shown that a technique known as cepstral mean normalization (CMN) increases performance while reducing the amount of enrolment data required.

In telephony applications it is well documented that recognition performance can be degraded as a result of variability in the properties of communications channels [28, 29]. The variability produces a mismatch between speakers' models produced by enrolment data, and the subsequent test utterances for which the models are supposed to be representative. The problem can be particularly acute in automatic speaker recognition systems where the enrolment data is collected over the telephony network in a single session, and hence inter session variability is not modelled.

In order to compensate for the channel effect two main approaches are commonly used. The first is in the use of channel robust features such as cepstral time matrices (see section 10.2.3) or RASTA filtering [30]. The second involves the removal of the channel distortion, as in the application of CMN. The work presented in this section focuses on the latter. An experimental assessment of five variations of CMN is performed and the results compared.

10.7.8   Speaker recognition — an overview

Automatic speaker recognition is the process of recognizing individuals from characteristic acoustic features of their voice. The subject area is divided into two categories. Speaker verification and speaker identification. Speaker verification is the process of authentication of a person's identity claim by comparing a sample of their speech to that of the claimed identity. Speaker identification is the process of attributing the speech produced by an unknown speaker to one of a population of known speakers for whom speech samples are available.

10.7.9   The channel effect

Many factors affect the properties of a communication channel such as line length, handset model and acoustic environment. To an approximation, these handset and channel effects can be modelled as a linear filtering operation, where the time-domain signal $y(t)$ is given by:

$$y(t) = x(t)*h(t) \qquad \qquad ... (10.11)$$

where:

$x(t)$ is the speech signal prior to filtering,

$h(t)$ is the overall impulse response of the channel.

In the frequency domain the convolutional channel distortion becomes multiplicative, that is:

$$Y(f) = X(f)H(f) \qquad \qquad ...(10.12)$$

In the log domain the channel distortion becomes additive and because of the linearity of the DCT, the effect of the channel on cepstra is also additive:

$$c_y(m) = c_x(m) + c_h(m) \qquad \qquad ...(10.13)$$

where:

$c_y(m)$ is the observed mth MFCC vector,

$c_x(m)$ is the mth MFCC vector, prior to transmission,

$c_h(m)$ is the mth channel distortion vector.

10.7.10   Channel normalization

The principle of CMN is to estimate the channel distortion vector over an observation sequence and subtract that estimate from every vector in the sequence. The effect of CMN is illustrated in Fig. 10.24.

The left hand column shows a trace of MFCC coefficient value over the time duration of an isolated utterance. In each case two traces are shown; these correspond to the same recorded isolated utterance transmitted via two different transmission channels. The channel effect produces a constant offset between the two traces, corresponding to the difference in transfer characteristic in the frequency bands between the two channels.

Referring to the right hand column, the plots of the normalized MFCCs are shown. It can be seen that the normalization has brought the traces into closer alignment.

The global estimate of the channel distortion vector L is given by averaging over the vectors in a sequence:

$$L = \frac{\sum\limits_{m=1}^{N} c_y(m)}{N} \qquad \qquad ... (10.14)$$

where N is the number of vectors in the sentence.

Each MFCC vector can be thought of as being made up of a speech component and a channel component as given in equation (10.13).

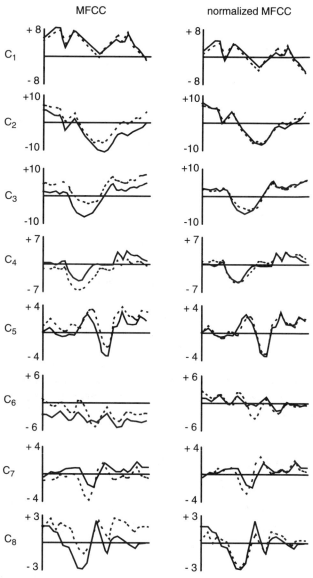

Fig. 10.24   Comparison between MFCCs and normalized MFCCs for the same speech transmitted through two channels.

Substituting equation (10.13) into equation (10.14), the global estimate L can be written in terms of its speech and channel components; that is:

$$L = \frac{\sum\limits_{m=1}^{N} (c_x(m) + c_h(m))}{N} \qquad \text{... (10.15)}$$

$$L = c_x + c_h \qquad \text{... (10.16)}$$

where:

c_x is the average of the speech over the observation sequence, and
c_h is the average channel vector over the observation sequence.

Equation (10.16) shows that the CMN technique does not actually estimate the true channel distortion vector c_h, but instead provides a value that is offset by the mean of the speech over which the estimate is calculated. The effectiveness of the estimate L consequently depends on the validity of two assumptions. The first is that the channel does not vary significantly over the observation period, and the second is that the average of the speech is the same for subsequent channel estimates.

In order to test the validity of these assumptions, the following normalization schemes were investigated. In all cases normalization was performed over both training and test utterances.

10.7.10.1   Long-term cepstral mean normalization (LCMN)

For LCMN [28], it is assumed that there is little or no variation of the channel distortion vector with time. The global estimate L is calculated over as long an observation period as possible in order to get the best estimate of the speech average c_x and the channel distortion vector c_h. Both speech and non speech segments are included in the calculation.

10.7.10.2   Speech-only long-term cepstral mean normalization (SLCMN)

An alternative to LCMN is to perform normalization over speech segments only. In this case the global estimate is calculated over the duration of the entire call, but non-speech portions are ignored on the assumption that they will give a poor channel estimate [31].

10.7.10.3   Non-speech long-term cepstral mean normalization (NLCMN)

In order to test the assumption that non-speech segments give poor estimates of the channel, CMN is performed over these segments only.

10.7.10.4   Word-level cepstral mean normalization (WCMN)

An alternative to LCMN is to normalize over individual words. It is anticipated that the average of the speech over individual words will be as consistent as that obtained over longer observation sequences. It is also expected that variations of the channel characteristic, over the relatively short observation sequence, are likely to be minimized.

10.7.10.5   Sub-word level cepstral mean normalization (SWCMN)

Finally, as with word level, CMN is performed over sub-word units.

10.7.11   Experiments

A set of speaker verification and speaker identification experiments was designed to compare the performance of the CMN techniques described above.

10.7.11.1   Speech database

The experiments used the BT Brent telephony database. This consists of 56 versions of the isolated digit set ('one' to 'nine' and 'zero') from 100 talkers. The first 12 versions are reserved for training and the remaining 44 versions (recorded 1 per week from at least 2 different locations) are for testing. The training sets are divided into the categories shown in Table 10.7.

Table 10.7   Training set categories.

Training Set	Description
1	Three utterances in a single call
2	Three utterances in three calls in a single week
3	Three utterances in three calls in three weeks
4	Eight utterances in two calls in two weeks (versions 1 - 4 from single call, versions 5 - 8 from separate calls)
5	Eight utterances in three calls in three weeks

The training set categories simulate different enrolment scenarios. Set 1, for example, represents a single enrolment session, which would be expected to give poorer performance than an enrolment over several days.

10.7.12   Signal processing

The experiments used an MFCC parameterization produced from speech sampled at 8 kHz. Following pre-emphasis, 32 ms frames (with a 16 ms overlap) were progressed as follows:

Hamming Window→ FFT→ Mel Filter (19 filters equally spaced on the Mel Scale)→ Log→ DCT.

10.7.13   Recognition system

Speaker recognition experiments were performed using a DTW classifier. This was chosen in preference to HMM or VQ techniques since it has been shown to give the best performance for the task [32].

Reference templates were constructed as follows. From the enrolment data the median-length utterance was designated as the initial template to which the others were time aligned using DTW. After alignment the utterances were averaged together into a single template.

The individual digits were segmented using Viterbi alignment with the correct phoneme-based isolated-digit models. This provided the appropriate annotation information required for the normalization schemes.

Recognition was performed using a 'city-block' distance measure between the template and the test token. Each test token consisted of the full isolated-digit sequence. A distance score was obtained for each of the ten digits and an overall score calculated by averaging the individual scores together.

Each recognition experiment consisted of 4400 test tokens (44 from each speaker). For speaker verification, speech from all talkers was matched against all individuals' models, and the resulting scores divided into true talker and imposter scores. True talker scores occur when the test tokens are taken from the same speaker as was used to generate the models (simulating true talker identity claims). Imposter scores occur when the test token is taken from a speaker other than that which was used to generate the models (simulating imposter identity claims).

The score values were used to calculate individual equal error rates (EERs) for all speakers, which were then averaged across the entire speaker set to give the overall performance figure. Speaker identification was simulated by presenting each test token to all models and choosing the best match.

10.7.14   Experimental results

For each of the training sets and CMN schemes, speaker verification and speaker identification performance is given in Figs. 10.25-10.34. Of the five CMN techniques investigated, long-term normalization over speech portions only (SLCMN) gives an overall improvement over conventional MFCCs. An overall improvement is also obtained for word-level normalization, the results being almost identical to SLCMN performance.

For speaker verification the most striking error reduction can be seen with training set 1. With SLCMN and WCMN, the EER decreases from the base-line figure of 5.1% to 2.5% and 2.7% respectively. Moreover, referring to performance across training sets, it can be seen that these techniques obviate the need for multiple enrolment sessions. Irrespective of the training set used (single or multiple enrolment), the EER is approximately 2.5%.

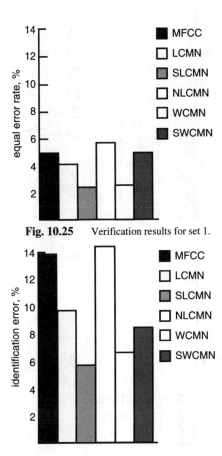

Fig. 10.25   Verification results for set 1.

Fig. 10.26   Identification results for set 1.

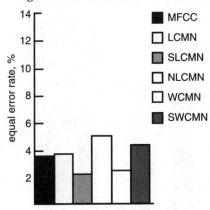

Fig. 10.27   Verification results for set 2.

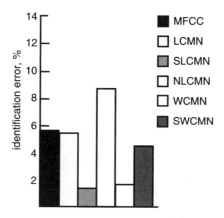

Fig. 10.28   Identification results for set 2.

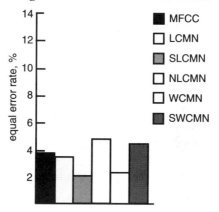

Fig. 10.29   Verification results for set 3.

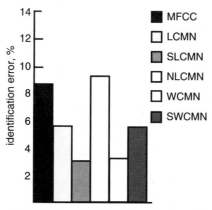

Fig. 10.30   Identification results for set 3.

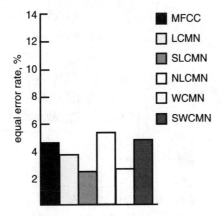

Fig. 10.31   Verification results for set 4.

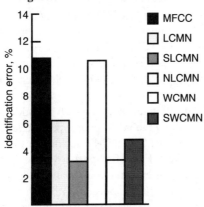

Fig. 10.32   Identification results for set 4.

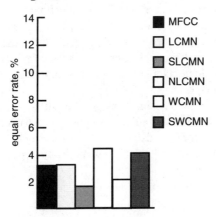

Fig. 10.33   Verification results for set 5.

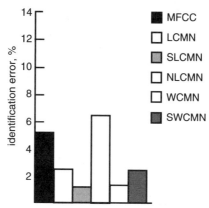

Fig. 10.34 Identification results for set 5.

For speaker identification the SLCMN and WCMN techniques substantially improve performance, indicating that speaker identification is very susceptible to inter-channel variations. This is not surprising since the identification process requires that test tokens are compared against all speaker models. If the models and test tokens are biased by different channel vectors, it is not unreasonable to expect additional errors.

The long-term normalization techniques which include non-speech segments, namely LCMN and NLCMN, give much poorer performance than that obtained for speech-only normalizations. The LCMN technique gives roughly the same performance as conventional MFCCs for speaker verification and slightly better performance for speaker identification. Normalization over non-speech segments only gives poorer performance throughout.

The favourable performance of the speech-only normalizations may be attributed to the fact that the speech averages for different instances of a word or words are likely to be consistent over different telephony channels (see equation (10.16)). Any differences in global estimates between the same words over different channels will therefore be due to variations in the channel term c_h and not the speech term c_x. If the normalization were to incorporate non-speech segments, the c_x term would be likely to be subject to more variation due to the random nature of the non-speech.

Although WCMN provides almost as good a performance as SLCMN, the trend does not follow for normalizations over phonemes. This could be caused by inconsistencies in phoneme boundary estimates, or it may be that there is greater variability inherent in the averages of phoneme instances compared with those found in whole word averages.

10.7.15   Discussion

There are two factors that affect CMN performance. Firstly, normalizations which include non-speech segments give poorer performance than those which do not. Secondly, normalizations over phoneme length intervals produce poorer performance than those obtained over longer observations. There is, however, a trade off between the length of the observation sequence over which the normalization is performed and the increase in accuracy achieved. Whereas normalization at word level gives a considerable improvement over that obtained at phoneme level, normalizations over longer observations give very little additional improvement. Word-level normalization is therefore the most appropriate technique for applications which require real-time or near-real-time capability. It also has the combined benefit of both increasing performance while reducing the amount of enrolment data required.

10.8   CONCLUSIONS

This chapter has shown new and powerful techniques which enable speech and speaker recognition technologies to operate successfully in a variety of environments. The key constituents of a robust recognizer have been considered and several technological advances have been demonstrated:

- a new feature-extraction technique, namely the cepstral-time matrix, gives higher recognition accuracy over the conventional cepstral vector augmented with higher-order time derivatives;

- a dynamic recognizer which provides an efficient solution to dealing with large vocabularies or long utterances;

- improved recognition accuracy via phoneme-based, context-dependent modelling techniques;

- user-selectable vocabulary using sub-word HMMs;

- increased OOV rejection via the use of sink models;

- noise and channel robustness via a combination of model adaptation and feature normalization techniques.

Although these techniques provide the basis for constructing sophisticated recognition systems there is still scope for improvement in many of the areas. Nevertheless this work is among the most advanced in its field and enables BT to provide state-of-the-art recognition systems.

REFERENCES

1.   Parsons T: 'Voice and speech processing', McGraw-Hill (1987).

2.   Hanson B A and Applebaum T H: 'Robust speaker-independent word recognition using static, dynamic and acceleration features: experiments with Lombard and noisy speech', Proc ICASSP, pp 857-860 (1990).

3.   Milner B P: 'Speech recognition in adverse environments', PhD Thesis (1994).

4.   Gales M J F and Young S J: 'An improved approach to the hidden Markov model decomposition', Proc ICASSP, pp 729-732 (1992).

5.   Woodland P C, Odell J J, Valtchev V and Young S J: 'Large vocabulary continuous speech recognition using HTK', Proc ICASSP, pp 125-128 (1994).

6.   Jelinek F: 'Continuous speech recognition by statistical methods', Proc IEEE, 64, No 10, pp 532-556.

7.   Katz S M: 'Estimation of probabilities from sparse data for the language model component of a speech recognizer', IEEE Trans ASSP, 35, No 3, pp 400-401 (March 1987).

8.   Kenn P, Hollan R, Gupta V N, Lennig M, Mermelstein P and O'Shaughnessy D: 'A* — admissible heuristics for rapid lexical access', IEEE Trans Speech and Audio Processing, 1, pp 49-58 (January 1993).

9.   Ney H: 'Search strategies for large-vocabulary continuous-speech recognition', in Rutio and Soler (Eds): 'Speech Recognition and Coding (New Advances and Trends)', Springer (1993).

10.   Schwartz R and Austin S: 'A comparison of several approximate algorithms for finding multiple (N_BEST) Sentence Hypotheses', Proc ICASSP, pp 701-704 (1991).

11.   Hwang M, Huang X D and Alleva F: 'Predicting unseen triphones with senones', Proc ICASSP, pp 311-314 (1993).

12.   Huang X D and Jack M A: 'Semi-continuous hidden Markov models for speech recognition', Computing Speech and Language, 3, pp 239-251 (1989).

13.   Lee K F: 'Large vocabulary speaker-independent continuous speech recognition', The SPHINX system, PhD Thesis, CMU-CS-88-148 (1988).

14.   Young S J: 'Benchmark DARPA RM results with the HTK portable HMM toolkit', Proc DARPA Workshop (September 1992).

15. Paul D B: 'The Lincoln continous speech recognition system: recent developments and results', DARPA Speech and Natural Language Workshop, pp 160-166 (February 1989).

16. Young S J: 'HTK version 1.3: reference manual', Cambridge University Engineering Dept, Speech Group (May 1992).

17. Svendsen T, Soong F and Purnhagen H: 'Optimizing baseforms for HMM-based speech recognition', Proc Eurospeech, pp 783-786 (1995).

18. Junqua J C and Angelade Y: 'Acoustic and perceptual studies of Lombard speech', Proc ICASSP, pp 841-844 (1990).

19. Van Compernolle D: 'DSP techniques for speech enhancement', Proc Speech Processing in Adverse Conditions (1992).

20. Munday E: Chapter 6 in Wheddon C (Ed): 'Speech and language processing', Chapman & Hall (1990).

21. Bridle J S, Ponting K, Brown M and Borrett A: 'A noise compensating spectrum distance measure applied to automatic speech recognition', Proc IOA, pp 307-310 (1984).

22. Klatt D H: 'A digital filterbank for spectral matching', Proc ICASSP, pp 573-578 (1976).

23. Milner B P and Vaseghi S V: 'Comparison of some noise-compensation methods for speech recognition in adverse environments', Proc IEE, 141, pp 280-288 (1994).

24. Wiener N: 'Extrapolation, interpolation and smoothing of stationary time series with engineering applications', MIT Press (1949).

25. Varga A and Moore R K: 'Hidden Markov model decomposition of speech and noise', Proc ICASSP, pp 845-848 (1990).

26. Gish H: 'Investigation of text-independent speaker identification over telephone channels', Proc ICASSP, pp 379-382 (1985).

27. Ariyaeeinia A M and Sivakumaran: 'Transitional features', IEEE International Carnahan Conference on Security Technology, pp 79-84 (1995).

28. Furui S: 'Cepstral analysis technique for automatic speaker verification', IEEE Trans Speech and Signal Processing, 29, No 2, pp 254-272 (April 1981).

29. Milner B P and Vaseghi S V: 'An analysis of cepstral-time feature matrices for noise and channel robust speech recognition', Proc Eurospeech, pp 519-522 (1995).

30. Hermansky H, Morgan N, Bayya A and Hohn P: 'Compensation for the effect of the communication channel in auditory-like analysis of speech (RASTA PLP)', Proc Eurospeech, Genova, Italy, pp 1367-1371 (1991).

31. Rosenberg A E, Lee C-H and Soong F K: 'Cepstral channel normalization techniques for HMM-based speaker verification', ICSLP, pp 1835-1838 (1994).

32. Yu K, Mason J and Oglesby J: 'Speaker recongition models', Proc Eurospeech, pp 629-632 (1995).

11

SPEECH RECOGNITION — MAKING IT WORK FOR REAL

F Scahill, J E Talintyre, S H Johnson, A E Bass, J A Lear,
D J Franklin and P R Lee

11.1 FROM ALGORITHMS TO SERVICES

Previous speech recognizers developed by BT concentrated on accurate recognition of small vocabularies across the telephone network, e.g. recognition of 'yes' and 'no' or the digits '0' to '9' using a whole-word model for each word. These recognizers suffer from the disadvantage that changes to the vocabularies require large collections of speech data to train models for the new words. Even with this drawback, these recognizers can still support many useful applications, such as BT CallMinder™.

Speech recognition algorithms have now progressed to the point where large-vocabulary and continuous-speech recognition are frequently demonstrated in the laboratory. The goal now is to make these algorithms available for use in real interactive speech services.

11.1.1 Flexible vocabularies

Developments in sub-word unit algorithms for speech processing mean that it is now possible to quickly create vocabularies for new applications or change vocabularies for existing ones, without additional data collections and model training. Some applications in fact depend on the ability to change the recognizer vocabularies on a day-to-day, or even minute-to-minute basis!

11.1.2   Large vocabularies

The accuracy and flexibility of recognition algorithms have improved to such an extent that it is now possible to offer services with active[1] vocabulary sizes of up to 1000 words. Such systems would have been impractical with previous speech recognition technology.

11.1.3   New services

The increases in vocabulary flexibility and vocabulary size have led to a huge range of trial automated services such as call-centre automation, e.g. bill payment by phone, automated database enquiries and advanced voice-mail services.

These services require a speech recognizer in which the latest speech recognition algorithms are implemented in a flexible, robust package of software that is portable to a wide range of platforms, including the BT/Ericsson interactive speech applications platform (ISAP) [1] and Unix® workstations.

This chapter describes such an implementation, known as Stap, and outlines how it was designed and how its capabilities meet the requirements of current and future interactive speech services.

11.2   SPEECH RECOGNITION — A BRIEF OVERVIEW

11.2.1   Vocabulary

A speech recognizer is designed to recognize one of a set of words or phrases specified in a vocabulary. The recognizer is presented with a segment of sound, usually several seconds long, which henceforth will be called an utterance. The recognizer attempts to distinguish the speech from the noise, and to label the utterance as matching one of the items in its vocabulary.

11.2.2   Hidden Markov models

Many speech recognizers, including Stap, use hidden Markov models (HMMs) to represent their vocabularies. An HMM is a statistical model [2]. Each HMM

---

[1] The active vocabulary is that which can be recognized at any one time. For example a recognizer could be capable of recognizing any of 10 000 town names, but be set to only expect one of 1000 in answer to a specific query.

represents a unit of sound, such as a word or a sub-word unit (e.g. phoneme) that is to be recognized. Each model is constructed using statistics calculated from examples of the speech unit that the model represents. This process is referred to as 'training' the models. The speech used for training should be representative of the speech that the recognizer will encounter when it is used. In some cases this is not possible and it becomes necessary to modify the models to suit the data, this is known as 'adaptation'. Some speech recognizers require that the speech data be provided by a single person, these are 'speaker-dependent' and will only recognize that person's speech. 'Speaker-independent' systems are those that combine the training data from enough speakers to work well for almost any person's speech.

11.2.3   Front end

To improve accuracy and efficiency, HMMs do not model the speech directly. Instead, they model a transformed version of the speech which is more compact and decorrelated. This transformation process is known as 'feature extraction' and is performed by the 'front end' (FE).

11.2.4   Parser

To perform the actual recognition match, a 'recognition network' of models (sub-word or whole word) is constructed to represent the entire vocabulary. The recognizer's task is to find the path, or paths, through the network which most closely match the utterance. The recognition component which searches for these paths is the 'parser'.

11.2.5   Rejection

Speech recognition is difficult — no speech recognizer, not even a person, is 100% accurate. There are many reasons why an utterance may be mis-recognized. For example, the person speaking may have an accent on which the recognizer has not been trained, or there may be a high level of background noise from a busy office.

Alternatively the person speaking may simply have said something that was not in the recognizer's vocabulary. An important attribute of any speech recognizer is therefore a 'rejection' capability. Rejection is the means by which a recognizer checks its confidence in the recognition match, so that the speaker can be asked to confirm the result if necessary.

11.2.6   Pause detection

Vocabularies can contain words or phrases of varying length. The user will speak the word or phrase an unknown amount of time after being prompted to speak. The speech recognizer cannot therefore rely on listening for a fixed period of time. It must accurately determine when the person has finished speaking so that the matching process can be stopped and the recognition result returned. Inter-word and intra-word pauses must be ignored. Henceforth this activity will be referred to as 'pause detection'.

11.3   WHAT IS STAP?

Stap is a package of software for performing advanced speech recognition. It comprises two main tools:

- StapRec — a real-time advanced speech recognizer;

- StapVoc — a recognition network creation tool.

A number of additional tools for more advanced users are also provided; these include:

- grok — a network creation tool which accepts any BNF style grammar;

- Xpressive — an X-window interface to Stap.

11.3.1   StapRec

StapRec is:

- speaker-independent, i.e. it works for any speaker without prior training or enrolment;

- able to accept input from the telephone network or a workstation's audio system;

- 'real time' (speech data is processed as it is received) — the recognizer will normally give out an answer as soon as it has detected that the speaker has finished speaking;

- multi-language — currently British English and American English are supported, other languages being added by creating new StapRec data files;

- flexible — the vocabulary is defined by a recognition network which may consist of isolated words, short phrases, connected digits, alphanumerics or any combination thereof, with the active vocabulary size varying from 1-1000 words, which can be subsetted or completely replaced very quickly.

11.3.2   StapVoc

StapVoc is:

- a vocabulary generator that converts textual vocabulary specifications into recognition networks and the associated files necessary to configure and control a StapRec;

- easy to use by application developers without the need for prior speech-recognition knowledge.

11.3.3   Key issues

In addition to the functional requirements for speech recognition, Stap was designed with a number of other key issues in mind.

- Portability and maintainability

  Computing hardware is changing at an increasing rate. New and more powerful processors are constantly becoming available. Similarly the requirements on and capabilities of speech recognition algorithms are constantly improving. It is therefore essential that Stap is portable and maintainable.

  To this end Stap was specified and designed using the Booch object-oriented design methodology [3], and implemented in C++ [4]. The Booch notation for expressing object-oriented designs is briefly described in the Appendix.

- Cost

  The commercial viability of an automated interactive speech service is largely dependent on the cost per channel. Speech recognition is one of the most expensive facilities that a speech service may have to provide. Therefore a primary goal for any speech recognizer is memory and computational efficiency. This allows useful speech recognition to be provided on low power hardware, yet permits more powerful hardware to make a trade-off between handling more channels simultaneously and performing more accurate speech recognition.

- Ease of use

  The rapid and widespread deployment of speech recognition can only be achieved if there is a simple application developer interface. Application developers should not require an understanding of hidden Markov models to add interactive speech to their services.

11.4   STAPREC

11.4.1   The recognizer architecture

Figure 11.1 shows a conceptual overview of the main components of StapRec.

11.4.1.1   Feature extraction

Speech arriving at the front end is already in digital format, for example 64 kbit/s A-law for ISAP based applications. The feature extraction block performs the signal processing required to reduce the speech signal to a more compact and easier to recognize parameterization. StapRec uses a combination of Mel frequency cepstral co-efficients (MFCC), delta MFCCs and delta log energy (see Chapters 9 and 10).

As can be seen from Fig. 11.1, the feature extraction block is separate from StapRec. This allows for easy replacement with alternative parameterizations for efficiency or accuracy improvements. It also allows for the different speech formats that can be encountered when Stap is run in a workstation environment. Furthermore the front end can be run on a separate processor, such as a digital signal processor (DSP), with better price/performance ratio for the type of vector maths processing required.

11.4.1.2   Variable frame rate

The parameterization process initially generates a frame of data every 16 ms. The number of frames is then adjusted to give variable frame rate (VFR) data.

So that timing information is not lost, the VFR process attaches a number to each retained frame, indicating how many real frames it represents. Using VFR data reduces the amount of processing required during recognition, and also emphasizes the more discriminatory changing regions of the signal.

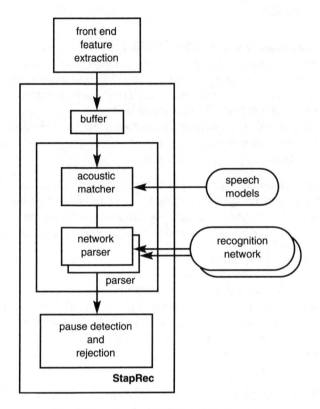

Fig. 11.1   Components of a speech recognizer.

11.4.1.3   Buffering

The rate at which StapRec can process the frames from the FE depends on a number of factors including the CPU hardware and recognition vocabulary size. The processing load also varies during the course of recognizing an utterance, depending on the proportion of the recognition network that is active. To allow for these variations StapRec maintains a separate buffer process between the front end and the parser. In principle, this buffering could be integral to the FE, but, in practice, this causes problems with digital signal processing (DSP) implementations of the front end due to memory limitations.

The separation of the buffer into a separate Unix process also allows StapRec to receive data from the FE before it has completed configuring. This can be important when switching between large recognition vocabularies on the fly.

11.4.1.4   Parser

StapRec extracts data from the buffer process and passes it, one frame at a time, to the recognition parser which performs the detailed pattern match. The parser matches the VFR feature data against the HMM speech models, and combines the outputs of this match with the recognition network which defines the vocabulary to be recognized. The parser will be described in more detail later.

The separation of the acoustic match and network parse enables StapRec to support efficiently[1] multiple parallel network parsers recognizing the same feature data. This leads to a number of advantages.

- Faster reconfiguration where vocabularies overlap — here, an applications vocabulary can be split into shared and independent regions. When the recognizer is switched between vocabularies then only the independent regions need to be reloaded. Thus, re-configuration time is reduced.

- Multiple independent matches available — normally a real recognizer must be able to detect silence and out-of-vocabulary (OOV) utterances in addition to finding the best matching vocabulary item. This is usually done through the use of silence and OOV recognition paths. If these were incorporated in the same network as the vocabulary then the network parser design would be complicated since it would have to treat these specially. The parallel network design enables a simpler parser design while providing a more generic functionality.

- Possible parallel processing — for very complex recognition tasks a single CPU may not be powerful enough for real-time recognition. Parallel networks will in the future facilitate the distribution of large recognition tasks across multiple CPUs.

11.4.2   Pause detection and rejection

After each frame has been processed by the parser, StapRec checks to see whether the speech has finished and, if it has, extracts the recognition results from the parser. A number of confidence measures are then applied to the results and, dependent on these, a decision is made — typically 'accept', 'reject' or 'query'.

To ensure good performance the pause detection and rejection software make use of both raw signal measures (e.g. SNR) together with the quality of match measures derived from the recognition results.

---

[1] Since the CPU-expensive parts of the HMM calculations are performed only once.

This mix of features means that the recognizer is capable of detecting errors caused by:

- out-of-vocabulary speech;

- noisy lines;

- poor pronunciation;

- confusable vocabulary words.

11.4.3   Advanced facilities

StapRec includes some advanced facilities which are not normally found in speech recognition software. These facilities have been developed to meet the requirements of BT's own advanced speech applications.

11.4.3.1   Vocabulary weighting and subsetting

For many large-vocabulary tasks, such as automated Directory Enquiries, there is a considerable amount of information regarding the *a priori* probability of a particular word in the vocabulary being used (see Chapter 16). Overall accuracy can be substantially improved if these probabilities are used to weight paths in the recognition grammar.

StapRec therefore provides a facility where a probabilistic weighting can be applied to items in the vocabulary. In its simplest form this provides a mechanism for subsetting the vocabulary by turning off some vocabulary items altogether, i.e. weighting them to zero. The advantages of this facility are clear — any reduction in the vocabulary size will increase accuracy and reduce the computational cost.

The application developer controls the vocabulary weighting by passing a simple text list of the vocabulary words and their weightings. The weightings are applied by modification of node penalties internal to the recognition network. It is performed at the start of each recognition and the network is reset at the end of the recognition. There is no significant increase in recognition time for performing this process.

Unlike other commercial speech recognizers which provide similar facilities, Stap provides the control down to specific words and can perform soft weighting (use of *a prioris*) or hard subsetting (setting some weightings to zero and others to one).

11.4.4   Homophones

StapRec, like all speech recognizers, is only able to recognize what a particular word sounds like. This is a problem when two or more words in a vocabulary sound the same. These are known as homophones, e.g. brake and break[1].

It is inefficient for the recognition network to contain separate copies of the homophones since this leads to unnecessary duplication of calculations. Instead, a single path in the grammar is used to represent the homophones and this path is expanded to include all the associated vocabulary items.

The mapping of network paths to the vocabulary items is produced when the recognition network is generated by StapVoc. Since StapVoc knows the phonetic transcription of the words, it is able to detect homophones and relabel those paths in the network with a 'homophone tag'. A separate mapping between these tags and the real vocabulary items is updated and stored.

The advantage of this approach is that the application developer needs to take no special measures to cope with homophones. Stap compresses and expands them automatically and invisibly.

11.4.5   Class diagram

Figure 11.2 shows how the basic components of StapRec translate into a class design using the Booch notation[2].

At the centre of the architecture is the RecogController. This co-ordinates and delegates the other components to recognize an utterance encoded as a sequence of frames.

The Recognizer class manages the multiple network parsers, ensuring that their recognition matches are combined in sensible ways and that all the parsers are synchronized. The ModelSet is the repository for all data relating to the HMMs that is not specific to their context in the network. To allow for integration of other types of network parser with Stap, the Network class is decoupled from the rest of StapRec through an interface class VegaIF.

StapRec detects when the caller has finished speaking as an integral part of the recognition process. Hence, after each frame has been processed by the Network objects, the PauseDetector must examine the current recognition result, together with some raw signal data, to decide whether the caller has finished speaking. When the end of speech has been detected, the results from the Networks are combined by the Recognizer and passed to the Rejector for checking. Finally the RecogResults are returned to the applications via the RecogController.

---

[1] This problem is most commonly found when automating access to large databases, where the application developer has little control over the contents of the database.

[2] See the Appendix for a brief explanation of Booch.

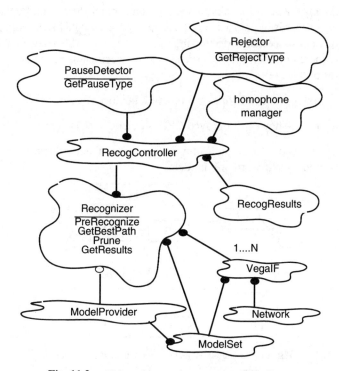

Fig. 11.2   Main architectural components of StapRec.

11.5   VEGA — THE PARSER

The heart of StapRec is the acoustic matcher and grammar parser. The collection of classes that perform these roles is known as Vega. As the principle of the algorithm-matching process is adequately covered by Chapters 9 and 10, this section concentrates on the functionality and design of the parser.

11.5.1   Networks, nodes and tokens

The parser design is based upon three main concepts — networks, nodes and tokens.

11.5.1.1   Networks

In Vega, the speech recognition vocabulary is represented as a finite network (or graph) of nodes defining the HMM interconnections. This representation allows considerable generality and flexibility.

For example, a network constructed by StapVoc, for the recognition of a constrained set of words or phrases, has the property that (roughly speaking) no node has two distinct successor nodes having the same model. This leads to a tree-like topology where common prefixes are shared between paths in the network. The advantages of this are:

- fewer nodes in the network — improved memory efficiency;

- shared calculations — improved CPU efficiency.

Figure 11.3 shows a simple example for the vocabulary duck, dog, calf and cat represented by sub-word unit models.

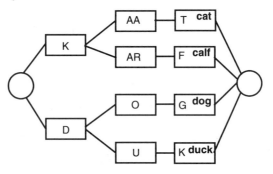

Fig. 11.3   Simple StapVoc generated network topology.

A useful feature of this network is that each vocabulary item corresponds to a single terminal (leaf) node. This has two advantages:

- the match scores for all of the vocabulary can be made available to the application (since all vocabulary items are represented by a unique path through the network);

- vocabulary items can be turned off (subsetting) or probabilistically weighted by the application of simple penalties to the terminal nodes.

Other recognition tasks require different network topologies. For unconstrained connected digit recognition a tree structure is neither practical nor desirable. Here a more sensible network is one where paths can merge and can loop back to a previous node (see Fig. 11.4).

Vega accepts arbitrary network topologies and importantly, arbitrary network sizes. In principle, the size of network that Vega can load is limited only by the memory available on the platform. This is typically up to 64 Mb which is sufficient for recognition vocabularies of 100 000 words containing more than a million nodes.

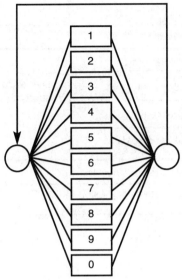

Fig. 11.4   Unconstrained connected digit network topology.

11.5.1.2   Nodes

A node is the basic processing unit in speech recognition. It normally has a model attached to it for recognizing a single phoneme or word, but alternatively can be just a connection point (null node — see below).

A null node is useful when handling many-to-many network connections. A null-node replaces these types of connections by a many-to-one and one-to-many connection. Thereby reducing token propagation from N^2 to $2N$ without loss in generality. The structure of a node is shown in Fig. 11.5.

As can be seen, the node is split into static and dynamic data. For memory-efficient recognition with network sizes of up to a million nodes, each node must occupy as little memory as possible. In Vega the static node data is represented typically[1] in 50 bytes of memory, while the dynamic data is allocated from a central pool only if the node becomes activated, and is returned to the pool if the node is deactivated.

The node design supports multiple node scratch spaces which, when combined with the node signature, provide an efficient mechanism for multiple candidate processing [5].

---

[1] Actual space used is dependent on the number of output edges from the node.

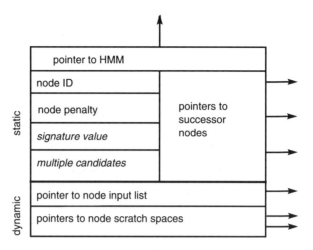

Fig. 11.5   Structure of a node.

11.5.1.3   Tokens

The network, and the nodes from which it is made up, describe the vocabulary of the speech recognizer. During recognition the input speech is matched against the network of nodes. Each frame that is processed causes a token to be emitted from each node. Before the subsequent frame is processed tokens move from a node to successor nodes in the network. For example, in Fig. 11.3 a token would be emitted from the 'D' node into the 'O' and 'U' nodes.

Figure 11.6 shows the structure of a token. The pointer to the node gives the identity of the last model matched and the time index indicates when this

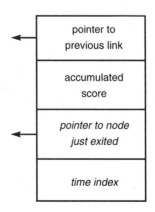

Fig. 11.6   Structure of a token.

occurred. By tracing back through the pointers to previous tokens a sequence of matched models is given.

11.5.2   Overview of the pattern matching process

For HMM-based speech recognition the matching process can be split into two areas:

- acoustic match;
- network parse.

11.5.2.1   Acoustic match

The acoustic match is given by calculating the HMM state output probabilities. This involves the evaluation of multi-modal, multi-variate probability density functions and so is particularly CPU expensive. Fortunately, the probabilities are dependent only on the current frame of data and the HMM parameters and so need only be calculated once per frame for each HMM. The state probabilities are then available to the HMM instances associated with each node in the network.

11.5.2.2   Network parse

The network parse can be viewed at two levels:

- inter-node parse;
- intra-node parse.

For the inter-node parse Vega embodies single-pass Viterbi recognition using the token-passing mechanism [6]. Information is passed between nodes using the tokens described in the previous section. A token contains information on the progress of the recognition up to a particular point in time. With each incoming frame of speech, tokens waiting at node inputs are passed into the node's model instance. Each model instance updates its state scores[1] according to the new frame and the node may then emit a new token. The emitted tokens are propagated along network edges to wait again at node inputs for the following frame.

The score in the token can be modified when it traverses the edges, as required by the vocabulary weighting and sub-setting described earlier. The score can also be modified on entry into a node, as required, to implement inter-word penalties such as bigram language models.

---

[1] Each state score in the model instance may have been derived from a separate input token. For example in an N-state model, the score in state N may be based on the token input at $t-N$, whereas the score in state 1 may be based on the token input at time $t-1$.

Figure 11.7 shows the state of a recognition network at the point that a new output token is created by the first node in the network and propagated to the successor nodes. It shows how the output token points to both the input token from which it is derived, and the node that created it.

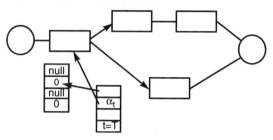

Fig. 11.7   Network state when first token is emitted from the first node with score α_t after T frames of data.

Tokens emitted from the final node in the network represent the best match between the data processed so far and the recognition vocabulary. The recognition result is obtained by tracing back through the linked list of tokens.

Intra-node processing requires the updating of the state scores in the node's model instance. The state scores are maintained in an array (node scratch space) that is dynamically allocated when the node is activated. Each state score is obtained by combining the acoustic match score for the current frame (state output probability) with the scores from preceding states in model instance together with the state transition probability matrix.

The scratch space has the form shown in Fig. 11.8. Since simple left-to-right (Bakis) HMM models are used within Vega, the scratch space is efficiently updated by processing the states in reverse order N down to 1.

State Index	1	2	3	..	N
Token Pointer					
State Score					

Fig. 11.8   Scratch space form.

Unlike the inter-node processing, neither the time index nor the identity of the best preceding state are recorded when state scores are updated.

11.5.3   Efficiency considerations

Configuration time is an important issue for speech recognizers running on platforms supporting multiple applications and multiple channels (e.g. ISAP). To ensure that the recognizers can be efficiently utilized it must be possible to quickly change recognition vocabularies. To improve configuration time, Vega supports:

- fast transfer between memory and file of the recognition network and HMMs in an internal binary format;

- model loading on demand — Vega loads new models only as required, with, in general, vocabulary changes involving new networks that contain arrangements of HMMs different from those that have already been loaded;

- recognition network pre-compilation — tools such as StapVoc and grok ensure that the recognition networks require no further processing on loading.

For efficiency during recognition, Vega employs a number of techniques, including:

- beam pruning — the deactivation of poor scoring tokens[1] in the recognition match is termed pruning [7], an extension used in Vega being 'variable beam-width pruning' where the pruning criteria are adjusted depending on the recognition network activity level;

- Vega-specific memory managers to provide efficient memory allocation, garbage collection and deallocation for the large number of temporary objects, such as tokens, that are formed during each recognition;

- node input lists implemented using ordered heap — to give reasonable CPU efficiency for the simple usual situation ($N = 1$) and for the multiple candidate situation ($N > 1$);

- small tokens — tokens usually take up 16 bytes, but in Vega most tokens in the system can actually be represented using half that storage.

---

[1]Relative to the best token.

11.6   STAPVOC

StapVoc constructs a recognition network from a textual list of vocabulary items (words or phrases to be recognized). The range and size of tasks with which StapVoc must deal is potentially very large.

It is envisaged that StapVoc will be used in the following scenarios:

- on-line (within a telephone dialogue) — for the generation of small, dynamically changing vocabularies, such as the names of race horses in an automated betting system;

- off-line — for the generation or update of large vocabularies, such as a list of surnames used by directory enquiries.

For on-line systems in particular, StapVoc is memory and CPU efficient. Small vocabularies can be created in fractions of a second using small amounts of memory and CPU. Large vocabularies may contain 100 000 surnames and require more memory (tens of megabytes), but can still be created within minutes.

Regardless of the specific vocabulary size or scenario, the main tasks for StapVoc remain the same, namely:

- phonetical transcription of each word — this involves creating a network of word or sub-word models representing the different possible ways of speaking that word;

- construction of the recognition network for the whole vocabulary from the individual word networks;

- adjustment of the recognizer parameters, e.g. rejection levels and time-outs according to the vocabulary.

The complete process is illustrated in Fig. 11.9.

To convert vocabulary items into sequences of models each phrase is treated as a network of words. Each word is processed by first querying a ModelServer to determine whether a whole-word model can be used to represent that word. If no such model is available a phonetic transcription for the word is required. This functionality is obtained by reusing the pronunciation component of the Laureate speech synthesis system (see Chapters 5, 6 and 7).

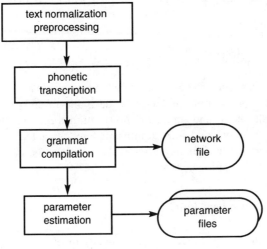

Fig. 11.9   StapVoc tasks.

11.6.1   Laureate pronunciation component (LPC)

The LPC software uses lexicons and text-to-speech rules. The lexicons store words and their phonetic transcriptions and facilitate rapid look-up of transcriptions, which becomes increasingly important as the size of the lexicon grows. If the transcription for a given word is not present in the lexicons, it is transcribed using rules that attempt to mimic the process used by a person when asked to pronounce an unfamiliar written word.

11.6.2   Advanced facilities

For most applications, simple translation of a text string into its phonetic equivalent is just part of the vocabulary generation process. The StapVoc input file format allows an application developer access to several other useful facilities.

11.6.2.1   Aliasing

Aliasing provides a form of shorthand within a vocabulary text file. For example, in a company name recognizer, there may be a number of companies with the suffix 'p.l.c.'. Clearly, a user may speak this in a variety of ways, including 'limited', 'plc' or 'public limited company'. An alias can be added to the input file, so that every entry which contains 'plc' will be associated with all the variants.

11.6.2.2   Tagging

Tagging enables a group of vocabulary items to return the same string when they are recognized. For example, it might be desirable to create a vocabulary in which the words 'help' and 'operator' both return the string 'help'. This can simplify the application code without imposing any unwanted restrictions on the recognition vocabularies. Tagging effectively overrides the default labelling of paths within the recognition network.

11.6.2.3   Spelt or spoken?

StapVoc will transcribe words either phonetically or by their spelling according to its built in rules. These rules can be overridden, for example, to specify that both phonetic and spelt transcriptions should be built into the recognition network allowing for recognition of either form.

11.6.3   Object diagram

Figure 11.10 outlines the sequence of messages that pass between objects in StapVoc during the construction of a recognition network.

The Vocabulary Generation process is controlled by the VocabGenController. Essentially the following process is repeated until all vocabulary items have been processed. The VocabGenController requests a Phrase from the VocabParser which has the responsibility of parsing the user-supplied text file of items to be recognized. Phrases are added to the PhraseVocabulary which calls upon the services of the PhraseExpander to deal with, among other things, the replacement of symbols with the words they represent. Each word in the expanded Phrase is then transcribed by the WordTranscriber (internally utilizing the LPC software). Transcribed Phrases are added to the VocabNetwork which builds up an optimized memory-resident recognition network of all the Phrases to be recognized. The VocabNetwork queries the ModelServer to identify the most appropriate models to use in each Phrase. This architecture allows the use of whole-word or sub-word models, and may also take into account the context of the model within the Phrase.

Once all the Phrases have been added to the VocabNetwork, the AnalysisServer scans the network in order to optimize various recognition parameters. Finally,

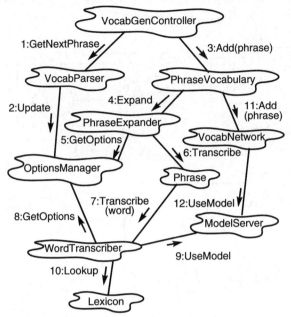

Fig. 11.10   Vocabulary generation.

the VocabNetwork object and recognition parameters are serialized to disk ready
for use by StapRec.

11.7   PLATFORMS

11.7.1   Stap and the ISAP

The BT/Ericsson ISAP [1] is a distributed processing system, containing a
variety of VME cards, which are responsible for telephony control, signal
processing and speech algorithm processing. The Stap component, StapRec and
StapVoc, are examples of ISAP speech service processes (SPSs) and reside on
the algorithm processor cards under the control of the ISAP process control
software (SPC).

Figure 11.11 shows how Stap residing on the algorithm processor (AG) card
interacts with a front end running on a signal processing (SP) card and with an
application running on an application processor (AP) card. All of the ISAP
speech services are ultimately under the control of an ISAP resource allocator

(SRA) which runs on a control processor (CP) card. A typical ISAP rack will contain a mixture of AP, CP, AG and SP cards, depending on the application's profile.

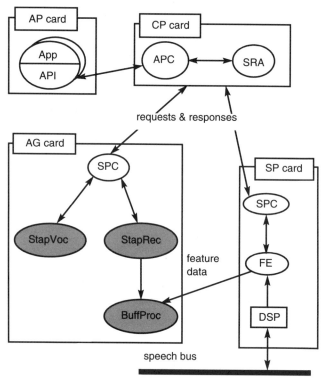

Fig. 11.11   Interfaces between StapRec and ISAP.

11.7.2   Separating Stap and ISAP

ISAP processes communicate via a proprietary TCP/IP inter-process communications library (ISAP-IPC), on top of which is layered an SPC-SPS protocol. This protocol is based around a small set of messages:

- Configure — reload the configuration files;

- Do — perform a recognition or create a vocabulary;

- Interrupt — stop the current Do and return any partial results;

- Delete — terminate the Stap process.

The intention in the design of Stap was for a clean separation between the speech services provided by Stap and the ISAP-SPS protocol, so that Stap could operate in other environments with other protocols, with little or no change. This separation was achieved by adopting the partitioning shown in Fig. 11.12.

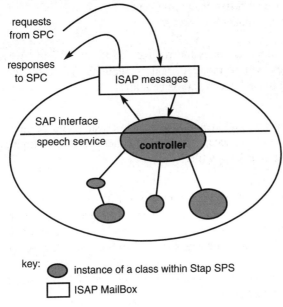

Fig. 11.12   Separation of ISAP and speech service layers.

The ISAP interface layer comprises an event loop which reads incoming requests from the SPC, invokes the appropriate actions on the controller, and conveys responses back to the SPC. The controller provides the public interface to the speech service (speech recognition or vocabulary generation), thereby shielding the components in the speech service from the SPC-SPS communications protocol and ISAP-IPC mechanism.

11.7.3   Stap for Unix® workstations

Although the ISAP is the primary target platform for Stap, it was designed to be easily portable to other environments.[1]

---

[1] To date, Stap is available for HP or Sun workstations and has also been ported to PCs running Consensus.

To drive Stap in a non-ISAP environment, software known as sip was developed which takes on the roles normally provided by the ISAP. sip is written in Tcl/Tk [8] and interacts with Stap using the ISAP-IPC mechanism such that Stap needed no modifications. In fact, the same Stap executables are used in the ISAP and stand-alone workstation environments.

The main components of sip are:

- sipstap, which assumes the role of the ISAP SPC by spawning the Stap processes when required and scheduling requests for both speech recognition and vocabulary generation;

- sipfep, which provides the signal processing and feature extraction for StapRec and can accept speech data either from the audio input on the local workstation or from a file.

Workstation applications communicate with sip through Tk 'send commands' (similar to remote procedure calls). These commands for speech recognition or vocabulary creation are converted into the corresponding ISAP-IPC message and sent to Stap. Response messages from Stap are translated by sip into the return value for the Tk send.

The use of Tcl/Tk also enables easy integration of speech recognition with graphical user interfaces.

11.8   TESTING STAP

Stap is a complex piece of software incorporating more than 60 000 lines of C++ source code. Given that this software is intended for a network platform, it has to be thoroughly tested.

Stap is functionally tested through component testing at the C++ class level and system/integration testing through a separate test harness which emulates the ISAP processes with which Stap must interface.

The system testing consists of:

- SPC-SPS protocol testing;

- recognition and vocabulary generation functionality;

- response to extreme loading;

- response to error conditions.

These tests are repeated for each platform to which Stap is delivered (currently SunOS and HP-UX).

In addition to confirming that Stap functions correctly, it must also be established that the accuracy and resource usage (memory and CPU) lie within expected ranges. A series of tests are performed using a database of speech recordings and a variety of vocabulary sizes and types to establish recognition accuracy and resource usage statistics. Failure to meet expected performance is treated as seriously as a major functionality failure.

11.9   STAP IN ACTION

Currently Stap is used in a number of trial speech applications, both internally to BT as well as with some of BT's customers.

11.9.1   PSTN applications

The majority of applications to date have been PSTN applications [9] utilizing the BT/Ericsson ISAP. They include the following.

- Advanced repertory dialling — users can store nine telephone numbers and names, the number being accessed by saying the associated name, e.g. 'Phone Home'. The user has complete control of the numbers and names and these are mixed with command words for accessing other facilities such as conference calling or voice mail. This application is an example of one in which recognition vocabularies are relatively small but change dynamically.

- Financial information services — here a caller can request information regarding their current investments. Vocabulary sizes are larger and more varied than in the repertory dialling application but tend to be less dynamic. Here the emphasis is on accurate recognition.

11.9.2   Workstation applications — 'What You Say Is What You Get'

In addition to the ISAP-based applications, a workstation-based application has also been under development. Stap has been integrated with the Mosaic [10] World Wide Web (WWW) browser to provide a means of navigating the World Wide Web via speech.

In this application, the hypertext links are extracted from a WWW page and passed to StapVoc to create the appropriate vocabulary configuration files for StapRec. Once StapRec has been configured, the user can drive the WWW browser simply by speaking the hypertext links. This demonstrates the speed and ease with which StapRec can be automatically reconfigured — StapRec is typically reconfigured and ready to recognize within a second or two of a WWW page being downloaded.

The architecture for the WWW application is shown in Fig. 11.13. As can be seen, this application makes use of the sip software described earlier which allows the Stap processes to work in a non-ISAP environment, and to receive speech through the workstation audio input.

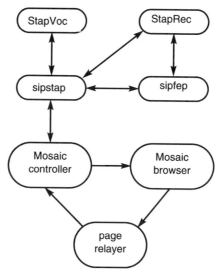

Fig. 11.13   Controlling a Mosaic browser via speech.

This application requires a slightly modified version of the WWW browser, Mosaic, which writes the received HTML page to a Unix fifo. An ancillary process, page relayer, monitors the output from the Mosaic browser. When a complete document has been received, the entire page is sent to the Mosaic controller. On receiving the new page, the Mosaic controller parses the page to extract the subject of each hypertext link and its associated URL. Some regularization is then performed on the link text to remove some of the unpronounceable characters and to account for the different ways in which people might speak it. The processed links are then passed to StapVoc via a file.

Additional words may be added to this file to allow voice control of some of the Mosaic browser commands such as 'home page', 'forward', 'previous', 'hot list', etc. StapVoc then creates the new vocabulary from this file and finally a StapRec is configured.

The user can now speak the text associated with any of the links on the page. Speech is captured through the standard audio input of the workstation and paramaterized by the sipfep process. The paramaterized speech is then passed to

StapRec in the same way as it would be in the ISAP. On completing recognition, StapRec presents the recognized text to the Mosaic controller (via sipstap) which determines the URL associated with the text and then instructs the WWW client to download the document. The process then repeats.

The major feature of this application is the close integration of the real-time vocabulary-generation and speech-recognition tools.

11.10   CONCLUSIONS

This chapter has given an overview of the design and capabilities of the BT real-time speech-recognition software suite — Stap. It has shown how the latest speech algorithms have been downstreamed into a portable, flexible product, that is being used with the BT/Ericsson ISAP as well as stand-alone workstations. With Stap, BT now has the capability to exploit a maturing technology in real interactive speech services (see Chapter 15).

APPENDIX

A brief introduction to the Booch notation

Booch has developed a notation for capturing and communicating the analysis and design decisions that are made during the development of complex object-oriented systems. These include, the taxonomic structure of the classes, the inheritance mechanisms, the individual behaviours of objects, and the dynamic behaviour of the system as a whole.

The Booch notation is expressive and quite detailed. However, a small subset is quite adequate for expressing most of the analysis and design issues. The notation defines the following views:

- class — show the existence of classes and their relationships in the logical view of a system;

- object — show the interactions that may occur among a given set of class instances;

- module — show the physical layering and partitioning of the architecture;

- process — show the allocation of processes to processors.

To illustrate this notation, a class and object diagram for an artificially simple scenario are shown in Figs. 11A.1 and 11A.2 respectively.

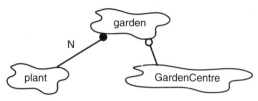

Fig. 11A.1   Class diagram.

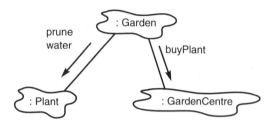

Fig. 11A.2   Object diagram.

Fig. 11A.1 shows that a Garden *has* one or more Plants and *uses* a GardenCentre. The *has* relationship denotes a whole/part relationship and appears as an association with a filled circle at the end denoting the aggregate. The class at the other end denotes the part whose instances are contained by the aggregate object. The *uses* relationship denotes a client/supplier relationship and appears as an association with an open circle at the end denoting the client.

Fig. 11A.2 illustrates an instance of Garden, invoking the water method on one of its Plant instances, and invoking the buyPlant method on its GardenCentre instance.

REFERENCES

1.   Rose K R and Hughes P M: 'Speech systems', in Dufour I G (Ed): 'Network Intelligence', Chapman & Hall, pp 75-94 (1997).

2.   Cox S J: 'Hidden Markov models for automatic speech recognition: theory and applications', BT Technol J, 6, No 2, pp 105-115 (April 1988).

3.   Booch G: 'Object oriented analysis and design with applications', (2nd Ed) Benjamin Cummings (1991).

4.   Stroustrup: 'The C++ programming language', (2nd Ed) Addison-Wesley (1991).

5.   Ringland S P A: 'Grammar constraints through signatures', in Rubio A et al (Eds): 'Speech Recognition and Coding: New Advances and Trends', Springer (1995).

6.   Young S J, Russell N J and Thornton J H S: 'Token passing: a simple conceptual model for a connected speech recognition system', CUED technical report (1989).

7.   Steinbiss V, Bach-Hiep T and Ney H: 'Improvements in beam search', ICSLP'94, $\underline{4}$, pp 2143-2146 (1994).

8.   Ousterhout J K: 'Tcl and the Tk toolkit', Addison-Wesley (1994).

9.   Talintyre J E and Ringland S P A: 'Isolated word recognition over the PSTN', IOA Autumn Conference (1989).

10. National Centre for Supercomputing Applications, NCSA Mosaic™ for the X-Window System: http://www.ncsa.uiuc.edu/SDG/Software/XMosaic/helpabout.html/

12

ARE SPEECH RECOGNIZERS STILL 98% ACCURATE, OR HAS THE TIME COME TO REPEAL 'HYDE'S LAW'?

R D Johnston

12.1 INTRODUCTION

In BT, automatic speech recognition (ASR) has already been applied to hands-free car phones, repertory dialling in networks and the control of answering services, and is now entering large-scale applications such as call centres and directory enquiries (see Chapter 1).

World-wide, over 70 companies [1] have speech recognition boards, algorithms or systems among their products and there are several hundred research institutes and laboratories actively developing advanced systems. Among these are the major telecommunications providers, for whom ASR is an essential enabling technology, providing customers with direct access to machine-based information — simply by picking up the phone.

BT is especially active in ASR (see Chapter 16) [2], but as a **user** as well as a developer. As such, BT needs to know not just what is available, but what is the very best. Only by searching out and deploying the 'best of breed' can customers be given the best possible service. BT is not alone in this, and in recent years much attention has been directed towards developing test methods. The US-based Advanced Research Projects Agency (ARPA) organization (whose motto is: 'Progress through evaluation') regularly organizes competitions to compare the various technical solutions developed within that programme. Within Europe the COST project 'Speech Technology Assessment', and two ESPRIT projects 'Speech Assessment Methodology' and 'Speech Quality Assessment in Language Engineering' [3] have provided a European focus for developing assessment methods. Most of these programmes have a broadly based remit and target both the large-vocabulary speech understanding systems and telephony-based applications. A major interest therefore is in evaluating speech recognizer performance in telephone networks and it is that which forms the basis of this chapter.

12.1.1 A change in emphasis

Only ten years ago commercial speech recognizer evaluation was dominated by issues such as power consumption, manufacturing cost, chip count and speed of response. At that time it was a major challenge to squeeze algorithms into the available processors and development effort was often directed at increasing code efficiency and 'cutting corners'.

The processing capacities of workstations have currently reached a stage where many speech recognition algorithms run in real time (see Chapter 1). As a consequence, two major new markets, one based upon telephony and one upon the 'voice activated typewriter', have emerged. Although both rely upon similar core technology (usually based upon hidden Markov models (see Chapter 9)), their realizations are very different. Contrasts are summarized in Table 12.1.

Table 12.1 Contrasts between telephony and office-based systems.

	Telephony	*Workstation/voice typewriter*
Vocabulary size	Usually 2-20 words occasionally up to 2000	5000 to 60 000 words
Flexibility in vocabulary	Preferably fixed — any changes may compromise performance	High — new words can be added quickly and easily
Adaptability	Little opportunity in most simple applications for continuous adaptation to talkers	Supervised learning strategies allow continuous improvement with use
Speaker variability	Speakers highly variable; wide range of accents, levels and vocal characteristics; often naive users; untrained	Only one speaker to whose precise characteristics and mannerisms the recognizer can adapt; opportunities for training
Signal quality	Extremely variable due to handsets, background noise and transmission interactions	Consistent — usually deploys a high-quality noise-cancelling microphone in a stable quiet environment; no transmission distortion
Word error rate	MUST be low to avoid invoking excessive error-correcting dialogues.	Higher rates tolerable as screen-based correction is quick and easy.

12.1.2   Background to performance measurement

Since the first demonstrable speech recognizer was constructed by Davis *et al* [4], engineers and scientists have tried to find reliable and realistic ways of measuring the performance both of the devices themselves and the systems into which they are built.

Until speech recording was possible, the only way to measure was using 'live' speakers. This was the method employed by Davis *et al* [4], where an accuracy of 97% (3% error rate) was claimed on the digits in a number of 'speaker-dependent' tests. Perhaps this set the standard as, over the following decades, most publications quoted similar results.

These consistent and surprisingly good results did not go unnoticed by researchers and in 1969 S R Hyde, of the Joint Speech Research Unit in the UK, proposed his 'law' stating that: 'The accuracy of speech recognizers is 98%', to be quickly followed by its first corollary that: 'Because speech recognizers have an accuracy of 98%, tests must be arranged to prove it' [5].

Although clearly tongue in cheek, 'Hyde's law' highlighted the fact that the results obtained from speech recognizers were so sensitive to test methods, that by judicious or nefarious use of test material almost any accuracy could be obtained for any recognizer! His figure was allegedly chosen because that was what his customers always said they required.

However, there was no deceit involved. 98% accuracy was genuinely being achieved in the laboratories and was even being publicly demonstrated [6] at the time. The reason for these impressive figures was the quality of the speakers. Sometimes they were specially trained standard speakers but often they were the researchers themselves. All spoke in a clear, well-articulated fashion, but the key to their performance was in the consistency with which the words were spoken. Provided that the speakers trained the system with speech patterns which they could reproduce later, even simple recognition strategies would work well. With such speech material (and highly motivated users who knew how to make the recognizers work) good performance could be virtually guaranteed.

This serves to highlight several key aspects of testing:

- any measured performance is the result of an interaction between the recognizer and the test material;

- high-quality or standardized speech is not appropriate for realistic testing;

- testing and training material must always be separated;

- speech recognizers must work despite the environment — not because of it.

12.1.3   Results from real networks

With telephony systems there are additional issues to be considered. As simple recognizers began to be used in trial telephony services, published accuracy figures deteriorated. In 1987 a paper from Nynex [7] published measurements made on several of the leading telephony-based commercial speech recognizers and reported that the best performance obtained on the digits was only about 80% — and this was after some 'bad' data had been removed. Within BT similar results were being obtained. Figure 12.1 shows those obtained in a number of measurements undertaken in 1990 and 1991 and illustrates two more important effects.

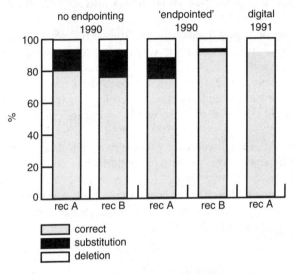

Fig. 12.1   Effects of database 'treatment'.

Firstly, even though the data was collected using substantial speaker populations (in these experiments there were over 500 talkers per test) the measured performance changed significantly when this data was 'endpointed', i.e. where the speech signal was examined and cropped to remove the waste/silence before and after the word — this process normally makes it easier for the recognizer to 'catch' the word.

The accuracies obtained before and after endpointing are as shown in the first four columns of Fig. 12.1. The results are unexpected, with one recognizer improving and the other deteriorating. This could possibly have been caused by this particular model trying to adapt to the now non-existent background noise.

The second effect was due to a change in the telephone network. The 1990 database had been collected over an analogue system (the 0800 overlay network), but the 1991 collection was over a fully digital network. At a stroke this single change in the BT network almost halved the measured error rate.

These examples serve to show that not only is there a strong interaction between the database content and the recognizers, but that it is sensitive to the collection processes and subsequent treatments.

Because of these strong interactions, for all practical purposes it is meaningless to refer to the 'accuracy' of a recognizer. This is why most experienced researchers only draw conclusions based upon 'head-to-head' comparisons and why, even then, results must be interpreted with caution.

12.1.4   Selection of speakers and speech material

At present, the only way to estimate the 'true' error rate of a telephony-based application is to create and run the service. In practice, logistical, ethical and legal reasons preclude this. Not only is testing normally required before a real service can be launched, but in a real service the time taken to collect sufficient data to estimate performance is often exceedingly long. It is also difficult to ensure sufficient coverage of those words which are crucial but appear only occasionally.

In practice, it is usually necessary to contrive a test database in the hope that it will be sufficiently close to the real application to allow comparisons to be drawn with regard to those factors (e.g. vocabulary type or recognizer type) which are important. This should contain data from customers prompted in a manner as close as possible to that of the genuine service. One way in which this may be done is illustrated in Fig. 12.2.

Fig. 12.2   Indirect prompt method.

Prompted in this way, effects due to speakers 'echoing' accents or manner-isms in the voice used for prompting are minimized and elements of 'thinking time' and spontaneity are introduced. Nevertheless, no matter how much care is taken in collection (and sometimes because of the care taken), such a database can never be exactly the same as that of a real application and it will not predict the absolute performance. All that can be expected is that such a database will allow recognizers to be compared and give confidence that the recognizer which does best in the laboratory will also do best in the field. Provided that it spans the range of speech qualities and collection methods found in real applications and that good statistical methods are used to analyse and interpret the results, then it is possible to predict which recognizer will be best despite the vagaries of popu-lations and collection methods.

How this analysis may be done is described later; before that, measurements must be made.

12.2   CHARACTERIZING PERFORMANCE

12.2.1   Classifiers, detectors and errors

Once the database has been obtained and the words presented to the recognizers to be compared, the errors can be counted. There are several types of error which must be considered and the terminology used is often confusing, having been extensively borrowed from other disciplines, e.g. radar and filtering. It is not uncommon, for example, to see terms such as deletion and rejection used synonymously in the literature. Generally context and common sense must be used as a guide. The following provides a self-consistent terminology which will be adopted in the rest of this chapter.

- A speech recognizer can 'get it right' by:

  — correctly classifying all 'valid' input words — this is true acceptance;

  — correctly rejecting (ignoring) all 'invalid' inputs — this is true rejection.

  For example, the set of 'valid' inputs to a digit recognizer would be 'zero..one.. two .. three... up to ...nine'. The set of 'invalid' inputs would be any other words or noise ....'sixty', 'fence', 'cough'.....etc. These are also known as out-of-vocabulary (OOV) words when used for testing.

- Also a recognizer can 'get it wrong':

— when a valid word is presented, but the match is not 'good enough' for the recognizer — this is a false rejection, or deletion error;

— when a valid word is presented but the classifier gets it wrong — this is a substitution error;

— when an 'invalid' input is presented and the recognizer fails to reject it — this is an insertion error.

These last two types of error are also sometimes known as false acceptances.

Although the performance of the recognizer is determined by its error profile the best performance is not necessarily achieved when the error rate is a minimum, as the occurrence of a false acceptance has different consequences from a false rejection. Usually the cost of a false acceptance is considered to be higher than that of a false rejection, as not only is the latter detectable but is amenable to recovery strategies using error-correcting dialogues.

All these error types may be analysed by considering speech recognition as a filtering and detection process, as shown in Fig. 12.3.

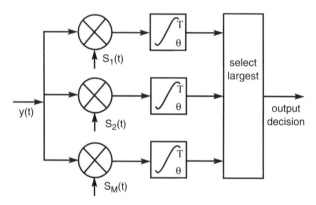

Fig. 12.3   Classical M-ary filter and detect process.

At its simplest, a speech recognizer receives signals which it labels as belonging to one of M classes. Viewed in this way a recognizer is a classical 'M-ary' filter [8] with an upper theoretical performance limit set by the extent to which probability distributions of the candidate signals overlap. For example if the words 'hut' and 'hat' are used and the speakers are drawn from a wide range of English accents there is an irreducible confusability as 'hut' spoken by some southern English speakers sounds just like 'hat' spoken by some northern English speakers. The costs of such vocabulary-dependent recognition errors are not symmetric, and this must be taken into account in the application. As an example of this asymmetry, consider the cost of confusing 'bombs away' and 'close bomb bay'!

12.2.2   Ways of representing performance of isolated words

Although most modern telephony-based speech recognizers are capable of connected-word operation, the formidable problems of error detection and correction means that this is not yet widely deployed in the simpler systems for public use [9]. Instead, such recognizers are constrained to operate in isolated-word mode with the 'connected' facilities used as a 'safety net', often wordspotting the occasional utterance where the talker fails to speak in an isolated fashion.

In principle, the process of measuring recognizer error rates of such systems seems straightforward. All that is necessary is to present the speech recognizer with isolated words spoken by number of speakers and count the errors. Such error counting forms the basis for most published results — often represented in terms of %correct or %accurate.

The usage of these terms (for isolated words) is discussed in Winski *et al* [10]. In this chapter the convention adopted is:

% correct = 100% − % substitution errors − % deletion errors
% accurate = 100% − % substitution errors

The distinction is made in this way because %correct is the figure which the user of an application experiences, whereas %accuracy is the figure from 'forced choice' tests — where the recognizer always responds to an input, such as is usually the case in database testing.

Less often a separate measurement is made using an out-of-vocabulary test set and the two figures combined to give a single figure of merit [11].

This average error-rate forms a useful single index, but a richer way of presenting performance is provided by the confusion matrix. An example based on the four words 'north', 'south', 'east', and 'west' is shown in Fig. 12.4. This shows the pattern which might emerge when 100 examples of each word are presented to a recognizer.

		word presented			
		north	south	east	west
word recognized	north	93	0	2	0
	south	4	98	0	0
	east	0	2	91	7
	west	3	0	7	93

Fig. 12.4   Substitution error confusion matrix.

Figure 12.4 shows, for example, that the word 'east' was correctly recognized 91 times, but was misrecognized as 'north' twice and as 'west' on seven occasions.

The matrix highlights inconsistency in word performance but in this form only shows substitution errors.

Other error types can be incorporated by augmenting this confusion matrix with an extra row and column to include the false acceptances and deletions (see Fig. 12.5). Here a separate out-of-vocabulary set of test words must be used.

		word presented				
		north	south	east	west	OOV words
word recognized	north	68	1	0	1	13
	south	3	89	1	1	12
	east	1	0	85	2	7
	west	0	0	4	84	8
	deletion	28	10	10	12	60

(a)   Results with a low threshold.

		word presented				
		north	south	east	west	OOV words
word recognized	north	54	0	0	0	3
	south	0	78	0	0	0
	east	0	0	75	0	1
	west	0	0	1	76	2
	deletion	46	22	25	24	94

(b)   Results with a higher threshold.

Fig. 12.5   Augmented confusion matrices — showing effects of different thresholds.

The practical advantage of the augmented confusion matrix is that trade-offs between the error types are apparent. For example, if the recognizer 'acceptance' criterion is tightened up (all other conditions being the same), then the number of substitution errors reduces and the number of 'out-of-vocabulary' words now being correctly deleted increases. However, these gains are at the cost of reducing the number of good matches of valid words, as may be seen by comparing the diagonals of the matrices.

12.2.3 The impact of altering thresholds

Figure 12.6 illustrates what happens as the criterion for accepting or rejecting words is altered.

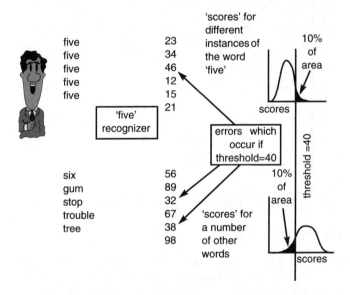

Fig. 12.6 Pattern matcher responses.

Here a 'single word' pattern-matching unit is shown responding to an ensemble of, firstly, the 'correct' valid words and, then, 'wrong' out-of-vocabulary words. A low number (distance score) corresponds to a close match. As the speaker never repeats a word identically, utterances, even of the correct words, vary, some matching better than others. With enough samples, distributions of the 'scores' for both curves can be derived as shown in Fig. 12.6.

The recognizer can only achieve perfect recognition if these two curves do not overlap. Unfortunately this is never the case in a real system and so, when a score is returned from a recognizer, a decision as to whether or not this is the valid word can only be made by comparing this score to a 'threshold'. The threshold setting of 40 shown in Fig. 12.6 is the point at which the number of the correct words rejected is equal to the number of invalid words accepted.

In Fig. 12.6, about 10% of the valid words are shown to be falsely rejected and about 10% of the invalid words falsely accepted.

The effect of moving the threshold upon the error ratio is illustrated in Fig. 12.7. This is analogous to the receiver operating curve (ROC) of filter theory. The sensitivity, which defines which curve is followed, is determined by the separation of the distributions.

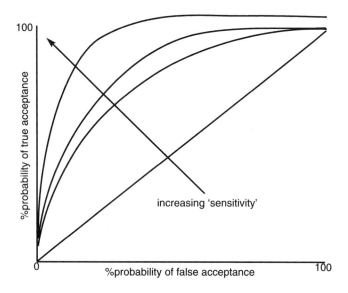

Fig. 12.7   Receiver operating curve (showing effect of increasing sensitivity).

For clarity, the above example was based upon a single-word recognizer. In a real recognizer the situation is complicated because every word undergoes multiple matching processes and all output scores compete with each other. In real systems, where each word recognition unit is independent, the means and variances of the scores from each class will differ as illustrated in Fig. 12.8.

However, provided that the outputs of all the matching units have similar covariance distributions (or can be post-processed to give this characteristic) it is only necessary to select the best candidate and ensure that it is above the threshold. In practice, this threshold may be based upon a score from the 'rejection' model. A comprehensive treatment of these issues is found in Devijver and Kittler [12].

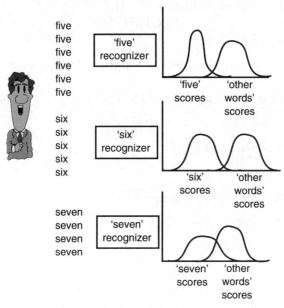

Fig. 12.8   Normal multi-word recognizer.

12.2.4   Practical considerations for selecting the best threshold setting

If an application can be tailored so that it is guaranteed that speakers will only utter 'valid' words (a condition often assumed when the recognizers are tested on a fixed set of words), then the lowest overall error rate is achieved when the recognizer threshold is set to 'accept all'. However, the lowest is not necessarily the best and, as an undetected substitution error is usually more 'fatal' than a false rejection which is always detected, it is usual to err on the side of caution.

A ratio of at least 3:1 false rejections/substitution error-rate is normally desirable for typical tasks involving menu selections. This means that a speech recognizer which has an underlying error rate of (say) 4% will be operated at a point where the total error rate is doubled to 8%, of which 6% are false rejections and 2% substitution errors. Although the number of 'fatal' undetected errors is halved users will now experience twice as many total errors and will have to endure error-correcting dialogues on 6 out of 100 occasions.

12.3 COMPARING THE PERFORMANCE OF REAL RECOGNIZERS

12.3.1 An alternative strategy

So far it has been shown that any measured error rate is determined by:

- the 'true' error rate of the recognizer;

- the threshold setting strategy;

- the vocabulary;

- the 'out-of-vocabulary' test material;

- the population of speakers;

- the collection process.

The problem is that none of these can be meaningfully quantified.

One way to deal with such unquantifiable variability is to resort to larger and larger 'random' test databases in the hope that the larger the test, the better the coverage. This has been one of the trends in speech recognizer evaluation in recent years — spurred on by the need for equally large databases for training statistically based recognizers. However, this is only a valid strategy if the sample data is fully representative of the real application. Unfortunately, as has been shown, this is not the case.

This kind of difficulty is not unique to speech recognizer evaluation. The same problem arises in many other areas — especially those concerned with the life sciences where the test environments can never be specified nor repeated and are often highly variable. Many statistical techniques have been developed to deal with such instances, all based upon the original work of Fisher [13], and comprehensively discussed in Cochran and Cox [14].

12.3.2 Balanced databases

The first step involves designing the experiment so that the sources of variation are separated and samples are taken over a small number of fixed conditions. In this way, the material from each speaker is subjected to various treatments determined by the fixed conditions (see Fig. 12.9).

Using these principles, a typical database consists of:

- 12 speakers — six male and six female;

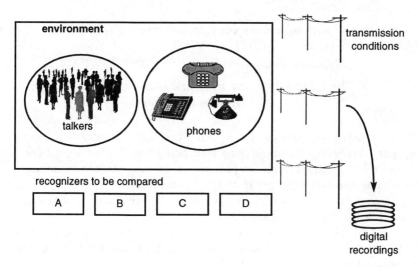

Fig. 12.9   Principle of a balanced data collection.

- 10 words (digits) — repeated several times by each speaker for each condition;

- 3 telephone lines — short (500m), medium (1.5 km) long (7km);

- 3 telephones (two electret and one carbon);

- 2 background noise levels.

Each speaker utters the same words into the three telephones over the three lines in two environmental noise conditions.

After the recorded material has been presented to the speech recognizers, the effects of the different treatments on error rates can be investigated. For example, as there are three sets of data containing speech from the same people speaking over each type of telephone line, the effects of the telephone lines can be compared. Similarly, as half the data is from males speaking over exactly the same wide range of conditions as the females, the results can be compared to see if there is a difference between male and female speech despite the variability of lines, phones, recognizers, etc. There are also three groups partitioned according to telephones and, as the speakers have all spoken at least twice over every condition, there are two sets of data which differ only in time of recording. Also, as the same data can be presented to different recognizers, it can be found not only if one is significantly better or worse than the others, but also over what range of conditions the best performance is obtained.

How this has been applied to speech recognizer testing is now illustrated with three specific examples.

12.3.3 The fixed telephone network — are all digit recognizers the same?

One such important question concerns the choice of recognizer technology or supplier. This is important as, if there is no performance difference between the various recognizers, decisions can be made on the basis of other functionality or price. The question may be answered by restating the question as a null hypothesis that 'all commercial recognizers are the same' and the data analysed to see if this assertion is sustainable.

Consider what conclusions might be drawn from the results shown in Fig. 12.10 where the data is partitioned into scores for male/female talkers, two different noise levels and a sequence effect[1].

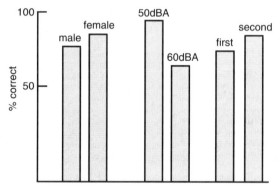

Fig. 12.10 Male/female, background noise and sequence.

The first observation is that background noise appears to be having a greater effect than the other two factors. Had the data not been derived from a balanced data set an appropriate test would have been a test of significance between the two noise conditions based on chi squared [15]. However, the balanced nature of the data means that better statistical tests may be made by comparing the various effects. For example, the male/female effects are so very much smaller than the noise effects that it would seem reasonable to use their small variability to 'benchmark' the relative importance of other respective factors. This basic principle underlies the technique of analysis of variance (ANOVA) which works by comparing variances determined in different ways — usually within and between

---

[1] Sequence effect results from each subject repeating each word. The first set of data is for the first utterance; the second is from the second set of utterances'.

the various classes, but often with a residual variance. This is the variance left after all those which can be accounted for are subtracted from the total, and is somewhat analogous to the background noise within the experiment.

This not only allows the significance of various factors to be estimated, but provides a means for estimating the overall reliability of the experimental process itself.

The partitioning of the data may be at many levels, and interactions between several factors may also be evaluated. For example, Fig. 12.11 shows how individual talkers fare when tested over all lines, all phones, all words and all recognizers.

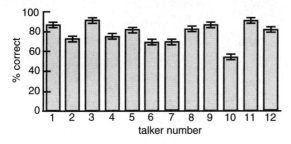

Fig. 12.11   Recognition by talker.

Much more variability is apparent in Fig. 12.11. This is not unexpected as there are now more degrees of freedom with more groups consisting of smaller numbers — which in itself will increase the between-group variation.

Figure 12.12 presents the same data — but this time showing performance by vocabulary item with each word averaged over all other factors.

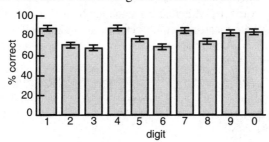

Fig. 12.12   Effect of vocabulary item.

Although the degrees of freedom are slightly different for these two cases (there are 12 talkers, but only 10 words), the variabilities are similar. Even greater detail can be found by examining the interactions between the words and the recognizer and Fig. 12.13 suggests that one recognizer is not only better, but is much more consistent than others — a result which was borne out when the data was analysed using ANOVA.

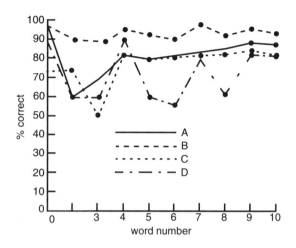

Fig. 12.13   Word-recognizer interactions.

Although derived from a compact database, statistical analysis of the results shows that it is more than adequate to separate out major effects between the factors. It is clear, for example, that there is a significant difference between recognizers and that the best one is consistently better despite everything else.

One way in which all this data can be compactly represented is by plotting the proportion of variance attributable to each factor as shown in Fig. 12.14. This shows that the single most important factor is the talker, followed by line, followed by recognizer. A more detailed description of these results can be found in Wyard et al [16].

In Fig. 12.14 the eta-squared measure, $\eta^2{}_X$, is defined as:

$$\eta^2{}_X = \frac{SS_X}{SS_T}$$

where SS_X and SS_T are the effect and total sums of squares respectively [17].

The beauty of this method is that many different analyses are made on the same data. For example, Fig. 12.15 shows the line-telephone interaction for each recognizer.

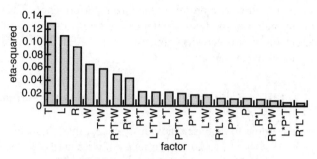

Fig. 12.14   Proportion of variance (real lines)
(T : talker, W : word, R : recognizer, P : phone, N : noise, L : line).

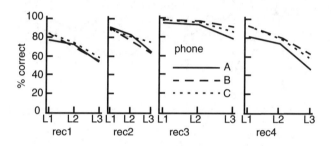

Fig. 12.15   Line-telephone interaction for each of four recognizers (lines L1, L2 and L3).

12.3.4   Do speech recognizers work with cell phones?

The previous experiment examined the main factors affecting the performance of fixed telephones and showed that talker variability was the most important single effect, but what about cellular telephones? Not only are cellular telephone users much more likely to access services based upon speech recognition, but buttons on a cellular telephone are generally much less accessible than those on a fixed phone and makes speech recognition a very attractive option.

From a technical point of view, however, there are good reasons for suspecting that radio links might impair speech recognizers. In particular:

- fading;

- speech clipping due to voice switching;

- spectral distortion due to coding and bandwidth compression techniques;

- echo control techniques for delay;

might all be expected to compromise recognizer performance. The last two of these are incorporated in the GSM digital cellular network and are therefore of special interest for future networks; but first it is necessary to establish if all cellular telephones behave similarly.

12.3.5   Cellular telephones — analogue system

The design was similar to that used previously except that it now contained three cellular telephones. There was no 'line' effect. All were used at full signal levels.

As before, the data was collected from 12 speakers. Figure 12.16 shows the performance by talker.

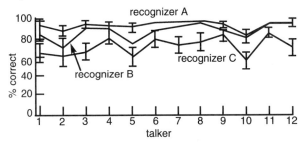

Fig. 12.16   Mobile phone — percentage correct by talker.

This is a very similar pattern to that found in the fixed telephone experiments. The ranked variances for all factors obtained are given in Fig. 12.17.

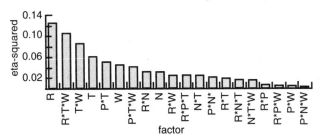

Fig. 12.17   Ranked variances for all factors
(T : talker, W : word, R : recognizer, P : phone, N : noise).

The overall conclusion is that users experience exactly the same performance from cellular telephones as they do over the fixed network — but only with certain recognizers. The recognizer performances were now more variable than the talkers, which was a different result to that found for fixed lines. They were also much more sensitive to background noise and in one instance there was a strong cellular telephone/recognizer/noise interaction. While a slight change in background noise level had little effect on the 'normal' phones, it had a much more marked effect on one of the recognizers when used with the cellular telephones.

12.3.6   GSM

The same principles were employed to compare GSM directly against the analogue phone set. The main results for most factors were found to be similar to those for the TACS system and the fixed network. Figure 12.18 shows talker variability for two recognizers.

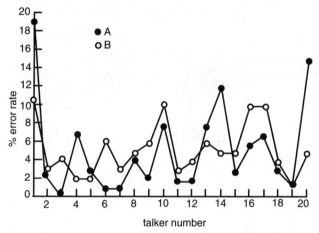

Fig. 12.18   GSM — talker number for two recognizers.

Table 12.2 shows the results from a direct comparison between GSM and the fixed network.

Table 12.2   Direct comparison of GSM with the fixed network.

	Recognizer B	Recognizer C
GSM (average)	5.3%	4.25%
Fixed (average)	5.56%	6.11%

12.3.7   International access

Finally, the same technique was used to investigate how international telephony access affects overall performance. Already a substantial number of calls to UK information systems originate from other countries and trends in 'globalization' mean that customers from anywhere in the world are now likely to be passed to a single call centre. If such call centres are to deploy speech recognition, they must be able to give comparable performance to speech originating from anywhere.

The data for this was gathered as part of the European Collaborative Project COST232 which collected a speech database from a range of participating European countries (see Fig. 12.19).

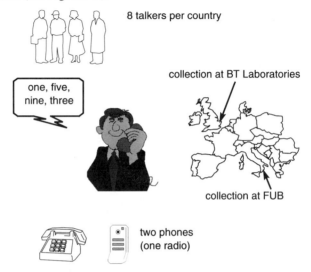

Fig. 12.19   COST232 data collection process.

In this case the balanced data consists of utterances of the (English) digits spoken by people in each of the participating countries. All speakers spoke over both fixed and radio networks and made separate, but identical, calls to both the UK and Italy.

The database contains 96 talkers spanning 12 accents and 24 transmission links. Half the speech is male and half is female. Every talker speaks the same words, over two phone types to two countries.

Figure 12.20 shows the general pattern of results obtained for the speech collected in the UK for three recognizers.

As before, this design allows extensive comparisons to be drawn. Many of the results have confirmed those trends found within the UK. For example, the differences between male and female speech and between fixed and mobile were again found to be very small and the ranking of the recognizers was the same.

As would be expected the 'local' performance, with native English speakers talking over national networks, is high — but so are several others. Whether this is due to the 'accent' effect or the 'lines' effect is not yet known — once the full set of data is available this can be determined. The lower line in Fig. 12.20, which shows the distance in 100s of kilometres from the UK, shows a trend which suggests the greater the distance, the worse the performance.

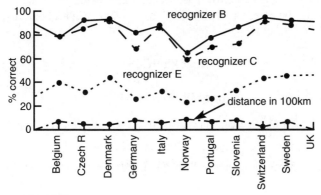

Fig. 12.20 Performance by country (three recognizers) and distance.

It is encouraging that a number of other 'foreign' countries appear to achieve a similar performance to that of the UK, suggesting that international boundaries need not be a major barrier to international service provision.

12.4 CONCLUSIONS

By undertaking experiments designed to compare the effects of specific factors, many of the problems traditionally associated with speech recognizer testing and evaluation can be avoided. Meaningful results can be obtained from small databases and the method also allows clear conclusions to be drawn from tests.

Some limitations, however, remain. In the main, the results presented here are those from one laboratory and are still limited in scope to the UK networks, the English language and a relatively small number of recognizers. To ensure that speech recognition continues to move forward other experiments must be done in other laboratories to replicate experiments, confirming or denying the generality of the methodology and the conclusions drawn.

Probably the most pressing need is for a suitable reference system to support the methodology so that experiments carried out in different laboratories and different languages can be compared. This could be based either upon human performance [18] or upon a software standard [19, 20]. Work towards such a reference has started, but, for such a reference to become an accepted standard, it needs to be tested and validated for several languages and environments.

But what of Hyde's law? By formulating his law (and like many before, Kant[1], and later, Deming [21]), Hyde drew attention to the folly of trying to apply absolute measurements to that which cannot be meaningfully measured. Perhaps it is here that the true value of Hyde's law lies. In the future, when methods based upon total quality replace those based upon 'attainment of numerical goals', Hyde's law may be quietly repealed. Until then, it has a rightful place.

REFERENCES

1.   'Voice + buyers guide', Advanstar Publications (1995).

2.   Brooks R M (Ed): 'Speech applications', BT Technol J, 14, No 2, pp 9-83 (April 1996).

3.   'Speech input/output assessment, multilingual methods and standards' Ellis Horwood (1989).

4.   Davis K H, Biddulph R and Balashak J: 'Automatic recognition of spoken digits', J Ac Soc of America, 24, pp 637-645 (1952).

5.   Hyde S R: 'Automatic speech recognition: a critical survey and discussion in the literature', in David E E and Denes P B (Eds): 'Human Communications: a unified view', pp 390-438, McGraw Hill, NY, reprinted in: 'Automatic speech and speaker recognition', IEEE Press (1979).

6.   Flanagan J: 'Voices of men and machines', J Ac Soc of America, 51, pp 1375-1387 (1972).

7.   Yaschin B et al: 'Performance of speech recognition devices — evaluating speech produced over the telephone network', ICASSP (1989).

8.   Mohanty N C: 'Signal processing', Van Nostrand (1987).

9.   Kose K R and Hughes P: 'The BT speech applications platform', Br Telecommunications Eng J, 13, pp 304-308 (1995).

10.   Winski R et al: 'EAGLES: spoken language working group overview document', Proc of Eurospeech, pp 841-845 (1995).

---

[1] "It is a great and necessary proof of wisdom and sasgacity to know what questions may be reasonably asked. For if a question is absurd in itself and calls for an answer where there is no answer, it does not only throw disgrace on the questioner, but often tempts an incautious listener into absurd answers, thus presenting as the ancients said, the spectacle of one person milking a he-goat and of another holding the sieve". 'Critique of Pure reason', Immanuel Kant (1724-1804).

11. Pallett D: Chapter 13 in Bristow G (Ed): 'Electronic speech recognition — techniques, technology and applications', Collins (1986).

12. Devijver R A and Kittler J: 'Pattern recognition — a statistical approach', Prentice Hall International (1982).

13. Fisher R A: 'Design of experiments', Oliver and Boyd, Edinburgh (1942).

14. Cochran W G and Cox G M: 'Experimental design', Wiley (1957).

15. Morony M J: 'Facts from figures', Pelican (1951).

16. Wyard P et al: 'The relative importance of the factors affecting recognizer performance with telephone speech', Proc of Eurospeech, Berlin (1993).

17. Cohen J: 'Eta-squared and partial eta-squared in fixed factor ANOVA designs', Educ and Psychol Meas, 33, pp 107-112 (1973).

18. Moore R K: 'Evaluating speech recognizers', Proc of ASSP, 25, No 2 (1993).

19. Chollet G F: 'On the evaluation of speech recognizers and databases using a reference system', ICASSP, pp 2026-2029 (1982).

20. EC COST232: 'Speech recognition over the telephone line', Final report (1994).

21. Deming W E: 'Out of the crisis', Cambridge (1986).

13

SPEECH ENHANCEMENT

B P Milner, A V Lewis and S V Vaseghi

13.1   INTRODUCTION TO SPEECH ENHANCEMENT

The past decade has seen an enormous growth in the use of speech technology. Together with this increase, public expectations of the technology have also grown. This can be attributed, at least in part, to the endless flow of television programmes and films which portray technically superior speech interfaces. Therefore it has been commercially vital to ensure that these new systems are able to operate effectively and reliably when placed into the network. One of the major factors affecting the reliability of speech systems is the quality of the speech itself. It is likely that the speech signal will be subject to various distortions as a result of the operating environment. Many phenomena act as speech contaminants, including additive acoustical and electrical noise, non-ideal communications channels, echo, and talker stress. This chapter reviews several techniques for achieving robust performance in such environments.

Most speech applications are initially tested using clean speech, which is ideally suited for a first assessment of the system. However, for deployment into the network it is often necessary to use enhancement to improve the quality of speech and hence increase the reliability of the system. It is the aim of this chapter to discuss techniques for the enhancement of speech for use in real applications.

Poor quality speech results in two distinct problems for a speech processing system. In the first case, it is unpleasant for users to be subjected to noisy or distorted speech when listening to a telephone conversation. In the second case, low quality speech may introduce errors into speech processing applications. For example, below a signal-to-noise ratio (SNR) of about 20 dB, additive noise severely degrades the performance of speech recognition systems. Also, for speech coding applications, it is important that the speech is relatively distortion-free to ensure reliable coding.

Many factors may contribute to speech distortion. These can be broadly categorized into the three areas discussed in this chapter — additive noise, communications channel effects, and echo. Normally the contamination of speech by noise is an additive process. In situations where either the speech or the noise is very loud this may cause amplifier saturation, making the distortion nonlinear. This chapter is concerned only with linear types of contamination. The distortion of speech by a non-ideal communications channel is a convolution process in the time domain. This means that the speech signal is filtered by the channel, and as a result may suffer from serious attentuation in parts of its spectrum. The most obvious example of a channel distortion is the telephone line. Telephony signals are bandlimited from 300 Hz to 3.5 kHz, but within this band the spectral attenuation may vary by as much as 40 dB. Additionally, the telephone channel frequency response may vary from call to call. On a telephone line, echo results from an impedance mismatch in the hybrid which connects the subscriber's two-wire line to the four-wire line in the exchange. This imbalance allows some of the talkers energy to be coupled back to the loudspeaker. This, combined with the time delay taken in passing around the loop results in an echo being perceived by the user that can be a serious barrier to fluent conversation.

13.2   NOISE REMOVAL TECHNIQUES

An enormous variety of noise-reduction techniques have been developed in the last 40 years. Early methods used analogue filters and were able to offer only limited improvements in restricted noise conditions. For example, banks of notch filters were used to remove periodic noise and its harmonics. With the advent of digital signal processing in the early 1970s, noise reduction has become much more effective.

In order to remove, or suppress, noise an estimate of the contaminating noise must be obtained. Two basic methods exist for obtaining this noise estimate — dual and single channel systems.

The dual channel cancellation system is attributed to Widrow [1] in 1975. In this scheme an adaptive filter is used to match the reference noise to the noise component of the speech. To work effectively it is important that there is a high degree of correlation between the reference noise and the component of noise which is contaminating the speech. In practice, it has been found that to cancel 90% of the noise energy in a car environment, the two microphones need to be as close as 5 cm apart. This makes it very difficult to ensure that no speech component enters the noise reference microphone. This can be avoided by increasing the microphone separation, but then the only coherent noise is that of the engine. Darlington, Wheeler and Powell [2] found that this lack of noise coherence was also apparent in a jet fighter cockpit environment.

Most work in noise reduction is now based on the single channel method, where an estimate of the noise is obtained during periods of speech inactivity. These techniques are, however, reliant on accurately locating periods of speech inactivity in which to estimate the noise. Techniques such as spectral subtraction and Wiener filters are now increasingly being applied to speech processing applications to improve their environmental robustness.

In the time domain, a noisy signal $y(m)$ is modelled as:

$$y(m) = x(m) + n(m) \qquad \qquad \text{... (13.1)}$$

where $x(m)$ and $n(m)$ are the signal and noise respectively. This can also be expressed in the frequency domain:

$$Y(f) = X(f) + N(f) \qquad \qquad \text{... (13.2)}$$

where $X(f)$ and $N(f)$ are the signal and the noise spectra. The recovery of a signal distorted by noise is a statistical estimation process, and can be formulated using the Bayesian estimation framework [3]. The *a posteriori* probability density function (PDF) of the signal $x(m)$, given a noisy observation $y(m)$, can be expressed using the Bayes rule:

$$f_{X|Y}(x(m)|y(m)) = \frac{f_{Y|X}(y(m)|x(m))f_X(x(m))}{f_Y(y(m))} \qquad \qquad \text{... (13.3)}$$

where $f_{X|Y}(y(m)|x(m))$ is the likelihood of the observation $y(m)$ conditioned on the signal value $x(m)$, and $f_X(x(m))$ is the prior PDF of $x(m)$. Assuming that the signal and the noise are Gaussian with PDFs $f_X(x(m)) = N(x(m),\mu_x\sigma^2_x)$ and $f_N(n(m)) = N(n(m),\mu_n\sigma^2_n)$, the maximum *a posteriori* (MAP) estimate, obtained by setting the derivative of the log likelihood function $\ln(f_{X|Y}(x(m)|y(m)))$ to zero, is given by:

$$\hat{x}(m) = \frac{\sigma^2_x}{\sigma^2_x + \sigma^2_n}(y(m) - \mu_n) + \frac{\sigma^2_n}{\sigma^2_x + \sigma^2_n}\mu_x \qquad \qquad \text{... (13.4)}$$

Note that the estimate $\hat{x}(m)$ is a weighted interpolation between the unconditional mean μ_x and the observed value $(y(m) - \mu_n)$. At poor signal-to-noise ratios (SNR) the noise variance is large compared to the signal variance — $\sigma^2_n >> \sigma^2_x$, and so $\hat{x}(m) \approx \mu_x$. Conversely at high signal to noise ratios $\sigma^2_n << \sigma^2_x$, so $\hat{x}(m) \approx y(m)$.

The remainder of this section describes three noise reduction techniques in increasing order of their utilization of the signal and the noise statistics — spectral subtraction, Wiener filtering, and a Bayesian signal restoration method based on hidden Markov modelling of the signal and the noise processes.

13.2.1   Spectral subtraction

In many situations, such as the handset of a telephone, there is no independent access to the contaminating noise. In these situations it is not possible to cancel out the random noise, but it is possible to reduce the **average effect** of the noise on the signal spectrum. Figure 13.1(a) shows that the effect of additive white noise on a time domain speech signal is to increase its variance and leave the mean unaltered. In the magnitude spectrum, Fig. 13.1(b) shows that the noise increases both the mean and variance of the magnitude spectrum.

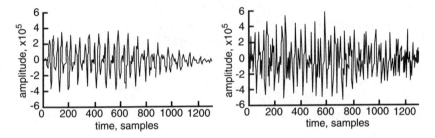

(a)   Time domain clean and noise-contaminated signals.

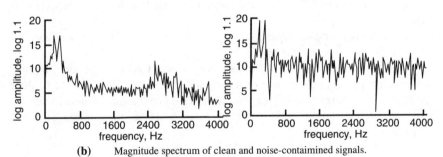

(b)   Magnitude spectrum of clean and noise-contaimined signals.

Fig. 13.1   Illustration of the effect of noise on a signal in the time and frequency domains.

Spectral subtraction is able to reduce the effect of noise by subtracting an estimate of the mean of the noise spectrum from the received signal spectrum [4, 5]:

$$\left|\hat{X}(f)\right|^{b} = |Y(f)|^{b} - \alpha\overline{|N(f)|^{b}} \qquad\qquad \ldots (13.5)$$

where $|\hat{X}(f)|^b$ is an estimate of the original signal spectrum $|X(f)|^b$, and $|N(f)|^b$ is the time-averaged noise spectra. It is assumed that the noise is a wide sense stationary random process. For magnitude spectral subtraction the exponent $b = 1$, and for power spectral subtraction $b = 2$. The parameter α controls the amount of noise subtracted from the noisy signal. For full noise subtraction $\alpha = 1$ and more typically $\alpha > 1$ for over-subtraction. The time-averaged noise spectrum is obtained from the periods when the signal is known to be absent and only the noise is present:

$$\overline{|N(f)|^b} = \frac{1}{K} \sum_{i=0}^{K-1} |N_i(f)|^b \qquad\qquad \text{... (13.6)}$$

where $|N_i(f)|$ is the spectrum of the ith noise frame and it is assumed that there are K frames in a noise-only period, where K is variable.

For enhancing a time domain signal the estimate of the magnitude spectrum $|\hat{X}(f)|$ is combined with the phase of the original noisy signal, $\theta_Y(k)$, and then transformed back into the time domain using an inverse discrete Fourier trans-from (IDFT):

$$\hat{x}(m) = \sum_{k=0}^{N-1} |\hat{X}(k)| e^{j\theta_Y(k)} e^{-\frac{j2\pi}{N} km} \qquad\qquad \text{... (13.7)}$$

The restoration is based on the assumption that the audible noise is mainly due to the distortion of the magnitude spectrum, and that the phase distortion is inaudible.

Evaluations of the perceptual effects of the human ear confirm the assumption that the ear is insensitive to phase.

Due to variations of the noise spectrum, spectral subtraction may produce negative estimates of the power or magnitude spectrum. This outcome is more probable as the signal-to-noise ratio decreases. To avoid negative magnitude estimates the subtracted output is processed using a mapping function $T[.]$ of the form [6]:

$$T\left[|\hat{X}(f)|\right] = \begin{cases} |\hat{X}(f)| & if\ |\hat{X}(f)| > \beta|Y(f)| \\ \beta|Y(f)| & otherwise \end{cases} \qquad\qquad \text{... (13.8)}$$

Typical values for β are between 0.1 and 0.01 which set the maximum attenuation of the spectral subtraction filter to be -10 dB to -20 dB in the power spectrum.

13.2.1.1   Processing distortions

The dominant processing distortion in spectral subtraction is due to the non-linear mapping of the negative, or small valued, spectral estimates. This distortion results in a metallic sounding noise, known as 'musical noise' due to its narrowband spectrum and the tone-like sound. The success of spectral subtraction depends on the ability of the algorithm to reduce the noise without introducing processing distortions. In the worst case, the residual noise can have two forms:

- a sharp trough or peak in the signal spectrum;

- isolated narrow bands of frequencies.

In the vicinity of a high amplitude signal frequency, the noise-induced trough or peak is often masked, or made inaudible, by the high signal energy. The main cause of audible 'musical' degradations is the isolated frequency components illustrated in Fig. 13.2. These musical noises are characterized as short-lived narrow bands of frequencies surrounded by relatively low-level components. In audio signal restoration the distortion caused by spectral subtraction can result in a significant deterioration of the signal quality. In particularly bad conditions the effect of spectral subtraction may result in a signal that is of a lower perceived quality, than the original noisy signal. A range of methods for compensating for these processing distortions may be found in Vaseghi [3].

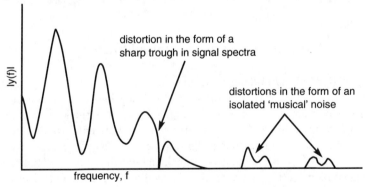

Fig. 13.2   Illustration of distortions that may result from spectral subtraction.

13.2.1.2   Effects of spectral subtraction on signal distribution

To illustrate the distorting effect of spectral subtraction on the magnitude spectrum of speech a simple two-dimensional feature space is utilized by dividing the signal into a low and high frequency band, f_l and f_h, which form the two axes of the feature space.

Figure 13.3(a) shows an assumed distribution of the spectral samples of a noise-free signal in the two-dimensional frequency space. The effect of the random noise, shown in Fig. 13.3(b), is to increase the mean and the variance of the spectrum. The increase in variance constitutes an irrevocable distortion. The increase of the mean in the magnitude spectrum can be removed through spectral subtraction. Figure 13.3(c) illustrates the distorting effect of spectral subtraction on the distribution of the signal spectrum. Due to the noise-induced increase in the variance of the signal spectrum, after subtraction a proportion of the signal becomes negative and has to be mapped to non-negative values. This process distorts the distribution of the low SNR part of the signal spectrum.

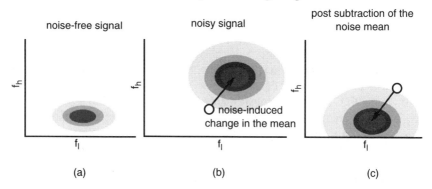

Fig 13.3   Illustration of the distorting effect of spectral subtraction on the space of the magnitude spectrum of a signal.

13.2.1.3   Implementation of spectral subtraction

Spectral subtraction is normally implemented as a block processing algorithm. The incoming audio signal is buffered and divided into overlapping blocks of typically 128 samples, as shown in Fig. 13.4. Each block is Hanning (or Hamming) windowed, and then transformed using a DFT to the frequency domain. After spectral subtraction and post-processing, the magnitude spectrum is combined with the phase of the noisy signal, and transformed to the time domain. Each signal block is then overlapped and added to the preceding and succeeding blocks to form the final output. The choice of the block length for spectral analysis is a compromise between the conflicting requirements of time resolution and spectral resolution. Typically a block length between 5 ms and 50 ms is used.

Figure 13.5a shows a speech signal contaminated by helicopter noise and in Fig. 13.5b spectral subtraction has been applied to reduce the noise.

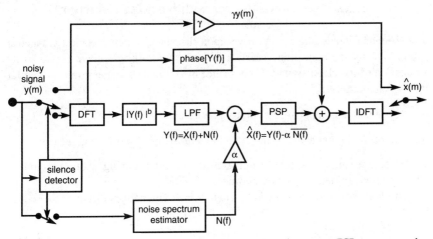

Fig. 13.4   Block diagram configuration of a spectral subtraction system (PSP = post spectral subtraction processing).

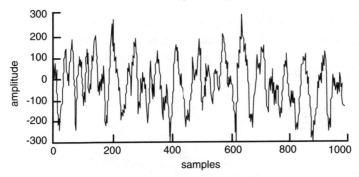

Fig. 13.5a   Noisy speech signal.

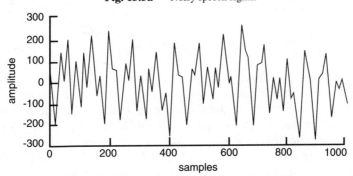

Fig. 13.5b   Restored signal after spectral subtraction.

13.2.2   Wiener filter for additive noise reduction

The Wiener filter is able to improve upon the enhancements made by spectral subtraction by incorporating statistics of both the noise and clean signal in its estimation of the clean signal. Wiener filter theory makes two assumptions about the signal and noise:

- the signal and noise are stationary processes;

- an *a priori* knowledge of the signal and noise spectra is available.

Of course speech is highly non-stationary, but by using a block adaptive structure this problem can be overcome.

The least squared error discrete-time Wiener filter for additive noise reduction [7] is given by:

$$w = (R_{xx} + R_{nn})^{-1} r_{xy} \qquad \qquad \dots (13.9)$$

where R_{xx} and R_{nn} are the autocorrelation matrices of the noise-free signal, and the noise respectively, and r_{xy} is the crosscorrelation vector of the noisy signal and noise-free signal. By Fourier transforming equation (13.9), the frequency-domain Wiener filter is:

$$W(f) = \frac{P_{XX}(f)}{P_{XX}(f) + P_{NN}(f)} \qquad \qquad \dots (13.10)$$

where $P_{XX}(f)$ and $P_{NN}(f)$ are the signal and noise power spectra. Dividing the numerator and the denominator by the noise power spectra, $P_{NN}(f)$, and substituting a signal-to-noise ratio variable $SNR(f) = P_{xx}(f)/P_{nn}(f)$ gives the filter gain at frequency f in terms of the SNR:

$$W(f) = \frac{SNR(f)}{SNR(f) + 1} \qquad \qquad \dots (13.11)$$

Figure 13.6 shows the Wiener filter gain as a function of the SNR.

The Wiener filter frequency response is a real positive number in the range $0 \le W(f) \le 1$. Considering the two limiting cases of a noise-free signal, $SNR(f) = \infty$ and an extremely noisy signal $SNR(f) = 0$. At very high SNRs $W(f) \approx 1$, and the filter applies little or no attenuation to the noise-free frequency.

On the other extreme at very low SNRs, $W(f) \approx 0$. Therefore for additive noise, the Wiener filter attenuates each frequency in proportion to an estimate of the signal-to-noise ratio.

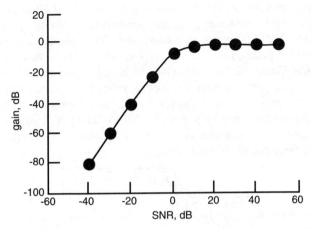

Fig. 13.6   Variations in gain of the Wiener filter response with SNR.

An alternative illustration of the variation of the Wiener filter frequency response with SNR is shown in Fig. 13.7. At a peak of the signal spectrum, where the SNR is relatively high, the Wiener filter frequency response is also high, and the filter applies little attenuation. At a signal trough, the SNR is low and so is the Wiener filter response. Hence, for additive white noise the Wiener filter response broadly follows the signal spectrum.

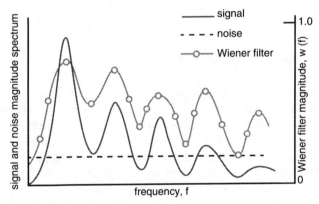

Fig. 13.7   The variation of Wiener frequency response with signal spectrum for additive white noise.

13.2.2.1   Implementation of Wiener filters

The implementation of a Wiener filter for additive noise reduction requires knowledge of the autocorrelation functions, or the power spectra, of the signal and the noise. The noise power spectrum can be estimated from inactive periods in the received signal, assuming that the noise remains stationary between the update periods. This is a reasonable assumption for many noisy environments, such as car noise emanating from the engine, aircraft noise, office noise from computers, machines, etc. The main problem in implementing a Wiener filter is that, as the desired signal is often observed in noise, the autocorrelation or power spectrum of the original signal is not readily available. Figure 13.8 illustrates the block diagram configuration of a Wiener filter for additive noise reduction. An estimate of the signal power spectra is obtained by subtracting an estimate of the noise spectra from that of the noisy signal.

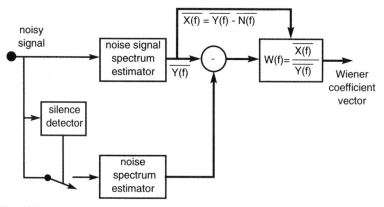

Fig. 13.8   Configuration of a system for estimation of frequency Wiener filter coefficients.

13.2.2.2   HMM-based Wiener filters

An alternative method for obtaining signal and noise statistics is to use the states of hidden Markov models (HMMs) [8], trained on clean speech and noise [9, 10]. This method is particularly suitable when operating with HMM-based speech recognition systems, although the technique is still suitable for speech enhancement provided clean speech models are available. Figure 13.9 illustrates an HMM-based state-dependent Wiener filter. To implement the filter, estimates of the state sequence for both the signal and the noise are required. In practice, for signals such as speech there are a number of different HMMs — one for each word, phoneme, or any other acoustic unit of the signal. In such cases, it is necessary to classify the signal to ensure that the state-based Wiener filters are derived from the most likely HMM. Assuming that there are V HMMs $\{\lambda_1, ..., \lambda_V\}$ for the signal process, and an HMM for the noise, the algorithm can be broken down into four stages.

Step 1 — calculate the maximum likelihood state sequence of the noisy speech for each HMM, λ_i.

Step 2 — for each HMM, λ_i, using the state sequence, produce a series of state-dependent Wiener filters.

Step 3 — for each HMM, λ_i, use the state-dependent Wiener filters to filter the noisy speech.

Step 4 — obtain a probability score for the filtered signal from its respective HMM, and output the signal from the highest probability score HMM as the restored signal.

This method is based on the hypothesis that when the speech is filtered by the Wiener filter associated with the **correct** HMM, a 'matched' filter sequence is produced, and hence the effect of noise is reduced. However, when filtering is performed by an **incorrect** HMM-based Wiener filter, an unmatched filter sequence is produced which merely distorts the speech signal.

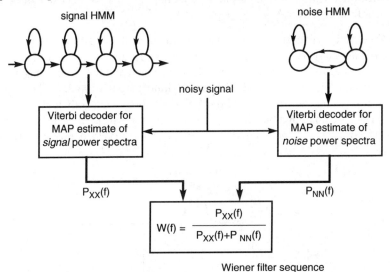

Fig. 13.9   Illustrations of HMMs with state-dependent Wiener filters.

13.3   ECHO CONTROL IN SPEECH COMMUNICATIONS

Humans have biological sidetone channels, because we hear the things we say. Sidetone has always been a natural part of telephony, because reflected signals were inevitable on connections that used, for economy, two wires rather than four. Sidetone is usually advantageous, since it helps regulate talker speech level [11]; but if these same reflected signals suffer a significant delay, we hear not sidetone but an echo of our own voice and that can be a serious barrier to fluent conversation. The subjective impact of echo depends strongly on the objective values of echo amplitude, frequency-response and delay [11]. Increased transmission delay, due to distance or to the use of speech and channel coding techniques, increases the audibility and unpleasantness of echo.

13.3.1 Echo control

The main sources of echo signals in telecommunications are:

- at impedance mismatches in 2-wire cables and at 2-to-4-wire converters;

- due to residual acoustic coupling inside 4-wire handsets in mobile and digital telephones;

- from loudspeaker-to-microphone coupling in hands-free telephones and conference terminals.

Historially, echo suppression was the main method of preventing audible echo in telephony. An echo suppressor is effectively an alternate-simplex or half-duplex connection, where speech signals in the inactive direction of transmission are blocked (Fig. 13.10). If both parties talk simultaneously (double-talk), a small loss is inserted in each direction of transmission, attenuating both double-talk and echo. Because unambiguous detection of double-talk is difficult, especially in the presence of echo and background noise, suppressors mutilate double-talk to some degree.

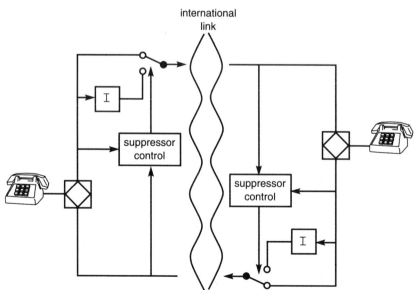

Fig. 13.10   Echo control in telephony, using two echo suppressors.

In the last decade, DSP technology has made echo cancellation [12, 13] economically feasible. An adaptive filter [1] is used to model the echo path and synthesize a nearly-identical copy of the expected echo, which is subtracted from the signal transmitted (Fig. 13.11). In this way, the echo signal is attenuated with little or no change to the speech signal from the local talker, so that a duplex (both-way simultaneous) connection is maintained. In principle, this transparency avoids the mutilation of double-talk and the modulation of background noise caused by suppressors.

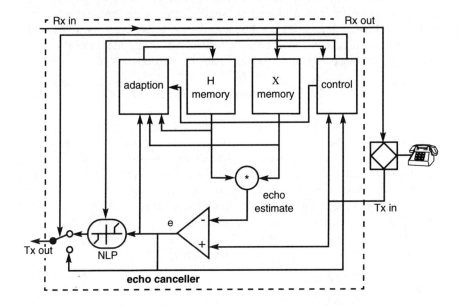

Fig. 13.11   Signal processing in a network echo canceller, with conventional terminology.

The advantages of echo cancellers are particularly clear when used to attenuate acoustic coupling in the more expensive types of hands-free telephones and loudspeaking conference terminals. Here the microphone location reduces the speech-to-ambient-noise ratio and the presence of multiple talkers makes double-talk more likely. Practical hands-free tests, in both office and payphone environments, have shown that a full-duplex connection is audibly superior to a half half-duplex link [14]. However, echo cancellers have a susceptibility that suppressors do not show. The echo replica is synthesized with a linear filter, meaning that cancellers are delicately sensitive to nonlinearity in the echo path, The weak nonlinearity of A/m-law telephony coding at 64 kbit/s (ITU-T G.711) limits echo attentuation to less than 35 dB — which is insufficient to conceal echo on long-delay connections. A kind of centre-clipper, called a nonlinear processor (NLP), is used to suppress the residual uncancelled echo. For this reason, all commercial implementations of network canceller are a hybrid mixture of canceller and low-level suppressor.

Sources of more severe nonlinearity include low bit-rate speech coders and energetic echoes outside the response window of the canceller — commonplace in acoustic canceller applications. Ironically, nonlinearity can also arise from the presence of other cancellers, with a similar convergence speed, in the same echo path. Practical tests with speech coders and cascaded cancellers have given results, dependent on implementation, ranging from the benign addition of a few dB to the noise floor to a disastrous amplification of the original echo. It is prudent to disable intermediate echo cancellers in a cascade connection. After recent improvements, the latest types of network echo canceller offer good echo control performance, especially on long-delay connections such as satellite links.

13.3.2   Echo canceller design

The echo response varies with each telephone call, depending on the length and gauge of cable in the 2-wire circuit as well as on the impedance of the telephone. Therefore the echo model must be individually adapted to match the echo response of each call. The choice of adaptation algorithm and its method of implementation are central to effective and affordable echo canceller design. The normalized least mean square (NLMS) algorithm, first proposed by Widrow and Hoff [15], has proved a robust and economic choice in many commercial designs. This algorithm (equation (13.12) is recursive, producing the desired echo response through successive approximations:

$$h_{j+1} = h_j + \frac{2\mu e_j x_j}{\sum\limits_{i=0}^{i=N-1} x_i^2} \qquad \qquad \dots (13.12)$$

where h is the echo impulse response of N coefficients, x is the signal input at time j, e is the echo residual after subtaction of the echo estimate and μ is a constant less than 1, influencing the speed of convergence of the successive approximations.

A white noise input signal, having an autocorrelation matrix with equal eigenvalues, gives optimum convergence of the NLMS algorithm; but a speech signal has significant eigenvalue disparity, since its spectrum is neither flat nor stationary. This behaviour is of minor significance in network echo control applications, where fast cancellation is needed of the most energetic components in the speech signal. Weaker components, that take longer to be attenuated by the

NLMS algorithm, are less audible as echo. However, acoustic cancellers are sometimes used to ensure loop stability and these have tougher requirements, since they need to accurately identify all the echo response — even at frequencies where the echo signal may have low energy.

Many other recursive algorithms have been proposed for echo canceller adaption, with a wide range of computational complexity [16-19]. Of these, the recursive least squares (RLS) algorithm (equation (13.13)) is probably the most commercially significant and is claimed to show faster convergence than the NLMS algorithm with a speech input signal. This improved behaviour stems from the replacement of the scalar constant μ by the matrix R^{-1}, which is the inverse of the autocorrelation matrix of the input signal. This matrix, which is also calculated recursively, reduces or removes the eigenvalue disparity of the received speech signal:

$$h_{j+1} = h_j + e_j R_j^{-1} x_j$$

$$\text{where } R_{j-1}^{-1} = R_{j-1}^{-1} - \frac{R_{j-1}^{-1} x_j x_j^t R_{j-1}^{-1^t}}{1 + x_j^t R_{j-1}^{-1} x_j} \qquad \text{... (13.13)}$$

Direct methods of calculating the echo response are also possible and those based on the fast Fourier transform (FFT), in sliding-block average form, are particularly computationally efficient. However, the simultaneous need for fast adaption and the avoidance of loss or adaptation during double talk means that such blocks must be calculated at a rate that is independent of block length (about one hundred times per second) so that the blocks overlap in most applications. The disadvantage of this approach is the increased numerical precision required, demanding the use of floating-point hardware or double-precision software in fixed-point hardware. This implies a significant increase in cost, so that almost all commercial network echo cancellers to date have used the NLMS algorithm. Table 13.1 gives indications of the typical computational complexity, excluding control overhead, of the NLMS and RLS algorithms in fixed-point form and of the sliding-average FFT approach in floating-point form, for a speechband echo canceller sampling at 8 kHz with echo model durations of 10, 60 and 300 ms.

Table 13.1 Numerical complexity of echo cancellers with the NLMS, RLS and sliding-block-average FFT algorithms, in millions of multiply-and-add operations per second.

	10 ms model	*60 ms model*	*300 ms model*
NLMS algorithm	1.20	7.68	38.4
RLS algorithm	6.67	38.7	192
Sliding FFT (float)	0.82	3.85	19.3

Accurate control of adaptation, in response to the flow of conversation in a wide range of background noise, is generally more important and more difficult to achieve than a better adaption algorithm. This is especially true of acoustic cancellers, which must rapidly and reliably distinguish double-talk from uncancelled echo, due to the possible changes in the echo response with user movement. Poor control of a superior adaption algorithm can easily result in worse performance than good control of a conventional NLMS algorithm. More complex signal classification and modelling techniques are likely to significantly improve the performance of the current control methods that are usually based on broadband level measurements. For example, improved signal classification techniques are already in use in the GSM voice activity detector (see Chapter 4).

A foreground/background architecture is known to offer better performance [14] with the existing adaption control methods, although at the price of about 50% more computation. Faster adaption, to reduce audibility of echo at the start of a call or after a change in the echo path, is promised by a multiband canceller architecture [20]. A similar benefit for network echo control might prove possible with a sliding-window technique, where fewer and more rapidly adapting filter coefficients are positioned around the echo, using a bulk delay configured during call set-up.

13.3.3   Echo canceller implementation

Each channel of a typical network echo canceller needs separate speech and tone detection circuits, to control the adaption and NLP functions and to disable cancellation on voiceband data and facsimile calls. Thirty-one such channels are grouped with a signalling channel for connection to 2-Mbit/s streams (ITU-T G.703), with a common diagnostic and configuration interface. Commercial designs have been implemented in fixed form with custom-integrated circuits or with the more fluid programmability of DSP software. A compromise route, recently emerged, uses semi-custom circuits for the fixed parts of the design and a DSP core, with associated software, to retain future flexibility.

In recent years, British Telecommunications has contributed strongly to revisions of the ITU-T recommendation for network echo cancellation, due to be published as G.168 [21].

The performance requirements of acoustic cancellers, used to reduce loudspeaker-to-microphone coupling, are harder to define than for network cancellers and are not yet the subject of ITU-T recommendation, although G.167 describes general aspects. Those requirements are usually are more stringent and occasionally conflicting. For example, the long duration of acoustic echo demands many adaptive filter coefficients which implies a slow rate of convergence. Yet user movement can rapidly change the echo response, demanding a fast rate of convergence.

13.4   CHANNEL EQUALIZATION

Historically, most work on channel equalization has been concerned with improving the reliability of data transfer over a speech link. Typically a modulator-demodulator (modem) carries digital signals over a telephony channel by converting the binary data to voice-frequency signals and back again. Due to the non-ideal frequency response of these channels the received signals may overlap, resulting in intersymbol interference (ISI). This distortion is a major obstacle to high-speed data transmission and requires the use of an equalizer to remove the effect.

However, this section is concerned with the equalization of distorted speech signals rather than modem signals. In a telephone system three areas of channel distortion can be identified — the acoustic environment in which the microphone is placed, the microphone itself, and the communications channel in the network. For most applications the effect of the acoustic environment can be ignored. However, for good quality speech it is necessary to compensate for the effects of the microphone and communications channel.

For most speech channels the distortion can be adequately modelled as a convolution of the channel input signal $x(m)$ with the impulse response of the channel $h(m)$, to give a distorted output signal $y(m)$:

$$y(m) = x(m)*h(m) \qquad \qquad \text{... (13.14)}$$

Additive acoustic noise, $n_1(m)$, can contaminate the speech prior to entering the channel. In addition, channel noise, $n_2(m)$, may also corrupt the speech inside the communications channel. Thus the speech output from the channel $y(m)$ is modelled as:

$$y(m) = \left\{ \left[x(m) + n_1(m) \right] * h(m) \right\} + n_2(m) \qquad \qquad \text{... (13.15)}$$

which is a convolution of the sum of speech and acoustic noise with the channel impulse response, followed by the linear addition of the channel noise. Figure 13.12 illustrates the communications channel model.

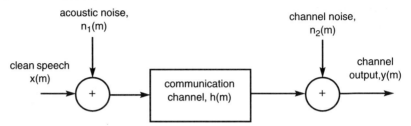

Fig. 13.12   Generalized convolution of a speech signal in a communications channel.

In the frequency domain, convolution becomes multiplicative, hence the channel output signal in the frequency domain is represented:

$$Y(f) = \left[X(f) + N_1(f)\right]H(f) + N_2(f) \qquad \text{... (13.16)}$$

If the acoustic and channel noises are ignored the problem of equalizing the channel is simplified to:

$$Y(f) = X(f)H(f) \qquad \text{... (13.17)}$$

Figure 13.13 illustrates the use of an equalizer $C(f)$ to deconvolve the channel and restore the original speech signal.

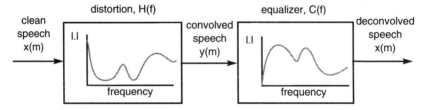

Fig. 13.13   Equalization of a channel-distorted signal.

In an ideal situation, the equalizer has a frequency response equal to the inverse of the channel. The ideal equalizer is able to fully restore the original speech signal from the convolutively distorted signal. In such cases the equalizer $C(f)$ is defined as:

$$H(f)C(f) = 1 \qquad \text{... (13.18)}$$

Real communications channels may not necessarily have invertible responses, hence $H^{-1}(f)$ may not be defined at all frequencies. Such a situation occurs in channels which have frequency regions which are heavily attenuated. For example, a telephone channel, which is essentially a bandpass filter, has two non-invertible regions, as shown in Fig. 13.14.

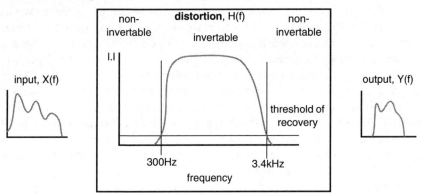

Fig. 13.14   Illustration of a non-invertible channel, e.g. bandpass characteristic.

The threshold of recovery indicates the lowest amplitude of signal which can be recovered. This is determined by any channel noise present and the quantization resolution of the sampling processes.

A second form of non-invertable channel occurs when the channel distortion has a maximum phase response which has zeros outside the unit circle [22]. The inverse channel, or equalizer, requires poles outside the unit circle and will therefore have an unstable impulse response. Channel equalization is a simple process when the distortion process is known and is invertable. In most applications the channel distortion is unknown and the only information concerning the channel is contained within the distorted speech signal itself. A possible method of obtaining the channel response in this situation is to use an equalizer training period. A known signal is transmitted from which the receiver is able to estimate the channel distortion. This approach has several drawbacks such as the increased time and energy required to perform the training phase. An alternative method is to use a 'blind' equalization technique to remove the channel distortion. Blind equalization (or blind deconvolution) requires some statistical knowledge of the clean signal in order to identify the equalizer. The remainder of this section describes three different methods of blind deconvolution, namely homomorphic deconvolution, all-pole modelling-based deconvolution and a Bayesian equalization approach.

13.4.1 Homomorphic deconvolution

One of the earliest attempts at channel equalization was that of blind deconvolution proposed by Stockholm [23]. This work concentrated on deconvolving the effects of the acoustic recording horn used in old gramophone recordings. In the 1920s, musical recordings were made using the apparatus shown in Fig. 13.15. This consists of a horn which is used to gather the acoustic energy and focus it on to the sound box, comprising a diaphragm which converts the acoustic energy to mechanical energy. This is linked to a stylus mechanism to cut the recording groove into the wax disk which rotates on a turntable. When played back the disk suffers from two types of distortion. Noise is introduced in the form of scratches, and clicks which arise from imperfections in the recording surface. The main distortion in the recording comes from the channel response of the acoustic horn and stylus assembly. The frequency response of the horn is typically limited to between 200 Hz to 4 kHz. In addition to this, resonances occur which can vary in amplitude from 10-20 dB (at frequencies between 100 Hz and 1 kHz). This introduces an unpleasant megaphone type quality to the recording and also produces loud bursts of sound when these particular resonant frequencies are prominent in the music.

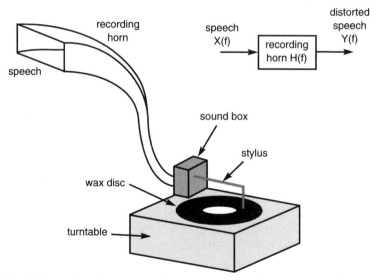

Fig. 13.15   Recording mechanism used in 1920s for recording music on to wax discs.

In this situation neither the original musical signal $x(m)$, nor the recording horn frequency response $H(f)$ are known. This means that the equalization of the horn becomes a blind deconvolution problem. In the time-domain the effect of the horn is convolutional which makes techniques such as Wiener filtering impractical as the channel is unknown. Using homomorphic processing developed by Oppenhiem [24], the channel can be made additive. Taking the Fourier transform and complex logarithm converts the channel distortion to an additive operation allowing conventional filtering techniques to be applied:

$$\log Y(f) = \log X(f) + \log H(f) \qquad \qquad \ldots (13.19)$$

The signal is divided into short duration blocks which are long compared to the channel impulse response using a suitable windowing function. Taking the time averages over the segments yields the relationship:

$$\frac{1}{N} \sum_{n=0}^{N-1} \log|Y_n(f)| = \frac{1}{N} \sum_{n=0}^{N-1} \log|X_n(f)| + \log|H(f)| \qquad \ldots (3.20)$$

Thus the channel is computed:

$$\log|H(f)| = \frac{1}{N} \sum_{n=0}^{N-1} \log|Y_n(f)| - \frac{1}{N} \sum_{n=0}^{N-1} \log|X_n(f)| \qquad \ldots (3.21)$$

This requires both the average distorted signal spectrum $\bar{Y}(f)$, and an undistorted average spectrum $\bar{X}(f)$. Stockholm used a modern recording of the same piece of music to obtain the average undistorted spectrum, which was recorded using modern hi-fi with a virtually flat frequency response. In practice, the equalizer is made to have a maximum attenuation of around 40 dB, and is typically smoothed in the frequency domain by removing sharp peaks in the equalizer spectrum.

13.4.2   Blind equalization using model factorization

Linear predictive models have been successfully used in speech modelling to estimate vocal tract filter parameters and the speech excitation signal from the lungs [25]. This is essentially a blind deconvolution problem as neither the input signal (excitation) nor the channel (vocal tract) is known. Two assumptions are made for the model — that the channel is minimum phase and that the input signal is random, with a flat spectrum. Spencer and Rayner [26] describe a method of blind deconvolution based on linear predictive modelling for separating time-varying and stationary systems — for example, the separation of a speech signal from a channel distortion. Figure 13.16 illustrates a cascade of a time-varying speech model and a stationary channel model, producing the distorted speech signal from the excitation signal.

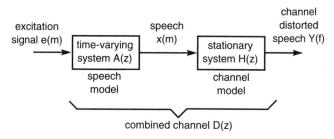

Fig. 13.16   Distorted signal modelled as a cascade of a time-varying and stationary system.

The linear predictive model for the speech model, $A(z)$, is written as:

$$A(z) = \frac{G_1}{1 - \displaystyle\sum_{k=1}^{P} a_k z^{-k}} = \frac{G_1}{\displaystyle\prod_{k=1}^{P} (1 - \alpha_k z^{-k})} \qquad \ldots (13.22)$$

where a_k is the kth FIR filter coefficient and α_k is the kth pole of the Pth order system. In a similar manner the stationary channel can be modelled using a linear predictive model:

$$H(z) = \frac{G_2}{1 - \displaystyle\sum_{k=1}^{Q} b_k z^{-k}} = \frac{G_2}{\displaystyle\prod_{k=1}^{Q} (1 - \beta_k z^{-k})} \qquad \ldots (13.23)$$

with b_k being the kth FIR filter coefficient and β_k the kth pole of the Qth order system. The combined system model $D(z)$, of the speech and channel model is given by the product of $A(z)$ and $H(z)$:

$$D(z) = \frac{G_1 G_2}{1 - \displaystyle\sum_{k=1}^{P+Q} d_k z^{-k}} = \frac{G_1 G_2}{\displaystyle\prod_{k=1}^{P+Q} (1 - \gamma_k z^{-k})} \qquad \text{... (13.24)}$$

which has filter coefficients d_k and poles γ_k. Clearly the $P+Q$ poles γ_k comprise the P poles α_k and the Q poles β_k. The poles from the speech model will be time-varying and the poles from the channel model will be stationary. Thus, if the channel-distorted speech signal is divided into N short duration blocks — of say 100 samples — and factorized into $P+Q$ poles, then a set of $N(P+Q)$ poles will be produced. The Q poles produced from the stationary channel will tend to be clustered together with a tight variance whereas the P time-varying poles from the speech model will be scattered around much more widely. By clustering these points using a furthest neighbour algorithm, the clusters with smaller variances can be identified as those poles resulting from the channel. Figure 13.17 illustrates these z-plane clusters and the identification of the poles of the channel.

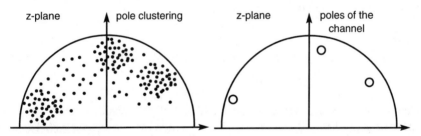

Fig. 13.17   Clustering of poles and identification of channel poles.

Once the poles of the channel have been identified, an equalizer can be used to filter the effect of the channel from the speech signal.

13.4.3   Bayesian channel equalization

The Bayesian inference method [27] provides a framework for the inclusion of statistical knowledge concerning the channel input signal and the channel itself, and is ideally suited for blind deconvolution. The channel distortion is modelled using homomorphic processing in the cepstral domain as an additive operation:

$$y_t = x_t + h \qquad\qquad \text{... (13.25)}$$

where y_t is a P-dimensional cepstral vector of distorted speech at time instance t, x_t is the undistorted cepstral input to the channel which is represented by h and is assumed to be time-invariant.

The joint estimation of the channel and channel input signal is a non-trivial problem. An alternative is to first estimate the channel $\hat{h}$ and then use this to estimate the channel input signal $\hat{x}_t$:

$$\hat{x}_t = \hat{y}_t - \hat{h} \qquad\qquad \text{... (13.26)}$$

Using Bayes rule the *a posteriori* probability density function (PDF) for a channel, given a sequence of T channel output vectors, $Y = \{y_0, y_1, y_2, ..., y_{T-1}\}$, which are assumed to be independent, is:

$$f_{H|Y}(h|y_0, y_1, ...,y_{T-1}) = \prod_{t=0}^{T-1} \frac{1}{f_y(y_t)} f_{Y|H}(y_t|h) f_H(h) \qquad\qquad \text{... (13.27)}$$

which is expressed in terms of the likelihood of the channel output given the channel, $f_{Y|H}(y_t|h)$, and the *a priori* distribution of the channel, $f_H(h)$. The reciprocal term $f_Y(y_t)$ is a constant and merely acts as a normalizing term so that the *a posteriori* density integrates to unity.

Using the equality:

$$f_{Y|H}(y_t|h) = f_x(y_t - h) \qquad\qquad \text{... (13.28)}$$

the *a posteriori* PDF can be written:

$$f_{H|Y}(h|y_0, y_1, ...,y_{T-1}) = \prod_{t=0}^{T-1} \frac{1}{f_y(y_t)} f_X(y_t - h) f_H(h) \qquad\qquad \text{... (13.29)}$$

Assuming that both the channel input vector and channel have Gaussian distributions with means and covariances, μ_x, Σ_x, μ_h, Σ_h respectively, the logarithm of the *a posteriori* PDF can be differentiated with respect to the channel h and the result set to zero to maximize the *a posteriori* probability:

$$\frac{\partial \log (f_{H|Y}(h|y_0 y_1,\ldots y_{T-1}))}{\partial h} =$$

$$\sum_{t=0}^{T-1} \left\{ (y_t - h - \mu_x) \Sigma_{xx}^{-1} - (h - \mu_h) \Sigma_{hh}^{-1} \right\} = 0 \qquad \ldots (13.30)$$

This gives the maximum *a posteriori* (MAP) estimate of the channel as:

$$\hat{h}^{MAP} = \frac{\Sigma_{hh}}{\Sigma_{xx} + \Sigma_{hh}} (\bar{y} - \mu_x) + \frac{\Sigma_{xx}}{\Sigma_{xx} + \Sigma_{hh}} \mu_h \qquad \ldots (13.31)$$

where $\bar{y}$ is the average of the distorted cepstral vectors y_t. The MAP estimate of the channel requires a knowledge of both the channel input signal statistics and the statistics of the channel. A suitable source of channel input signal statistics may be obtained from a vocabulary of HMMs which are used in speech recognition applications. This makes the technique ideal for working in parallel with HMMs to give channel robustness to speech recognition systems. However, this does not prohibit such a system from being used for channel equalization in other communications applications, provided that a suitable set of HMMs are available to provide the channel input signal statistics.

By collecting the cepstrum of many different channels, the channel mean and covariance can be estimated. If no prior knowledge of the distribution of the channel $f_H(h)$ is known, only the likelihood of the observation given the channel $f_{Y|H}(y_t|h)$ can be used to estimate the channel distortion. Maximizing the likelihood by differentiating the log of the likelihood with respect to the channel and setting the result to zero gives the maximum likelihood (ML) estimate of the channel as:

$$\frac{\partial \log (f_{Y|H}(y_0 y_1,\ldots y_{T-1}|h))}{\partial h} = \sum_{t=0}^{T-1} \left\{ (y_t - h - \mu_x) \Sigma_{xx}^{-1} \right\} = 0 \quad \ldots (13.32)$$

with:

$$\hat{h}^{ML} = (\bar{y} - \mu_x) \qquad \ldots (13.33)$$

The ML estimate of the channel does not use any prior knowledge of the channel. It is interesting to observe that the MAP channel estimate tends to the ML estimate when the variance of the prior channel distribution Σ_{hh} tends to infinity. This means that the PDF of the prior channel estimate becomes a uniform distribution and contains no information about the channel.

13.4.3.1   HMM-state-based Bayesian equalization

As with all blind deconvolution methods, some statistics of the channel input signal are required for the maximum likelihood or maximum *a posteriori* channel equalization techniques. One solution for obtaining this information, which is particularly suited to speech recognition, is to use the statistics found within the states of HMMs trained on undistorted clean speech.

A simple ML implementation may obtain the clean speech mean and covariance μ_s and Σ_{ss} from a single-state HMM trained on a range of speech sounds.

Alternatively a more sophisticated scheme may use a vocabulary of HMMs trained on either words, phonemes, digits, etc. In this case a decision must be made as to which HMM is providing the clean signal statistics. This is a time-varying process and will require the HMM to change as the speech sound changes with time. Ideally, this chosen HMM will model the sound being spoken. Thus the likelihood of the observation, given the HMM λ_w, the channel and the state sequence q of the HMM (q_t is the state of the HMM at time t), is:

$$
\begin{aligned}
f_{Y|\Lambda,H,Q}(Y|\lambda_w,h,q) &= \prod_{t=0}^{T-1} f(y_t - h|\lambda_w,q_t) \\
&= \prod_{t=0}^{T-1} \exp\left(\frac{-1}{2}(y_t - h - \mu_{\lambda_w,q_t})^T \Sigma_{\lambda_w,q_t}^{-1} (y_t - h - \mu_{\lambda_w,q_t})\right)
\end{aligned}
$$

$$\dots (13.34)$$

By setting the derivative of the log-likelihood to zero, the maximum likelihood estimate of the channel, based on the model λ_w and state sequence q, is given as:

$$\hat{h}^{ML}(Y, \lambda_w, q) = \frac{\displaystyle\sum_{t=0}^{T-1} (y_t - \mu_{\lambda_w, q_t}) \Sigma_{\lambda_w, q_t}^{-1}}{\displaystyle\sum_{t=0}^{T-1} \Sigma_{\lambda_w, q_t}^{-1}} \qquad \ldots (13.35)$$

This leads on to a hypothesized approach [28] where each HMM is used to give a unique channel estimate which then equalizes the speech signal. Each estimate of clean speech is then re-applied to its respective HMM and a probability score obtained. The restored signal which gives the highest probability is output as the equalized signal. In Fig. 13.18 the effectiveness of the ML equalization approach is demonstrated by comparing the frequency response to that of the actual channel distortion.

13.5   FUTURE DIRECTIONS

Signal enhancement technology for telecommunications is evolving in several distinct and important respects. Firstly, the location and deployment strategy of such equipment is changing, for example, echo control was historically viewed as a special function, independent of transmission technology or call-by-call routeing. A different approach is now required, due to the rapidly increasing variety in network technology and interconnection — including the effects of fixed/mobile convergence, the growth of virtual private networks, and the impact of various kinds of voice-over-LAN connections. If customer satisfaction in speech quality is to be maintained, signal enhancement must be treated as an

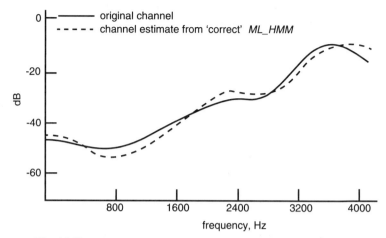

Fig. 13.18   Example of actual and estimated channel frequency responses.

integrated and indivisible part of transmission and switching, with an overall customer-to-customer viewpoint.

The second and ultimately more significant factor in the evolution of tele-communications signal enhancement is the dramatically decreasing cost of digital signal processing, in both semi-custom, DSP-core and native PC software form. This is likely to make new applications affordable and change the face of telephony. In future, economic echo control may be possible on the exchange line card, before the nonlinearity of A/μ-law conversion, or as part of the telephone or customer equipment. Computer-based telephony and conferencing products, with direct-to-customer digital links and multichannel echo control, might allow acoustic image conferencing, where speech communication would be hands-free in an acoustically-realistic sound-stage. These products will also be more susceptible to noise and channel distortions and it is vital that compensation is included to maintain quality. Many systems in the future will involve direct connection to speech recognition systems for enquiries and control. This will make the quality of speech an even more important issue to ensure the reliable response of the system.

In the near future, the spread of customer-to-customer digitization will offer a four-wire connection to the ordinary customer for the first time in the history of public telecommunications. This connection will be virtually free of the impairments of noise, echo and loss. Yet signal enhancement technology will be needed more than ever, because the full channel in communication reaches beyond the telephone socket on the wall. Acoustic propagation often forms the 'first and last' metre of this channel, especially in hands-free connections, and introduces noise, echo and fluctuations. The subjective impact of these degradations depends on how they interact with what goes on inside our heads — the psychoacoustic signal processing of human speech and hearing. For these reasons, the authors believe that noise reduction, echo control and equalization techniques will become increasingly important as telephony evolves, offering greater naturalness and realism in a widening range of environments for both person-to-person and person-to-machine conversations.

REFERENCES

1.   Widrow B: 'Adaptive noise cancelling — principles and applications', Proc IEEE, 63, pp 1692-1716 (1975).

2.   Darlington P *et al*: 'Adaptive noise reduction in aircraft communication systems', ICASSP-85, pp 716-719 (1985).

3.   Vaseghi S V: 'Advanced signal processing and digital noise reduction', John Wiley (1996).

4.   Boll S: 'Suppression of acoustic noise in speech using spectral subtraction', IEEE Trans ASSP, 27, No 2, pp 113-120 (April 1979).

5.   Lockwood P and Boudy J: 'Experiments with a non-linear spectral subtractor', Speech Communication, 11, pp 215-228 (1992).

6.   Milner B P and Vaseghi S V: 'Comparison of some noise compensation methods for speech recognition in adverse environments', Proc IEE, 141, No 5, pp 280-288 (1994).

7.   Wiener N: 'Extrapolation, interpolation and smoothing of stationary time-series, with engineering applications', MIT Press (1949).

8.   Rabiner L R: 'A tutorial on hidden Markov models and selected applications for speech recognition', Proc IEEE, 77, No. 2, pp 257-285 (February 1989).

9.   Ephraim Y *et al*, 'Speech enhancement based upon hidden Markov modelling', ICASSP-89, pp 353-356 (1989).

10. Milner B P and Vaseghi S V: 'Noise compensation methods for hidden Markov model speech recognition in adverse environments', IEEE Trans SAP, 5, No 1, pp 11-21 (January 1997).

11. Richards D L: 'Telecommunications by Speech', Butterworths, London (1973).

12. Duttweiller D L and Chen Y S: 'A single-chip VLSI echo canceller', Bell Sys Tech J, 59, No 2, p 149 (February 1980).

13. Murano K, Unagami S and Amano F: 'Echo cancellation and its applications', IEEE

14

ASSESSING HUMAN PERCEPTION

M P Hollier and G Cosier

14.1 INTRODUCTION

Intense competition and rising customer expectations are ensuring that price differentials are minimized in the telecommunications market-place. When price is desensitized in this way, other factors such as quality become critical differentiators. A key requirement, therefore, is the need for a reliable means of assessing the quality of performance experienced by the customer. Not only must acceptable standards be met consistently, but this must be done at minimum cost to the operator. Objective assessment of parameters, such as speech quality, will enable standards to be efficiently maintained together with effective assessment of systems during design, commissioning and operation.

The network increasingly contains complex and nonlinear processes such as low bit rate coding schemes for data reduction in mobile communications (see Chapter 3). Such processes are designed to operate on a speech signal and may deliberately exploit psychoacoustic phenomena such as masking [1] to minimize the audibility of error artefacts [2]. Conventional engineering performance metrics are inadequate for predicting the subjective performance of such systems where objective assessment requires an analysis which can predict the response of the human senses.

In response to this requirement for a new measurement approach, a technique has been developed which attempts to model the human senses in order to predict the perceptibility of any errors. Performance prediction based on models of the human senses can be referred to as perceptual analysis. This chapter discusses an audio-perceptual analysis method which has been developed at BT Laboratories and indicates the main applications. Future work towards the objective assessment of multimedia products and services is also noted.

14.2 A NEW MEASUREMENT APPROACH — PERCEPTUAL ANALYSIS

Perceptual analysis models the human senses which are the final arbiter of system performance. Because the analysis is independent of the complexity of the distorting process and the distortion characteristic, it is potentially a highly robust measurement approach and can be used to assess the full range of single and multiple nonlinear processes. It is this potential robustness in real-world applications, where the distortion type will not generally be known in advance, which makes a sensory modelling approach preferable to a purely empirical approach. Purely empirical approaches have been described [3-5], where a variety of objective measures such as coherence and cepstral distance are used to train a neural network. However, the performance of the analysis with an unknown distorting process is uncertain. The performance of a human sensory model approach is potentially independent of distortion type. Perceptual measurement techniques are a rapidly emerging new area and several models based on this approach have appeared in the literature [6-10]. Some of these models, e.g. Paillard B *et al* [7] and Stuart [10], are intended for the evaluation of hi-fi (music) codecs and are only suitable for predicting the onset of audibility of errors and not the subjectivity of errors which are audible. It is this subjectivity of error that is pertinent to network performance.

The requirements for a perceptual analysis process are a test stimulus which will maximally exercise the device under test, and an analysis which can predict the perceived performance. Requirements for a speech-like test stimulus are discussed in Hollier *et al* [6] (see Fig. 14.1). The perceptual analysis consists of two parts, a sensory layer and a perceptual layer.

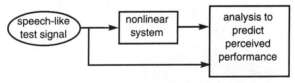

Fig. 14.1   Performance analysis with speech-like test stimulus.

14.2.1   Sensory layer

The sensory layer model reproduces the main psycho-physical transformations of human hearing including representation on a critical band or equivalent rectangular bandwidth scale, level-to-loudness transformation, and an approximation to simultaneous and temporal masking. Reproducing these transformations allows an audio stimulus to be represented with perceptually relevant dimensions. The BT Laboratories analysis represents the perceived audio signal as loudness per unit of auditory bandwidth, i.e. a specific loudness pattern. The subjective loudness of a signal, in sone, is then the area under the loudness curve.

Figure 14.2 shows the application of a subset of these psychoacoustic mappings to calculate the subjective loudness of a 2 kHz tone at a level of −16.6 dBV when heard via a telephone handset.

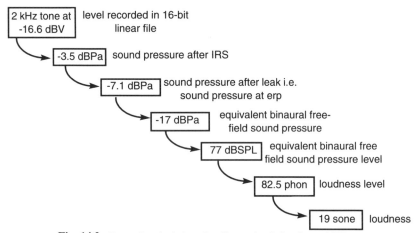

Fig. 14.2   Example calculation of auditory stimulation due to 2 kHz tone.

Original and degraded signals can be compared in this perceptual space in order to predict audible error. Visualization of audio signals and audible errors in perceptual space can be used as a diagnostic tool. Figures 14.3 and 14.4 show the

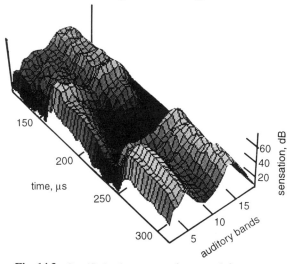

Fig. 14.3   Specific loudness pattern for a speech fragment.

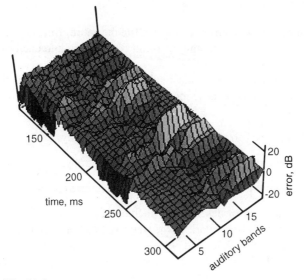

Fig. 14.4   Audible error pattern for speech fragment in Fig. 14.3.

specific loudness pattern and corresponding audible error surface for a fragment
of speech processed by a regular-pulse-excited linear-predictive codec with long-
term (pitch) prediction (as used in GSM) and a net bit rate of 13 kbit/s [11].

Careful examination of this audible error surface, with reference to the specific loudness pattern, shows that audible errors occur for:

- voiced sounds — particularly where the rate of change of amplitude is high;

- unvoiced sounds — particularly during onset and transition.

However, in many cases diagnostic interpretation by expert users is not sufficient and a typical requirement is to interpret the audible error surface in order to
predict an overall user opinion of performance. This requires the subjectivity of
any errors to be algorithmically assessed which is performed by a perceptual
layer model.

14.2.2   Perceptual layer

An important and novel feature of the audio-perceptual analysis developed at BT
Laboratories is that it is able to automatically interpret the subjectivity of audible
errors in order to predict user opinion using a perceptual layer algorithm. This is
a key requirement for a measurement system suitable for assessing the speech
quality of a wide range of real telecommunications systems.

In order to algorithmically assess the subjectivity of audible errors it is necessary to model the low-level psychoacoustics which determine this subjectivity. It
is apparent that the total audible error is not a sufficient indicator of error signifi-

cance, e.g. a given total error could permeate the entire error surface (perhaps due to a frequency response shaping) or could be collected into one or more discrete features.

Extensive examination of audio subjective test data resulted in the hypothesis that three criteria can be described which determine the subjectivity of audible errors:

- total amount of error — referred to as error-activity;

- distribution of error — referred to as error-entropy;

- correlation of error with original signal — referred to as error-correlation.

Inspiration for error descriptors which describe the first two of these criteria was taken from published work on the adaptive transform coding of video images [12]. Parameters are described which control the dynamic allocation of bits depending on the distribution of energy within the image. The application of these parameters to audible error surface interpretation is discussed in more detail elsewhere [13-15].

In summary, and with reference to Fig. 14.5, the total audible error may be referred to as the error activity. Based on the expression in Mester and Franke [12], for block activity, error activity can be defined as:

$$E_a = \sum_{i=1}^{n} \sum_{j=1}^{m} |e(i,j)| \qquad \ldots (14.1)$$

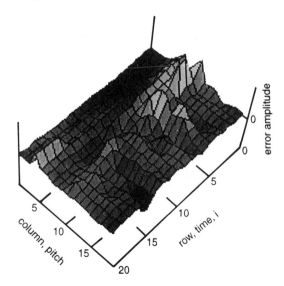

Fig. 14.5 Error-surface fragment.

where n and m are the dimensions of the error surface or error-surface fragment and $e(i,j)$ is the audible error in the ith row and jth column. The distribution of the audible-error may be referred to as the error entropy. Again an expression may be based on that given in Mester and Franke [12] to define a suitable quantity:

$$E_e = -\sum_{i=1}^{n} \sum_{j=1}^{m} a(i,j) \cdot \ln(a(i,j))$$ ... (14.2)

where $a(i, j) = |e(i,j)| / E_a$

The behaviour of the error-activity and error-entropy quantities is readily visualized and is illustrated in Figs 14.6 to 14.8, using hypothetical error-surface fragments which show the same error-activity distributed in different ways; these are summarized in Table 14.1.

Table 14.1   Error descriptors for hypothetical error-surface fragments.

	Error activity	*Error entropy*
response shaping (Fig. 14.6)	250	4.63
two peaks (Fig. 14.7)	250	3.78
single peak (Fig.14. 8)	250	0.87

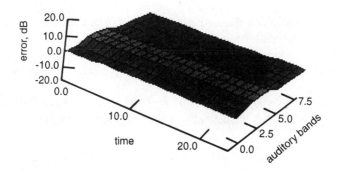

Fig. 14.6   Error-surface fragment — response shaping.

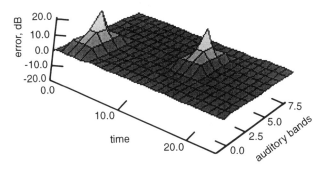

Fig. 14.7   Error-surface fragment — two peaks.

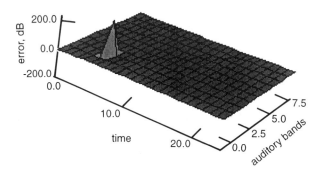

Fig. 14.8   Error-surface fragment — single peak.

Descriptors based on these error parameters can be calculated and their final combination calibrated using the results of a subjective experiment. Predictions by the resulting model can then be compared with further subjective test results. The combination of error descriptors and the prediction performance of the model are reported in Hollier and Hawksford [16].

These error-activity and error-entropy parameters were averaged across the sentence pairs for a subjective experiment and fitted to the subjective test data using a sigmoid function. The resulting combination equation is shown below and can be used to predict the subjective test results from further subjective experiments. The error parameters and opinion score predictions from an international subjective test to select the new half-rate GSM codec [17, 18] are shown in Figs. 14.9 and 14.10.

The combination function was:

$$\text{logit } Y = -132.18342 + 0.00307181 * E_a$$
$$- 0.00001425 * (E_a)^2 + 50.7525376 * E_e$$
$$- 6.4591057 * (E_e)^2 + 0.27230122 * (E_e)^3 = w$$

where:

$Y = (\text{MOS} - 1)/4$, and
$\text{logit } Y = ln\,[\;Y/(1 - Y)]$

The predicted score MOS' is then given by:

$$\text{MOS}' = 4 * Y' + 1,$$

where $Y' = [\exp(w)/(1 + \exp(w))]$, as described in Hollier and Hawksford [13].

The resulting function gives a mean squared error of 0.11 of an opinion score, with a correlation coefficient of 0.92. The prediction performance has also been tested across a variety of nonlinear distortions and example results are shown in Table 14.2.

Table 14.2   Average error and predicted MOS for different degradations.

Degradation	Subjective ranking	Average Log_{10} (error)	MOS prediction
DCME	Best 1	7.288	4.04
Speech codec A (EPX)	2	7.232	3.12
Nonlinear amplifier	3	7.810	2.97
Speech codec C (EPY)	Worst 4	7.555	2.36

It is apparent that the average error (total error averaged across test sentences) does not indicate the correct subjective ranking, while the MOS prediction, which takes account of the distribution of the error, does predict the correct subjective ranking.

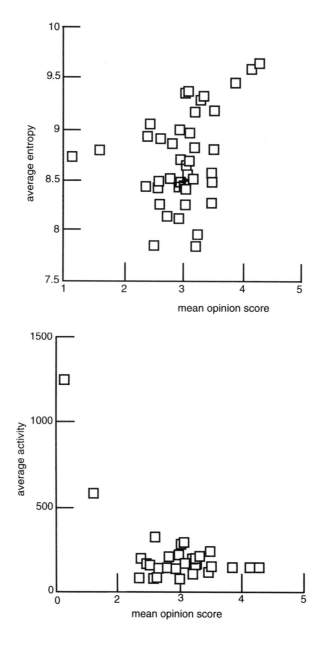

Fig. 14.9   Error parameters.

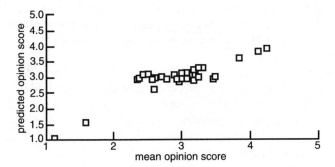

Fig. 14.10   Predicted opinion score.

14.3   HIGHER LEVEL PERCEPTUAL CONSIDERATIONS

The psychoacoustic properties modelled by the perceptual layer are low level, in terms of their neurological order, and describe the basic subjectivity of audible errors. These low-level properties do not account for higher level psychology and cognition which also contribute to the formation of subjective opinion. For example, the low-level psychoacoustic model may predict a particular degree of error subjectivity, but this may be strongly biased by the degree of attention paid to that sensory modality. Further, subjective opinion expressed by populations of users are dependent on the expectation of those users in connection with a particular product, service or activity. Aspects of speech cognition are also significant — in particular, utterance onsets are generally more critical for speech recognition than the relative steady-state portions of speech sounds.

14.3.1   Expectation factor

The actual opinions expressed by a group of subjects for a given audio quality will vary according to both the quantity to be judged (e.g. intelligibility versus quality) and the expectations of that user group. The quantity judged, e.g. Y_{LE} listening effort, is accounted for directly in the subjective data used to calibrate the analysis predictions. However, user expectation has also been found to be significant. An example of this was apparent during the course of this research — namely the variation in expectation between users of fixed, mobile and other terminals.

The performance of competing data reduction algorithms for a mobile application was assessed using conventional fixed network telephone handsets. The opinion scores from the subjective experiment were then used to select the codec which performed best over the required range of noise and tandeming conditions. Later, the performance of the selected codec was subjectively judged in a mobile communications context, resulting in significantly higher subjective ratings. The reason for this variation was found to be that users of mobile equipment expect lower audio quality and had unconsciously adjusted their subjective ratings to match this expectation.

The objective perceptual analysis described above will provide an opinion prediction which does not include this user expectation effect. It is proposed that an absolute objective performance metric is established which can be accurately predicted by measurement algorithms. The expectation factor appropriate to a given product or market sector can then be explicitly applied, as discussed by Hollier and Sheppard [19].

14.3.2   Cognitive aspects

Moving into cognitive function and task dependency, simple rules which describe the subjectivity of error in controlled conditions become less relevant. For example, in an emergency the task of obtaining assistance would dominate the cognitive process and minor audible errors in the communication channel would be irrelevant. Similarly, generalized perceptual rules cannot account for an individual's state of mind (happy, sad, angry) at the time of the subjective judgement. It should be recalled that the variable nature of human subjective judgement is one of the motivations for developing an objective measurement system. If an 'average' listener's opinion of subjective audio quality can be predicted, this provides the required objective measurement allowing system performance to be compared and improved.

There are aspects of cognition which are consistent and which can be included in a simple rule-based model. The main example is the recognition of language where certain words and phrases are more critical to the cognitive task of recognition than others. Certain parts of language such as consonant onset (consonant plus second formant transition) are consistently critical to recognition and can be weighted more highly in an analysis of subjective audio quality. This is a practical proposition since the test-signal must be known in advance for this kind of *a priori* analysis. To date only this simple linguistic cognitive rule has been implemented in a practical system. There is clearly potential here for further development.

14.4   APPLICATIONS FOR AUDIO PERCEPTUAL ANALYSIS

The main application areas for audio perceptual analysis are listed below:

- measurement tool
  — optimization,
  — iterative design aid,

— performance specification,
— procurement selection,
— refinement of proprietary equipment for service differentiation;

- measurement tool
  — speech quality assessment,
  — simple acceptance tester,
  — performance tester for commissioning and monitoring;

- perceptually motivated diagnostic tool, for speech and music
  — codec design,
  — expert fault finding,
  — psychoacoustic training aid;

- as part of a novel nonlinear network planning tool;

- analysis tool for opinion score prediction;

- research model, as a starting point for further development.

14.4.1   Measurements tools

An important requirement for many applications is the provision of a metric for speech quality which indicates that adequate performance is achieved, as perceived by the customer. When systems contain nonlinear processes this metric can best be provided by perceptually motivated analysis techniques. A prototype analysis based on a simplified perceptual model was developed to demonstrate a speech quality 'go/no-go' indicator for a mobile application.

When new equipment is installed it must be tested to ensure that performance is acceptable to the customer, and monitored to ensure that performance remains acceptable, particularly for private networks where new equipment increasingly contains dynamic data reduction of speech channels. In these cases a measurement tool is required which can provide a repeatable, graduated measure of speech quality, providing the following important functions:

- supplying commissioning engineers with easily interpretable evidence that systems are performing correctly — the objective assessment of multiple channels will be quicker than a combination of informal listening and conventional measurement;

- establishing a reference for speech-quality performance which can be monitored over a period of time, and in particular can be used to demonstrate to a customer that an agreed level of performance is maintained, perhaps as part of a service level agreement (SLA) in terms of a subjectively relevant metric.

14.4.2   Nonlinear network planning tool

Existing linear network planning tools rely on performance parameters for individual network components being combined to provide a prediction of overall performance. When one or more of the network elements are significantly nonlinear it is no longer possible to determine adequate parameters which are additive in the same way.

A new concept for a network planning tool capable of predicting the performance on multiple nonlinear network components has been proposed. The planning tool concept is based on perceptual analysis + network emulation as shown in Fig. 14.11.

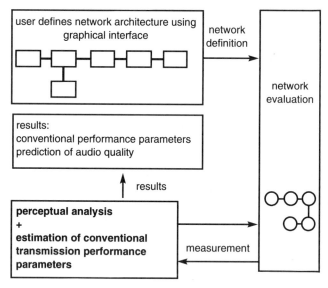

Fig. 14.11   Planning tool concept.

The graphical user interface would allow the target network architecture to be specified as a block diagram. The controlling software would then appropriately configure the network emulation ready to be assessed by the perceptual analysis. The results showing both a prediction of subjective audio quality and a range of conventional transmission performance parameters would be displayed via the user interface.

As part of a recent feasibility study, nonlinear network configurations were emulated, using the BT Network Emulator [20], and assessed with the perceptual analysis. The resulting prediction of their subjective performance was compared with that determined by subjective test with good correlation. While this does not represent an exhaustive study, which must consider the performance of a wide range of emulated network configurations, it does provide direct evidence of the feasibility of the new network planning tool concept.

14.4.3   Perceptual analysis to complement subjective experiment

Another important application for an established perceptual analysis will be to replace or complement subjective testing.

It is interesting to note that one of the earliest applications of the prototype auditory model was to help identify the test conditions for a large subjective test when doubt was cast on which condition subjects had actually experienced during a test. The prototype model was used in elementary fashion to show which subset of sentences had been subjected to particular degradations.

The main prerequisites for perceptual analysis as a replacement for subjective testing are:

- to compare many sets of subjective test results with perceptual analysis predictions in order to refine and verify its performance over a wide range of conditions and to establish the necessary criteria for an expectation factor, e.g. European Telecommunications Standardization Institute competition for codec selection;

- to pilot (alpha trial) perceptual analysis in parallel with subjective tests in order to show that its predictions were satisfactory — this is very important to gain credibility and confidence in the system prior to its use in contributing to high-value commercial decisions.

Further work is required to process a variety of subjective test data sets through the model in order to establish its credibility sufficiently to replace some subjective tests. Initially it is anticipated that the perceptual analysis could be used to complement a subjective test. The perceptual analysis could objectively assess a set of test conditions with a subset of the conditions being subsequently confirmed by human subjects in a reduced subjective experiment. The objective verification of test conditions and the reduction in size of the subjective test together represent a substantial potential cost and time saving.

14.4.4   As a starting point for future research

The final application for the auditory perceptual analysis is to provide the starting point for future research into:

- wideband audio models;

- models for non-speech audio applications, e.g. music;

- the philosophical starting point for related work on visual perceptual models;

- objective multimedia assessment.

14.5   MULTIMEDIA ASSESSMENT

Technology is opening up many new market opportunities and the way that we work, play and interact is changing [21]. Communications systems have historically been conducted in a single medium, but, with the advent of multimedia, this is now no longer the case. Multimedia, as the term suggests has the ability to stimulate the auditory, visual and tactile senses together. To address the perceptual understanding of a multimodal stimulus, a 'cognitive' framework based on multisensory perceptual analysis (see Fig. 14.12) is starting to be developed. (In order to build this 'cognitive' model that can predict perception, when more than one of the senses are being stimulated, it is essential to understand many aspects of human behaviour previously not investigated.) People are good at identifying similar patterns, in pictures, sound, or human behaviour — but, how they do it remains something of a mystery.

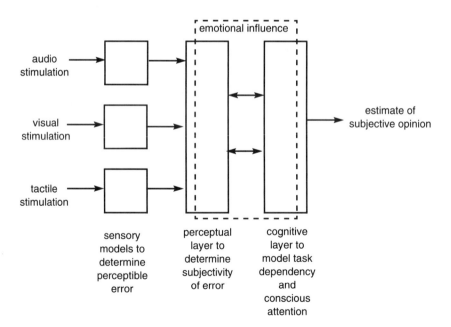

Fig. 14.12   Concept for multisensory perceptual analysis.

Objective assessment of multimedia products and services will require multi-sensory perceptual analysis. A visual perceptual model will be required to complement the audio model. Further, the relative contribution of each sense to the overall opinion of performance will depend on the task undertaken. A 'cognitive' layer model will therefore be required to provide the perceptual weightings relating to the task undertaken. Recent neurological evidence indicates that emotions are not a luxury — they are essential for 'reason' to function normally, even in rational decision-making. Furthermore, emotional expression is a natural and significant part of human interaction. Whether it is used to indicate like/dislike or interest/disinterest, emotion plays a key role in multimedia information retrieval and user preference modelling, and will form part of this 'cognitive' framework.

Only an elementary cognitive model is initially envisaged which appropriately weights the perceptual-layer descriptors for a small number of well-defined task types, such as a videotelephony scenario.

The BT approach to perceptual modelling has an advantageous perceptual-layer structure for this kind of sensory combination, since the task weighting can be applied to one or more of the error descriptors for each sensory model.

14.6   CONCLUSIONS

The economic necessity of providing adequate subjective quality at a given price has been described together with the increasing presence of nonlinear processes, such as data reduction in practical communications networks. Data reduction is already widespread in telecommunications, is also appearing in broadcasting (e.g. Digital Audio Broadcasting) and professional audio, and has emerged in domestic electronics in both Digital Compact Cassette (MPEG) and MiniDisc (ATRAC). The need for a new generation of objective measures to provide adequate characterization of such systems is apparent.

Objective performance assessment using an analysis based on a model of the human senses offers a potentially powerful and robust means of characterizing the perceived performance of nonlinear systems. An audio perceptual analysis has been developed and was introduced together with a description of its main applications:

- speech quality assessment;

- perceptually motivated diagnostic tool;

- complement to subjective tests;

- a component of a novel nonlinear network planning tool;

- a starting point and enabler for further research.

The potential extension of perceptual analysis to include other human senses has also been described. This is expected to be a key area for the communications industry since the bandwidth requirements for multimedia and virtual-reality applications are substantial and certain to contain data limitations or data reduction. It is argued that the perceptual implications of this data reduction in multi-sensory systems will require a perception-based performance analysis in order to provide a meaningful objective measure for:

- efficient optimization of systems;

- specification and monitoring of a minimum perceived performance.

The algorithmic evaluation of perceived performance of a multisensory task must be related to the task undertaken by the user. An outline model for a multisensory perceptual analysis is proposed, including a cognitive layer which will initially weight the combination of the senses according to a limited range of task scenarios.

The ultimate goal of this work is to view information, not in terms of Shannon's notation, but as it is described and understood by the humans who perceive it.

REFERENCES

1.   Moore B C J: 'Frequency selectivity in hearing', Academic Press (1986).

2.   Davis M F: 'The AC3 multichannel coder', Presented at the 95th Convention of the Audio Engineering Society (October 1993).

3.   Halka U and Heuter U: 'A new approach to objective quality measures based on attribute matching', Speech Communications (1992).

4.   NTIA, CCITT SG XII Contribution: 'Effects of speech amplitude normalization on NTIA objective voice quality assessment method', DOC. SQ-74.91 (December 1991).

5.   Irii H, Kozono J and Kurashima K: 'PROMOTE-A system for estimating speech transmission quality in telephone networks', NTT Review, 3, No 5 (September 1991).

6.   Hollier M P, Hawksford M O and Guard D R: 'Characterization of communications systems using a speech-like test stimulus', J Audio Eng Soc, 41, No 12 (December 1993).

7.   Paillard B, Mabilleau P, Morissette S and Soumagne J: 'PERCEVAL: perceptual evaluation of the quality of audio systems', J Audio Eng Soc, 40, No 1/2 (January/ February 1992).

8. Wang S, Sekey A and Gersho A: 'An objective measure for predicting subjective quality of speech coders', IEEE J on Selected areas in Communications, 10, No 5 (June 1992).

9. Beerends J and Stemerdink J: 'A perceptual audio quality measure based on a psychoacoustic sound representation', J Audio Eng Soc, 40, No 12 (December 1992).

10. Stuart J: 'Psychoacoustic models for evaluating error in audio systems', Procs Inst of Acoustics, 13, No 7 (November 1991).

11. Mouly M and Pautet M-B: 'The GSM system for mobile communications', Published by the authors, ISBN 2-950 71 90-0-7 (1992).

12. Mester R and Franke U: 'Spectral entropy-activity classification in adaptive transform coding', IEEE Journal on Selected Areas in Communications, 10, No 5 (June 1992).

13. Hollier M P, Hawksford M O and Guard D R: 'Error activity and error entropy as a measure of psychoacoustic significance in the perceptual domain', IEE Proc Vis Image Signal Process, 141, No 3 (June 1994).

14. Hollier M P, Hawksford M O and Guard D R: 'Algorithms for assessing the subjectivity of perceptually weighted audible errors', Presented at the 97th AES Convention in San Francisco, preprint No 3879 (November 1994).

15. Hollier M P, Hawksford M O and Guard D R: 'Objective perceptual analysis: comparing the audible performance of data reduction schemes', Presented to the 96th AES Convention in Amsterdam, Preprint No 387 (February 1994).

16. Hollier M P and Hawksford M O: 'A perception based speech quality assessment for telecommunications', IEE Digest No 1995/089 (May 1995).

17. ETSI/TM/TM5/TCH-HS: 'Second GSM half rate selection test plan', Doc No 92/14 (April 1992).

18. ETSI/TM/TM5/TCH-HS: 'Selection tests — basic data', Doc No TD92/39 (December 1992).

19. Hollier M P and Sheppard P J: 'Objective speech quality assessment: towards an engineering metric', The 100th AES Convention, Copenhagen, (Preprint No 4242) (May 1996).

20. Guard D R and Goetz I: 'The DSP network emulator for subjective assessment', in Westall F A and Ip S F A (Eds): 'Digital signal processing in telecommunications', Chapman & Hall, pp 352-372 (1993).

21. Whyte W S: 'The many dimensions of multimedia communications', in Whyte W S (Ed): 'Multimedia Telecommunications', Chapman & Hall, pp 1-27 (1997).

15

INTERACTIVE SPEECH SYSTEMS FOR TELECOMMUNICATIONS APPLICATIONS

S J Whittaker and D J Attwater

15.1 INTRODUCTION

The convergence of telecommunications and computing in the latter half of this century has probably had a greater impact upon society than almost any other technological change within that period. It has lead to an explosive growth in the availability and use of telecommunications and unprecedented demands and opportunities for sophisticated new products and services. Communication has become a global business, with the potential to involve almost everybody on the planet. This growth offers both challenges and opportunities, as a broadening base of users from diverse backgrounds demand increasingly varied products and services of a high quality, at a reasonable price and which are straightforward to use, and as this expansion drives a need for increased automation.

Speech interfaces build upon a basic human skill. The evolution of spoken language has given us an effective mechanism both for giving instructions and for expressing and interpreting ideas. Therefore speech technology can provide a flexible and intuitive approach to the control of advanced services and interface design can help to produce products and services which are straightforward to use and require little user training.

As developments in processor and storage technologies continue to drive down the cost of computing power in the network, on the desktop and in consumer and telecommunications equipment, speech will increasingly become the primary mode in many user interfaces across a wide range of market segments:

- in the office, voice control of personal computers and voice dictation is set to become more widespread;

- in the home, speech may become commonplace for the control of multimedia systems such as advanced interactive television;

- on the move, the convergence between mobile telephones and personal digital assistants (PDAs) will drive the development of ubiquitous voice and graphics driven communication assistants.

However, perhaps most dramatically, continued and rapidly accelerating growth will be seen in the use of speech technology within and at the periphery of the telecommunications network. Speech provides a method by which nearly everyone can control an increasingly feature-rich network, access sources of information and interact with customers and suppliers. Interactive voice response (IVR) systems can enable companies to re-engineer the way in which they do business with their customers, and it provides an approach to handling the explosive growth in the telecommunications and information sectors [1, 2]. It is likely that, despite the rapid growth in on-line access, in the medium term the majority of enquiries and transactions will occur by phone and much of it will rely on IVR technology. For example it has been estimated that within the US, 70% of call-centres employ IVR systems.

This chapter concentrates on the development of speech applications within the telecommunications area. It discusses relevant aspects of the technology and issues of integration and service design.

15.2   TELECOMMUNICATIONS APPLICATIONS OF SPEECH TECHNOLOGY

Speech technology is particularly relevant within the current telephone network as it can be used to:

- provide an intuitive interface;

- provide access to services without the need for additional customer equipment, with relevance both to customer acceptability and mobility;

- replace or supplement services which currently require the use of human operators;

- reduce the cost of service provision — new services can be inexpensively developed and made available through the international network.

Across a broad range of businesses, including telcos, the primary drivers for the introduction of automated services are:

- cost reduction;

- new revenue opportunities;

- increased customer quality and flexibility such as 24- hour operation.

Specifically within operator and call-centre applications, automation through the use of speech technology can allow people to handle individual calls more efficiently and be freed from many routine tasks to concentrate on high-value activities or problematic calls. This in turn allows flexibility in handling market growth and the ability to respond rapidly to new market opportunities. Similarly, new information services can be developed and deployed with reduced requirements for costly operator infrastructure, and existing manual services can be expanded to take advantage of prevailing market conditions or new customer groups.

IVR technology can enable a wide variety of applications, for example:

- advanced call-handling facilities:

  — voice controlled dialling;

  — auto-attendant/call routeing;

  — enhanced network services;

- messaging:

  — voice-mail;

  — voice access to electronic mail;

  — network message delivery;

- information and entertainment:

  — timetables;

  — directories;

  — music browsing and ordering;

- transaction processing:

  — home banking/home shopping;

  — work-force management;

  — automation and enhancement of operator services;

  — call-centre automation;

  — advertising tie-ins;

  — ordering of entertainment services.

IVR systems may be categorized by the primary mode of user input, typically either TouchTone® or speech recognition. The following sections describe their use in more detail.

15.2.1   TouchTone services

For some years TouchTone has been used as the primary method for controlling network-based speech applications, providing inexpensive solutions to a wide variety of applications. Often, a menu structure is implemented and the caller is guided through the structure with prompts of the form:

For News, press 1,
sport 2,

TouchTone can provide a good interface for motivated users and conventions for best practice in the design of dialogues have been established. Standards are available for the assignment of keys to commonly used functions and are discussed later in section 15.3.3.1. However, the mapping of system functions on to digits is arbitrary and not necessarily intuitive for complex tasks.

For many applications, TouchTone provides an appropriate and effective form of input. For example, the entry of a security PIN is well known to a public used to automated telling machines, and may give an increased sense of security to some users. Where long digit strings must be entered, for example card or telephone numbers, the greater speed and accuracy of TouchTone, may mean that it is preferred.

TouchTone can be used for the entry of alphabetic information, for example entering surnames to select a mailbox on a corporate Voice-Mail server. However, within the UK, few phones are marked with the letters and internationally not all labelling is consistent. Some applications rely upon more than one key press for each letter, which can be difficult and cumbersome for callers.

In some countries, penetration of TouchTone-capable telephones and even the proportion of capable telephones which are not correctly configured can restrict the availability of TouchTone-based services.

Other issues, such as the unavailability of the '*' key in some countries, may need to be taken into account.

15.2.2   Speech recognition for telephony services

For some applications, speech recognition can be used to provide a more user-friendly interface. For example, as discussed above, once alphanumeric or large vocabulary input is required, TouchTone becomes less appropriate.

In many cases, system designers may decide to offer speech and TouchTone in parallel, allowing the user to select the method with which they feel most comfortable. This may or may not be explicit in the outgoing prompts. In general, simply overlaying a dialogue based upon TouchTone menus with spoken digits is unlikely to result in an improved interface. For example:

For News, say or press 1,
Sport 2,

If possible, the natural vocabulary of the application should be reflected in the spoken vocabulary of the system. It is now possible to rapidly generate arbitrary recognition vocabularies which are intuitive within the particular application and to allow the entry of large-vocabulary information such as names. Where possible, alternative responses which may be reasonably expected from users should be catered for, e.g. 'football' and 'soccer'.

Please say the name of the service you require:
news, weather or sport

Speech recognition may be preferred for hands-free and eyes-free applications, such as those within cars — also for use by cellular phone users, when buttons are awkwardly positioned for use in interactive systems.

15.3   SPEECH TECHNOLOGY

This section discusses the main components of IVR systems which use speech recognition.

TouchTone systems typically contain many of the same basic components. Interactive speech systems comprise:

- dialogue controller — responsible for controlling the interaction with the caller, for example, maintaining the dialogue logic, sequencing prompts, selecting the recognition vocabulary, interpreting the semantics of user responses and determining the appropriate system operation;

- speech output — responsible for output messages and prompts;

- speech input — responsible for interpreting the caller's responses, often supplemented with TouchTone input for digit string entry and with 'speaker verification' for enhanced security;

- information manager — interfaces with the host application or application database and acts as a mediator for high-level dialogue requests, for example, building recognition vocabularies based upon database contents.

The separation between dialogue and information management may be trivial in a simple application and both activities may be handled by a single agent. This issue is discussed in more detail for advanced applications, such as directory searching, in Chapter 16. Figure 15.1 shows this diagrammatically, and the following sections discuss the speech and dialogue components separately and then issues of integration.

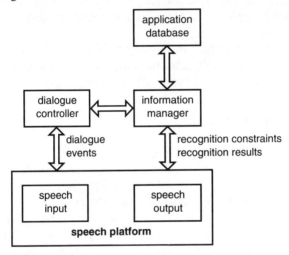

Fig. 15.1   Key components of speech applications.

15.3.1   Speech input technology

Most speech recognition systems currently in use are based upon statistical classification techniques such as hidden Markov modelling (HMM) (see Chapter 11). This is based upon building mathematical models from an explicit training phase. At run time, the task of the recognizer is to determine which model or sequence of models as defined by some lexicon or grammar (finite state or statistical) best explains the observed speech.

The performance of such systems depends upon how representative the training data and grammars are of the utterances which will be encountered in live use. Ideally, therefore, the initial training set should be as representative as possible of speech from the types of applications for which the recognizer is going to be used, for example telephony versus microphone speech, isolated word versus continuous speech.

The effective use of speech recognition is based upon the ability to predict and model the responses which are expected from the user. Hence the performance of an application which uses speech recognition is highly dependent upon the ability of the dialogue to steer the user to give responses in a predictable fashion.

Speech recognition is not a precise process. A certain error rate must be expected and should be catered for in the user interface design. This is particularly the case in telecommunications applications where the user population, background noise conditions, telephones and channels can vary widely and be difficult to predict. The dialogue should generally be designed to identify and repair such errors through the use of recognition confidence measures, confirmation and re-prompting.

Speech recognition systems can be categorized in a number of ways, and in general, increased functionality results in higher computational and storage requirements and hence, in the network context, per-line costs.

15.3.1.1   Types of recognizer

Speaker-dependent recognizers are trained specifically to the characteristics of a particular speaker. This involves collecting training data for that speaker. This technology has been largely superseded for telephony applications, with the notable exception of speaker adaptation described below.

Speaker-independent recognizers are trained on a corpus of data collected from a variety of speakers and can be used by any previously unseen speaker (given that the training speakers are representative of those expected in the live application). Speaker-independent systems form the basis of the majority of new services.

Speaker-adaptive recognizers adapt the characteristics of a speaker-independent recognizer as more information becomes known about the user. This technology has been used in systems where a single user uses the system extensively, e.g. voice dictation, although it is not yet commonplace in telephony applications.

15.3.1.2   Fluency

Systems can be designed which guide users to speak in different styles, for example as isolated words or in fluent speech. Different recognition techniques and components are required dependent upon the fluency of the speech expected and the detail of the interpretation which is required:

- grunt detection — the recognizer only identifies that something has been spoken; it does not attempt to identify the sound as a word or utterance; this can be used to interrupt long messages or to indicate a desired option from a menu;

- isolated-word entry — a single word or well-defined phrase is spoken in response to a prompt;

- word spotting — a single word or well-defined phrase is identified in a string of words;

- fluent-string entry — a sequence of letters or numbers is spoken in a stream; this is useful for entering spellings and alphanumeric account numbers, PINs and telephone numbers;

- fluent speech — complete phrases or sentences may be spoken.

At present most applications are implemented using isolated-word entry. Although fluent-string entry is increasingly used, fluent speech systems are only now becoming a realistic proposition for use in network applications within tightly restricted domains.

Fluent speech places very different requirements on other components of the system particularly the dialogue controller. For example, more than one piece of information can be given in a single response, or the user may refer to a piece of information given in a previous response. In addition to the basic recognizer, a parser is required to both identify words and extract the meaning from the sentence or phrase recognized.

In reality, these modes form a continuum. For example, if an isolated-word recognizer is to be robust to the variety of responses which users will actually give in a live application, it has to embody some degree of word spotting just to isolate the word from the surrounding noise and carrier phrases. It can be more helpful to consider the mode of entry as being more an issue of dialogue design, comprising the estimation of predicted responses, than one of recognition technology.

Good dialogue design can ensure that the majority of users can be primed and guided to respond in a constrained fashion. However, a proportion will always respond in unexpected ways, and it is therefore important that the recognizer is robust and can either ignore extraneous portions of a response or can detect and reject unexpected responses. It can be as important in an application to know if a response has been effectively recognized, as to know what the interpretation was. Dialogue decisions such as re-prompting can be made dependent upon such confidence criteria.

15.3.1.3 Vocabulary creation

Within an application, the vocabulary which will be used by the caller must be specified and used to configure the recognizers. This can be done in one of two ways:

- **whole-word** recognition training relies on the collection of multiple (tens or hundreds) examples of each word within the application;

- **sub-word** recognition training is based upon the construction of new vocabulary words from a pretrained vocabulary-independent set of phonetic (or other) sub-word units, using a pronunciation dictionary or lexicon.

Except where large pre-existing databases exist for a particular vocabulary as in the case of digits and 'yes/no', sub-word training based on an application-independent training database is now the preferred approach for the development of new applications. This is because the approach allows vocabularies to be defined and recognizers to be configured either by the designer or by reference to some application database, without the need for a specific speech data collection. It is therefore suitable for the construction of large vocabularies such as those required for directory access, catalogue shopping, etc.

An approach referred to as 'rapid vocabulary generation' allows new recognition vocabularies to be specified by the dialogue designer with little expert knowledge of the speech recognition component, simply by typing in the words of the vocabulary, using a GUI or service creation tool.

In practice, application-independent training databases may not optimally represent either the vocabulary of a particular application, or the way in which users respond. Therefore, once an application has been running for some time, application-specific training data should be collected and fed back into a retraining cycle, building either improved (application-specific) sub-word or whole-word models.

Sub-word recognition relies on the availability of high-quality phonetic lexicons which specify the pronunciations associated with potential vocabulary items in terms of phoneme sequences. These must capture both regional and international variations and the variations in pronunciation which may be reasonably expected within the application. This is particularly the case with real names, which show wide variations. Pronunciations can be generated in a number of ways:

- 'hand-crafting' based upon linguistic expertise;

- automatically by rule;

- automatically by analogy with similar vocabularies;

- automatically by auto-transcription of spoken exemplars using a speech recognizer.

Typically, the vocabulary of an application can be specified as a part of the application and dialogue design. There are, however, other instances where this is not the case and alternative approaches have to be taken, for example:

- user-selectable vocabulary, where a user can specify their own vocabulary items for use in, for example, voice-controlled dialling applications (e.g. 'call Bob'), typically requires the user to repeat the word or phrase a small number of times, under the guidance of the dialogue; alternative approaches to implementation allow the resultant vocabulary to be speaker-dependent or speaker-independent;

- database access — in applications where the vocabulary is changing rapidly, such as directory or customer identification applications, the system must be able to dynamically update its active vocabularies (database applications are discussed further in Chapter 16).

15.3.1.4   Vocabulary size

The size of vocabulary which can be supported within an application depends upon factors including:

- channel and terminal characteristics;

- confusability of vocabulary;

- available computing power;

- quality, quantity and applicability of training data;

- user population.

Although vocabulary requirements will vary from application to application, the data in Table 15.1 is indicative of the size of vocabularies which are currently available for speaker-independent telephony recognition supported on platforms such as the interactive speech applications platform [3, 4] (ISAP) (see also section 15.7.1).

Table 15.1 Example vocabulary sizes.

Mode	Vocabulary size
isolated yes/no	about five
isolated digits	twelve
isolated command words	tens
large vocabulary (e.g. names)	hundreds — thousands
fluent spelling and alphanumeric strings	tens of thousands of words
fluent digit strings	unconstrained

In general, increasing the size of a recognition vocabulary will reduce the accuracy of the recognizer. The similarity of the words also has a significant effect.

Chapter 16 discusses approaches which can be taken to combine the results of a number of different large vocabulary recognizers, arbitrarily spoken or spelt, to raise system performance in applications where each individual vocabulary is too large to achieve adequate accuracy in isolation.

Alternatively, recognizers can be configured to focus on the most common subset of responses expected at a particular point in a dialogue and to exhibit aggressive rejection or utterances outside of that subset. Such 'partial vocabulary recognizers' can be employed in applications where a small subset dominates what is, overall, a large vocabulary.

15.3.1.5   Speaker recognition

Speaker recognition is the term used to refer to both speaker verification and speaker identification. The technology is closely related to speech recognition.

Speaker verification is the process of validating or rejecting that a person is who they claim to be. For example, in a financial or calling-card application, where a caller has attempted to identify themselves by the entry of a card or account number, that claim can be validated using speaker verification. Speaker verification can have an impact on the perception of the level of security of a system. In some cases, TouchTone entry of a security PIN may be inappropriate (for example some hotel PBXs can record a calling-card PIN and display it on the bill) and so spoken security may be preferred. In this case, speaker verification adds an additional layer of security which can detect if an impostor has obtained the PIN.

Speaker identification is the process of identifying an individual from a set of speakers without any prior identity claim. This can be used, for example, to identify a family or team member using a shared system.

Just as applications using speech recognition have to cope with a certain error rate, so speaker verification has a corresponding error rate in terms of incorrect rejections of the real user and incorrect acceptances of impostors, and speaker identification will have a certain level of false categorization.

Typically, speaker-recognition systems require a separate enrolment dialogue in which the characteristics of the caller are obtained. In general, the performance of both technologies is dependent upon the amount of training material that is available during enrolment and on the amount of information available in the verification/identification pass. A balance has to be struck, dependent upon the particular application, between the amount of intrusion which the procedure causes within the dialogue and the performance required. In some cases, it may be possible to perform recognition and verification of the same utterance, for example, when an identification number is being entered.

Speaker recognition is not yet as mature a technology as speech recognition, and is only beginning to ease itself into applications. It is worth bearing in mind the difficulty of this task for people and how easily they can be fooled.

15.3.2 Speech output technology

Traditionally most IVR systems have used pre-recorded messages for dialogue prompts and information recordings. Often professional speakers are used to give a high quality of output and the same speaker or set of speakers may be used for a number of applications both to maintain a cor-porate identity and to ease the transition between services where one service gateways into another. In some cases, more than one speaker may be used within a single application to distinguish different activities or phases. For example, help messages may be spoken by a different voice with the purpose of indicating the structure of the dialogue to the caller.

As well as providing prompts, pre-recorded digits, words and phrases can be concatenated to enable the output of information such as telephone numbers as in the automatic announcement subsystem used within the BT directory assistance system. Care is needed in the selection of component phrases to ensure that the resultant concatenation is of a suitably high quality. For example, in the case of times or phone numbers, each word or number may be recorded in a number of contexts and the most appropriate unit selected at run time.

In situations where it is impractical to use concatenated recordings, a more flexible output system is required, capable of dealing with arbitrary output phrases. Such applications are those where:

- the information is changing rapidly,

- the information is dynamic and varied across users, such as reading out electronic mail,

- the vocabulary of the system is just too large to be practically recorded or stored such as directory applications.

Text-to-speech (TTS) synthesizers which are capable of speaking arbitrary sentences have been available for some years, but until quite recently, the quality has been very limited. While the available quality has been acceptable for high-value applications, such as those for the disabled or closed user group financial services, it has not been appropriate for the majority of telephony applications. However, the new generation of synthesizers, such as BT's Laureate system (see Chapters 5, 6, 7 and 8), have achieved a much higher level of performance and may be considered for use in advanced applications.

The combination of high-quality recorded messages with TTS-generated speech in another voice should be used with care. However, BT's Laureate TTS system can be trained to sound like a particular individual, which allows the combination of recorded messages from a given speaker with inserted segments generated in the same voice to cater for variables within an output phrase. For example:

'The number for' *<John Smith>* *'is'* *<01473 ...>*

---

where recorded messages are shown quoted and TTS variables bracketed.

15.3.3   Dialogue management

15.3.3.1   Dialogue design

Dialogue design is often the key to successful speech system design. In any IVR application it is the core of the system and specifies the user interface which the caller will encounter.

A significant amount of work has been undertaken at BT on the development of best-practice guidelines for the design and implementation of TouchTone and speech dialogues. Internally, BT designers have access to the BT voice application style guide which is based on extensive research and field experience and a number of external guides [5] have been produced to assist designers. In addition, BT is a partner in the Dialogues 2000 project [6] and funds a significant amount of research work at the Centre for Communication Interface Research at the University of Edinburgh [7]. A number of other standards projects, such as the EU-funded EAGLES project, are also under way. However, only a small number of international standards exist, such as ETSI/ISO standards for key assignment of standard functions within TouchTone dialogues, for example '1' for yes and '2' for no.

Generally, a successful dialogue will be one in which the user feels:

- comfortable;

- that they are in control, but being guided where necessary;

- that they know how to respond to any question;

- that they can get help at any point;

- that they can exit conveniently.

In brief, it must be clear and consistent.

The approach taken to the design of the structure of a dialogue will depend upon the characteristics of the users. In some call-handling applications, the caller will encounter the system unexpectedly and infrequently, in others they will have made a deliberate decision to access a value-added service or may be regular users.

Typically, TouchTone dialogues are based upon a menu structure, with dialogue prompts identifying options available. Complex services with a large number of options may require a hierarchical menu structure. The style guidelines discussed provide recommendations on issues such as the maximum size of individual menus and the order in which entries should be placed. A key requirement is that it is clear to the user both where in the dialogue they are and how to navigate around it. Usually the user can interrupt or pre-empt an outgoing menu.

There is a temptation when developing speech dialogues in domains for which TouchTone dialogues have already been developed to simply duplicate the menu structure and produce a 'speech enabled dialogue'. However, being able to select a recognition vocabulary pertinent to the application reduces the need to constantly explain the mapping between the user interface and system functions and can allow the development of more natural 'speech-centric' dialogue structures.

The majority of speech applications currently under development or deployed are 'system directed' and use finite state-based controllers. Here the system 'steers' the caller by imposing a clearly structured dialogue in which the current focus of the interaction is always clearly under the control of the system, and ensures that the caller uses an appropriate vocabulary and a form of speech which is compatible with the recognition technology used. For example, the dialogue may explicitly use prompts such as:

> *'Speaking after the tone, please say the name of the*
> *service which you require: balance, transfer or*
> *cheque-book <tone>'*

Such a message tells the user both when they can speak and what the valid responses are. However, such messages can become tiresome and in services where the caller has to navigate through a number of such prompts, a simpler message may initially be offered falling back to more direct phrasing only when necessary.

Typically, a dialogue would have to know how to respond to cases in which the user remains silent or an unexpected response is encountered. For example, a segment of dialogue might be:

> *Which service do you require?*
> <Silence>
> *I'm sorry I missed that. Please say the name of*
> *service which you require.*
> Help
> *The following three services are available, balance, transfer and checkbook.*
> *Which service do you require?*
> Balance

Although 'system-directed' dialogues have proved effective in applications using isolated-word speech recognition, they are less appropriate in the case of fluent speech entry. Here users are more likely to use complex linguistic forms and will tend to directly steer the dialogue. In these 'mixed initiative' dialogues, the system must keep track of the purpose of the conversation as well as the current active focus, allowing the user more flexibility in taking control of the interaction. In the extreme, the system takes on the role of an assistant, prompting for clarification and confirmation when required. For example [8]:

> *Which train do you require?*
> I want to go to London this morning

In order to allow such interactions, a more general model of dialogue is required which deals with high-level dialogue processes such as goal seeking, focus shifting and confirmation. The dialogue would typically be expressed in terms of this model, rather than the individual states of an application-specific finite-state dialogue (see Chapter 17).

Alternatively, where a large number of example user interactions with a particular automated application are available, a statistical model of the dialogue can be generated. This offers the potential for an interesting unification of representations for speech understanding from low-level signal processing to high-level pragmatics.

Special consideration needs to be given in the case of 'computer mediated dialogues' where the automatic system acts as an agent between two people, for example, in the reverse charge system described below or in simultaneous translation systems.

Ultimately, it would be desirable to develop dialogue systems which provide a broad 'world' knowledge rather than a closed model of a single application domain, as well as historical and contextual knowledge of the user. This would enable more 'intuitive' interfaces.

Coupling advanced dialogue front-ends with flexible agent-based back-ends, which can search, index and catalogue distributed information sources, may offer promise in this area (see section 15.8.3).

Hi Steve
Sort me out some tickets for a concert on Tuesday after I get back

---

Although much work continues worldwide, arguably no adequate general model of human/machine and machine-mediated speech dialogue has yet been developed.

15.3.3.2 Dialogue implementation

For even relatively straightforward dialogues, a dialogue control module can become complex, particularly if application and data manipulation functions are not separated from the main logic. As a result, much work has been done in the area of language and tool support for speech dialogues. The aim of such activities is to define a suitable representation scheme for the dialogue which allows user interface design to be separated from other system functions.

Tool support can significantly simplify and improve the development and in-life support. The use of 'service creation tools' (SCTs) provides a number of key benefits:

- separation of skill sets — specialist service designers can concentrate on important aspects of service design without the encumbrance of working with conventional programming languages;

- reduction of development time — SCTs permit interactive speech services to be implemented quickly and inexpensively in order to respond to a growing market-place where time to market is often critical;

- parameterization of services — established dialogue structures can be reused or reconfigured, e.g. for an additional language;

- establishing standards — emerging standards with regard to dialogue interactions can be incorporated into the tool, encouraging consistent 'look and feel' and the propagation of best practice;

- an aid to understanding the service — services represented in a graphical form can be appreciated by a wider range of stakeholders than with a textual representation.

SCTs support a number of roles within an iterative development process:

- service design;

- service review;

- service validation;

- documentation;

- message preparation;

- service simulation;

- in-life maintenance.

SCTs can take a number of forms including 'point and click' or 'drag and drop' graphical user interfaces, menu-based tools and high-level scripting languages.

The BT VISAGE (voice interactive service application generation environment) [9, 10] was developed as a rapid development tool for use across a number of platforms. VISAGE provides an intuitive, user friendly, graphical interface which supports the following tool-set:

- a flowcharting tool (see Fig. 15.2);

- message recording and editing tools;

- a service simulation tool for true service simulation using a local telephony interface, a screen-only mode or a 'Wizard-of-Oz' capability — see next section;

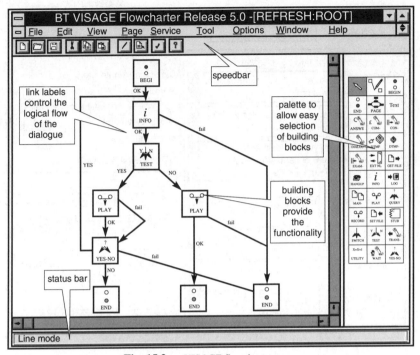

Fig. 15.2   VISAGE flowchart canvas.

- a code generator which produces code for execution on the target platform.

The flowcharting tool was developed using cross-platform development software and is available for use on PCs under the Windows family, and on Unix platforms (currently HP and SUN) under MOTIF and X.

VISAGE was designed around the concept of code reuse of a family of generic building blocks which cover telephony functions, dialogue interaction, control flow and miscellaneous functions. Standard dialogue blocks have been implemented which use standard, proven caller interactions and have been extensively tested [5]. It is also straightfoward for designers of complex applications to provide additional functionality.

Figure 15.3 gives an example of how a standard building block can be configured or parameterized to suit a particular service.

Many of the applications described in this chapter were developed using VISAGE.

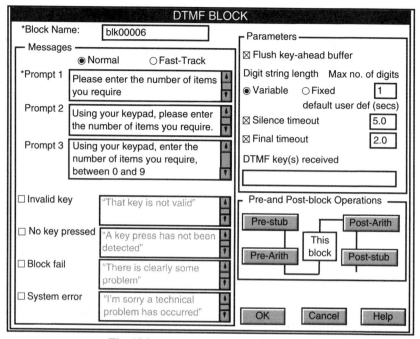

Fig. 15.3   VISAGE DTMF parameter form.

15.3.3.3   Dialogue design cycle

Developing an effective dialogue is not a straightforward process and an iterative approach is normally required, taking advantage of service creation tools such as VISAGE and rapid vocabulary creation facilities.

Early prototypes may be evaluated in small scale experiments with potential users, before proper system field trials take place.

In some cases where the requirements of the final system are not well understood or if a suitably accurate prototype of the recognition subsystem is not available, 'Wizard-of-Oz' experiments may be used in which the place of the recognizer is taken by a trained operator or 'wizard'.

The aim of this approach is to allow users to interact with an accurate simulation of the system. Varying portions of the system may be simulated, and appropriate constraints applied to the ways in which the 'wizard' can behave. For example, if the wizard is used solely to simulate the speech recognizer, it will only be able to hear and interpret the users' responses and enter that interpretation into the system, it will not be able to respond directly. As the recognizer will have a restricted vocabulary or grammar, the system may restrict the valid inputs which can be accepted. The sequencing of the dialogue and the selection and output of messages would typically be the responsibility of the system.

Care must be taken to ensure that any aspects in which the simulation may differ from the final system, such as the level of robustness to unexpected responses, are taken into account in the interpretation of results.

15.3.3.4   Trialling

Good dialogue design is best achieved by the application of best practice and the use of iterative trials concentrating on the human factors aspects of the service [11].

Typically, trialling will comprise both objective and subjective aspects. Subjective analysis may involve data collected from a variety of sources depending upon the size and scope of the trial, for example help-desk information, questionnaires used during follow-up interviews by market research agencies [12] and on-line World Wide Web questionnaires [13]. It will concentrate on:

- usability — the effectiveness and efficiency with which customers can use a service;

- acceptability — whether a service meets customer expectations and the level of satisfaction resulting from the service.

Figure 15.4 shows the results of subjective analysis within a trial of a Touch-Tone service [12]. This shows a number of standard dimensions against which the service was judged. In addition service-specific dimensions will typically be identified.

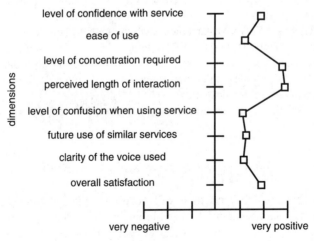

Fig. 15.4   'Third-party retailer' — subjective analysis.

Objective analysis will focus on measures such as transaction completion time, transaction success and speech recognition accuracy.

15.3.3.5 Dialogue maintenance

Usually, the design of a dialogue will be adapted during trials and after deployment to take into account the observed behaviour and responses of callers and their demographics. This process may include the modification of messages which are causing unexpected problems for callers, the extension of the recognition vocabularies or the statistical weighting of some words within them. Also, as described above, the speech recognizers can be optimized by using task-specific data.

In this respect, the dialogue design cycle can be viewed as extending beyond the initial deployment.

15.4 VOICE TECHNOLOGY IN CALL-HANDLING APPLICATIONS

The rapid growth in telesales, telemarketing, helpdesks and other services, such as telephone banking, have raised the importance of customer call handling for many companies. Call-centres have become a major business tool to focus telephone-based customer contacts and aid corporate process re-engineering. There is a strong similarity between call-centres and telco operator service bureaux.

Human agents and IVR systems are used in combination to provide the customer contact function [12]. The related technology of computer telephony integration (CTI) [14, 15] is a key enabler, enabling calls to be intelligently routed between agents or operators and IVR systems. The voice (or multimedia) call, with related call and customer information can be handled in a co-ordinated fashion throughout a complex call-flow, potentially involving IVR and multiple agents, departments or physical sites.

Figure 15.5 shows the architecture of a typical CTI and IVR-enabled call centre.

15.4.1 Call delivery and distribution

Calls are typically delivered to a call centre by the network provider in conjunction with number translation and billing services, such as those provided by BT Telemarketing services. This allows companies to take advantage of call stimulation and enhanced tariffing options.

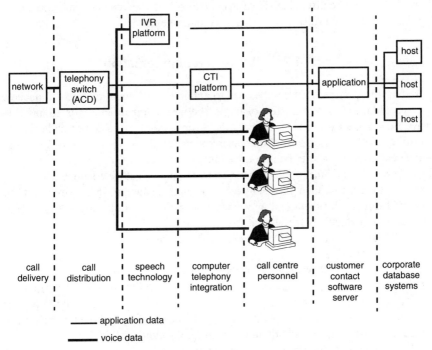

Fig. 15.5   IVR and CTI within call centres.

Traditionally an automated call distributer (ACD) is employed to distribute calls between and among groups of agents in order to balance agent loading, caller wait time and, in the case of 'blended call-centres', the mix of in-bound and out-bound calls. For example, queues can be established to segment applications into different agent skill groups or campaigns, or to select calls for IVR treatment, and used to correctly route calls based upon factors such as call origin or original number called.

Increasingly much of the intelligence is handled by the CTI controller, leaving the ACD with more basic call-handling functions. CTI allows the co-ordinated control of voice and data, and can use telephony-related data, such as calling line identity (CLI), information gained through IVR or agent transaction and application or corporate data, such as customer records obtained from back-end databases or mid-tier business process servers, to provide complex functionality.

For example, using information gained during an introductory dialogue, a corporate data resource can be searched and an agent's screen re-populated with the relevant call and customer information. This process of 'screen popping' ensures that, when agents or IVR systems either receive new calls, or are passed one by another entity, they have access to all of the relevant information to handle the call and fulfil the customer's requirements.

15.4.2   IVR for agent automation

Within this context, IVR can be used to automate fulfilment of selected queues, to provide call steering (for example, between departments) or to provide overflow capacity during peak periods. A proportion of customer calls can be fulfilled with a minimum of human intervention allowing increased flexibility and cost-effectiveness to be introduced into call processing.

Typically customer contacts handled by telephone can be conducted at significantly lower cost than those by other manual means, as information can be entered directly by the agent and many transactions completed and fully documented immediately.

Similarly, contacts fulfilled through IVR can be a further order of magnitude more cost effective.

In addition, IVR provides a mechanism for handling peaks in call rate effectively. For example, in many financial applications, call densities can be dramatically concentrated in relatively short peaks at predictable times of the day. If many of these can be handled through IVR, the existing call-centre agents can manage traffic variations without the need for an additional tidal workforce.

Developments in speech recognition technology will continue to allow progressively complex applications to be automated. Typically the benefits achieved through the appropriate use of IVR include:

- skilled agents can be focused on high value, rewarding activities;

- agents can be moved from reactive in-bound call handling to proactive out-bound;

- 24-hour, 7-days per week working;

- ability to handle traffic peaks;

- flexibility to target new market opportunities;

- new commercial channels are enabled for services which would not otherwise be cost-effective;

- services delivered at lower cost.

15.5   APPROACHES TO AUTOMATION

There are two primary approaches to the introduction of speech technology into new or existing applications, and each places different emphases on the underlying speech technology, service creation and effective dialogue design.

- Full automation

  A discrete portion of the service or business is identified and fully automated on a unique access number. Handover to a help-desk or operator may be supported, but this is secondary to the main mode of customer interaction. This is often an appropriate approach to the development of a new service, or the repackaging of an existing service (for example, for market expansion or re-pricing). In both of these cases, the application can be explicitly positioned.

- Progressive automation

  A framework is developed in which speech technology and operators are used co-operatively, the aim being to enable skilled staff to be deployed and operate as effectively as possible.

In general, there are a number of approaches to progressive automation:

- IVR front-end

  In this case, IVR is used to handle the early stages of all calls. In the simplest case, an 'auto-attendant' may be responsible for routeing calls to the correct individual, agent, business process or automated system. Alternatively, customers may be engaged in some preliminary dialogue phase, such as customer identification or verification, before being passed to an operator.

- Human front-end

  Here all calls are handled by a human and those suitable for automation are manually selected and passed to an IVR system.

This distinction can become blurred. For example, in many directory assistance applications, the initial personalized greeting is played to the caller automatically during the operator's inter-call rest, often without the caller being aware. Similarly, play-out of the telephone number is handled automatically.

This has led to the adoption of 'stealthy' approaches to progressive automation where those calls suitable for full or partial automation may be filtered automatically, according to the user's response to an initial prompt or dialogue.

Partial vocabulary recognizers (PVR), as described above, can be useful in applications of this sort. For example, if a user is to be prompted to provide a city name, the PVR may be trained to detect that subset of expected responses which identifies calls suitable for further automation. Those calls falling outside the subset can be routed directly to the operator and the initial response played. In this case, an automated dialogue may be used to hide the play-back process. This allows the maximum proportion of calls to be automated given the capabilities of the available recognition components.

15.6   ISSUES IN SPEECH APPLICATION DESIGN

In the design of any speech application a number of issues must be taken into account, for example the role of user priming material and consistency among related applications. In the case of complex automation applications a number of additional factors should be considered.

15.6.1   User knowledge

The amount of domain information known by the caller may be limited and inaccurate. The designer may wish to concentrate on information which is likely to be correctly known by the caller, depending on less reliable information only when necessary or during confirmation. In addition, some information may be intrinsically or culturally variable such as job titles and business functions.

15.6.2   Normalization of application databases

Many application databases are extremely irregular, with multiple spellings, spelling errors, variable punctuation and inconsistent abbreviations. Some normalization may be required.

15.6.3   Confidence and confirmation

In many applications the caller is asked for a series of pieces of information which are then used by the agent to conduct some database search. Often this will result in the operator being presented with several screens of information which may be intelligently browsed. Replacing this behaviour within a speech system is a complex process, which may typically result in more complex confirmation strategies and the acquisition of additional pieces of information.

15.6.4   Operator or agent interworking

The role of the operator or agent within an automated system is very important. The design of the system, and hence the dialogue and style of the application, may be biased to emphasize either or both. For example, the operator may be used as a fall-back 'helpdesk', in a 'pop-up' mode monitoring the progress of multiple calls, or as the prime focus of the application with the automatic system portrayed as a distinct phase. This leads to a spectrum from 'stealthy' automation in which the caller is unaware that the call is being fully or partially automated to more explicit systems which may be deliberately styled to emphasize the point, perhaps by some handover signal, such as ring tone or a change of voice.

15.6.5   Vocabulary processing

Particular care must be taken in the use of certain vocabularies. Name information requires additional modelling of homophones (words which sound the same but have different spellings, e.g. Miller and Millar) and synonyms (alternative words with the same meaning, e.g. Jim and James). Typically 30% of names may be homophones or synonyms of others. Place names may require complex locality modelling, as separate users may use different and overlapping terms for the same places.

15.6.6   Partial versus full spelling

Whereas it may be reasonable to expect a caller to spell their own surname, it is less likely that a caller entering the spelling of another individual's name will do so successfully. Partial spelling of the first N letters may be more appropriate, being less error prone, less variable and less time consuming. Partial spelling raises issues as to performance, resolving power and human factors.

15.7   SPEECH PLATFORMS

Speech applications vary significantly in scale and complexity, from single line systems residing in a home PC, through small office-based servers with some tens of lines and call-centre systems with perhaps hundreds of lines, to embedded network systems capable of supporting many hundreds or thousands of simultaneous calls.

In order to support this variety of services a range of platforms have become available, both at the periphery of the network and within it.

Large switches and some PABX products offer restricted TouchTone facilities. IVR platforms aim to provide a scalable, reusable and flexible component tailored to the delivery of multichannel voice solutions.

Such platforms usually provide the core components of TouchTone, speech input, output and storage and a graphical service creation environment. In addition, they support facilities for integration with application host computers and telco functions such as billing, alarm handling and service management. The BT interactive speech applications platform (ISAP) [3, 4] is such a platform now deployed within the BT network. Many of the applications discussed below are implemented on the ISAP.

15.7.1   ISAP

ISAP was designed and developed at BT Laboratories as a flexible and scaleable platform for a wide variety of services. It is now being deployed widely within the BT network supporting services such as CallMinder. It has been licensed to Ericsson Ltd for manufacture and sale. Key design criteria included:

- flexibility;

- functionality;

- scaleability;

- cost effectiveness;

- network readiness;

- support for future enhancements.

In order to achieve this, a number of design decisions were made, including:

- use of industry and open standards where available and practical;

- modular design within both hardware and software;

- common hardware base allowing highly efficient dynamic allocation strategies to be used for speech/signal processing resource management.

These decisions allow individual components or algorithms to be upgraded for enhanced performance, functionality, channel density or ongoing cost reduction without extensive system or sub-system redesign.

The ISAP is implemented using industry standard CICS, RISC and DSP components with Unix and Unix-like real-time operating systems as appropriate. The system is based upon a VME backplane architecture with an additional 120-

channel PCM speech bus providing high-bandwidth switching of speech data. Figures 15.6 and 15.7 show the high level architecture of the ISAP.

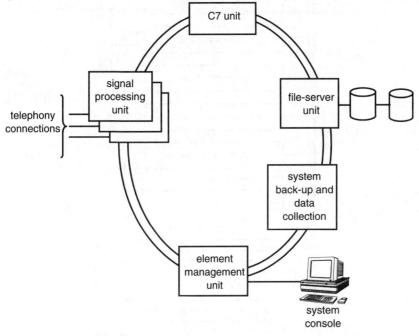

Fig. 15.6   The ISAP.

The ISAP makes extensive use of dynamic speech resource allocation in which the costly need to provision the platform *a priori* for 'worst-case operation' is reduced. For example, a platform may be running a variety of applications with vastly differing signal processing requirements, and a given application will typically vary its requirements through a dialogue. Thus it is not necessary to accommodate the most complex algorithms in the most resource-hungry combination which may be encountered (e.g. simultaneous speech recognition, recording of the user response and parallel TouchTone detection) for all channels at all times.

Resource control is achieved by a combination of dynamically switching incoming channels to processors where a required process (e.g. a signal processing function, such as a speech recognizer) is loaded but not in use, and by dynamically managing the loading of software-based processes into the processor pools to match the total requirements of the platform at any particular time.

The ISAP system software is implemented using object-oriented principles. This provides an effective approach to dealing with the complexity of large systems, and allows a high degree of abstraction, for example, to enable the choice

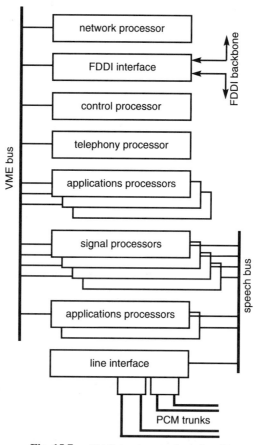

Fig. 15.7 ISAP signal processing unit architecture.

of network protocol to be changed without modifying other components, e.g. from CCITT No 7 to DPNSS.

ISAP provides a number of programmer's interfaces including:

- application programmer's interface (ISAP-API) — this provides the functionality required by the application developer;

- mangement programmer's interface (ISAP-MPI) — this provides facilities for the management of the platform and its components as well as the raising of alarms.

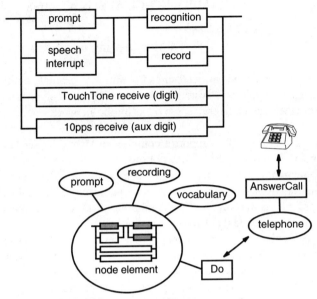

Fig. 15.8   A typical node element template.

Much of the complexity of the ISAP-PI can be encapsulated within 'node elements'. These provide templates for composite behaviour regularly needed by application developers n given domains (see Fig. 15.8).

In turn, the dialogue building blocks supported within VISAGE (see section 15.3.3 above) can be implemented as combinations of node elements.

15.7.2   Emerging standards

Standards are emerging for use in the integration of call and speech-handling functions with desktop and call-centre applications — for example Dialogics SCSA, Microsoft's TAPI and SAPI, and TSAPI from Novell and the Versit consortium. Along with the availability of low-cost speech technology and customer premises platforms, these will encourage the growth in the use of speech technology within applications at the periphery of the network as well as within it.

In many cases a designer has the choice between a customer premises-based solution and one based on a managed service provided by the network or other service provider.

15.8 EXAMPLE VOICE APPLICATIONS

15.8.1 Voice messaging

Voice-mail systems have been widespread within the corporate environment for some years now. Typically, systems have been operated either as a corporate resource or as a managed service provided by the network operator or service provider.

These systems normally support a complex structured TouchTone dialogue and are aimed primarily at the business or mobile phone user.

Speech recognition offers the potential both to simplify the user interface and to make the functionality available to a wider audience. BT's CallMinder™ service [16] provides a mass-market residential network-based call-answering system using speaker-independent speech recognition.

This is an example of how a major new service can be introduced using speech technology. Through the use of an easy-to-follow directed dialogue and a relatively small vocabulary, the system can be used by any customer without the need for a complex instruction card or TouchTone phone, although TouchTone can be operated according to user preference.

CallMinder provides a rich set of features.

- Call-answering

  One of the benefits of a network-based system over answering machines is the ability to take multiple simultaneous messages and to handle engaged lines as well as 'ring tone no reply' — an increasing issue as more residential customers use their lines for access to on-line services.

- Message retrieval

  Mailbox owners can gain access from their own phone (via a shortcode) or remotely through PIN access.

- Mailbox configuration

  Owners are, for example, able to:
  — manage their outgoing message;

  — manipulate stored messages;

  — select a fast-track message retireval dialogue for experienced user;

  — select options for the dialogue experienced by callers;

  — change PIN numbers.

- Message waiting indication

  In order to advise a customer that they have messages waiting, CallMinder changes the user's dial-tone to a distinctive 'stuttered' tone and attempts to call them at preset intervals.

  Figure 15. 9 shows the CallMinder platform. Beacham and Barrington [16] give an insight into some of the engineering and project management issues involved in launching a large, multi-site voice service such as CallMinder. Among the factors to be considered are:

- network dimensioning;

- platform dimensioning;

- network connection and resilience — each ISAP is connected to two or three main trunk exchanges via single C7 routes and multiple physical links;

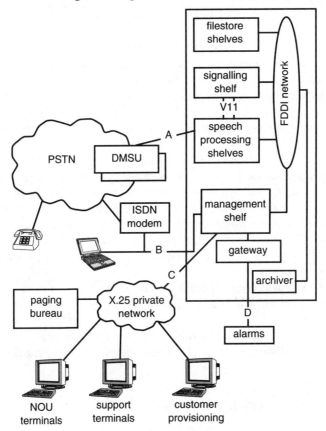

Fig. 15.9   CallMinder platform architecture.

- network operations — platforms are controlled remotely from BT's network operations units (NOUs); alarm handling is also extended remotely via the BT network alarm management system (NOMS1);

- operational management — customers are automatically loaded on to the system after requests are received on BT's customer services system (CSS), through the operations and maintenance centre systems (OMC);

- security;

- data and statistics collection;

- inter-working with other services and network features;

- the trial, pilot, delivery and launch processes.

15.8.2   Call delivery

Uncompleted calls, either through number busy or no answer, are a cause of frustration to callers and represent a loss of revenue to the operator. Systems such as CallMinder tackle this problem from the perspective of the called party; message-delivery systems approach it from the other direction by offering the caller the ability to leave a message for later delivery or distribution. Typically the system may attempt to deliver the message a number of times over a set period or at a particular time.

Payphone voice messaging is an example of this and provides a delivery facility from payphones. A system has been successfully trialled in a number of locations, including motorway service stations and major London railway stations, to assess the impact of such a service. The very nature of the service means that all calls are made from pay-phones many of which are sited in high ambient noise locations, providing a specific challenge to speech recognition. In this case, the dialogue was extremely straightforward and was implemented using TouchTone.

This application is currently being developed as a facility for BT's new cash-less services platform which is a replacement for the BT Chargecard (calling card) platform.

This is a simple example of how speech technology can begin to convert the telephone network from a seemingly simple system to a more intelligent assistant.

15.8.3   Operator assistance

BT's operator assistance centres receive over 100 million calls per year from customers requiring assistance with making telephone calls, enquiries of a general nature, or requests for a wide range of telephony services such as alarm and reverse change calls.

Orrey and Adams [17] describe the results of a number of trials investigating the use of speech technology which provides many opportunities for the streamlining of the operator assistance service. Typically, once the nature of many calls has been identified, they can be completed automatically, either through a simple message, call redirection or an automated dialogue. Examples include:

- calls to operator assistance which are normally chargeable services provided on another access number;

- calls which require the operator to act as an intermediary in a routine dialogue such as that associated with reverse charge (call collect) calls;

- advice duration and charge (ADC);

- alarm calls.

In this case, the operator selects those calls which are suitable for automatic treatment. The following dialogue fragment shows the automated response given to a caller in the case of a referral:

> *I'm sorry, this service is not available on 100. Directory Assistance is a chargeable service and can be contacted directly on 192. If you wish me to connect you, please hold on; otherwise please replace the handset.*

The handling of reverse charges is an interesting example of computer-mediated dialogue. In this case, the system dials the third party and holds an additional dialogue with them before connecting the calling and third parties.

To the calling party:

> *This is an automated service. Please state your name <long beep like an answering machine>*
> Bob
> *Thank you. After ringing there will be silence while I try to connect you. Any problems and you will be reconnected to an operator.*

To the third party:

> *Good <morning/afternoon/evening>. You have a BT automated reverse charge call from <recorded name> from <the geographic area>. If you will pay for the call, please say yes after the tone, otherwise please say no.*

To the calling party, dependent upon the response from the third party:

> *Go ahead please*

or:

> *If you wish to be connected to an operator please hold the line, otherwise please replace the handset*

or:

> *I'm sorry, the number you are calling will not pay for the call. If you require further assistance, please hold for reconnection to an operator, otherwise please replace the handset.*

The human factors trials showed that automation of such services is acceptable and appropriate. In addition, the savings in operator time shown in Table 15.2 were identified compared to manual handling.

Table 15.2   Operator time savings.

operation	saving
referrals and onward connect	10 seconds
alarm call delivery	38 seconds
advice of duration and cost	68 seconds
reverse charge calls	23 seconds

Overall, given the observed call mix, this represents an average of 3 seconds of potential savings per call over more than 100 million assistance calls per year.

15.8.4   Account management

A significant number of BT's business transactions with residential customers are conducted over the telephone by customer services advisers based in regional call centres. These transactions include telephone account management, product ordering and fault reporting. A significant proportion of these calls are low-complexity. Whittaker et al [12] describe trials investigating the application of voice automation within BT's Personal Customer Division in support of customer service operations.

The ACE (automated customer access) platform is aimed at automating such calls, enabling advisers to concentrate on work of higher value, both for customers and BT. Figure 15.10 shows the platform architecture and can be compared with Figure 15.5.

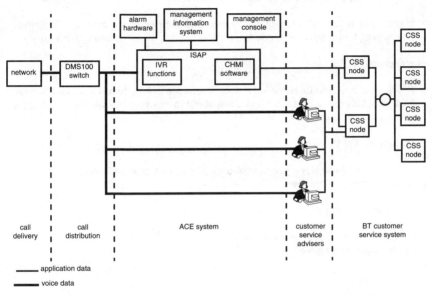

Fig. 15.10   ACE platform architecture.

Many of the activities carried out by the customer service advisers involve interaction with BT's customer service system (CSS) database. CSS maintains information on every BT telephone installation in the UK and incorporates processes for transactions such as ordering of new network products and the retrieval of customer installation details.

The ACE system integrates with CSS to perform many of these functions without human intervention.

A number of applications were trialled including 'Paid-My-Bill'. There are many reasons why payment of a bill may not have been received by the due date. In such cases, a reminder letter is issued with a telephone number for account queries. A filtering system based on TouchTone was implemented which prompts the caller as to whether they have paid their bill or if they wish to speak directly to an adviser. If they elect for the latter option, they are transferred to the appropriate regional customer service centre, which is identified from the number initially dialled. For other callers, the system checks the status of the customer's account on CSS. If the account is clear, the customer is informed, otherwise the details are processed and the account handled in line with normal business processes. Figure 15.11 provides a typical dialogue fragment.

Good morning, welcome to BT's payment service.

If you have a tone dialling phone, I can deal with your call quickly. Otherwise, please wait for an adviser. If you have a star button, please press it twice now.

<caller enters '**'>

Thank you. You now have three choices. If you are calling to say you have already paid your bill please press '1'; if you are about to pay your bill, please press '2'; to speak to an adviser, please press '3'.

<caller enters '1'>

Again, using the buttons on your phone, please key in the full telephone number including the dialling code.

<caller enters a telephone number>

I recognized the phone number as <number>. Using '1' for yes, and '2' for no, is this the correct number?

<caller enters '1'>

Please wait a moment while I look up your account details. Please wait a moment longer. You are not being charged for this call.

I have checked your account, and it is clear. I have entered your call on our records.

Thank you for calling BT. If you would like to know more about our bill payment options and discount schemes, please call us free on 150.

Goodbye.

Fig. 15.11   Example 'Paid-My-Bill' interaction.

15.9   NEXT GENERATION SYSTEMS

This section describes some of the prototype systems which have been developed at BT Laboratories, and which give some indication of the way in which speech technology can be used to approach the development of advanced applications.

15.9.1   PaymentLine

'PaymentLine' was developed as a prototype system on which to develop technologies for advanced applications such as home shopping and customer handling. It demonstrates some of the generic aspects of automated call centres and illustrates the automation of customer bill payment. Customer identification is a problem common to many applications. For each customer, the database contains:

- an identification number, name and address;

- payment methods used to date;

- financial information relating to the account.

The system aims to:

- fully automate all calls (i.e. no operator involvement in the majority of calls);

- allow the caller to select a new method of payment and make specific payments;

- record 'problem' calls for later off-line transcription.

A sample dialogue is shown below:

Good evening. Welcome to BT's PaymentLine. First of all, I need to find out who you are. Please say the last 8 digits of your reference number:
Double-8-3-7-double-3-double-7
After the tone, please give your surname:
Smith
Is your surname Smith?
Yes
Thank you. Please wait while I check your details. The full balance of your account is £43.71
...
Please let me know how much you would like to pay by saying full balance, monthly payment or other amount.
...

15.9.2   Directory automation

Access to directory information is a key issue for many companies, both corporate directories within companies and directory assistance information in the case of network operators. Speech technology can be used to supplement existing methods for accessing directory information.

A series of demonstration and trial systems have been developed at BT Laboratories to investigate the feasibility of various approaches to the full and partial automation of directory assistance (DA) [1].

In the UK, DA is a chargeable service with a database of 20M residential business and government entries involving half a million distinct surnames and nearly 30K distinct localities of differing sizes. One such prototype dialogue is shown below and shows the use of large-vocabulary spoken and spelt recognition over the telephone network to access a subset of the database:

Directory Enquiries. Which town please?
Martlesham
Do you know the road name?
Yes
Please say the complete road name
Viking Heights
And now say the surname
Foster
And now spell the surname
F-O-S-T-E-R
Is the surname Foster?
Yes
The number is ...

Due to the complexity of this application and the size of the vocabulary and application database, the automation of directory assistance will remain a significant challenge to speech recognition for some time.

Chester *et al* discuss the design and implementation of an advanced architecture for the development of complex large vocabulary application such as directory automation [13]. It takes as a study case the Brimstone corporate directory trial [18].

15.10   CONCLUSIONS

This chapter has discussed various aspects of the development of interactive speech systems with a particular emphasis on telecommunications applications. In addition a number of example services and advanced prototypes have been described.

Speech technology offers great promise for high quality user interfaces to a wide variety of applications and services. It is rapidly becoming a vital enabling technology for a wide range of business and consumer applications, enabling services to be made available rapidly to large markets.

Although, it is widely agreed that there have been few substantial theoretical breakthroughs in speech recognition technology within the last decade, the art of engineering high-quality speech systems has reached the level at which it now has a major impact on product and systems development.

REFERENCES

1.   Whittaker S J and Attwater D J: 'Advanced speech applications — the integration of speech technology into complex services', ESCA Workshop on Spoken Dialogue Systems — Theory and Applications, pp 113-116 (1995).

2.   Southcott C B and Whittaker S J: 'Advanced network based speech applications', AVIOS (1994).

3.   Rose K R and Hughes P M: 'The BT speech applications platform', British Telecommunications Eng J, 13, pp 304-31 (January 1995).

4.   Hughes P M et al: 'The interactive speech applications platform', BT Technol J, 14, No 2, pp 24-34 (April 1996).

5.   'Now You're Talking... Voice Services', BT Publication (1994).

6.   Schmidt M et al: 'Dialogues 2000. Towards the Introduction of best practices in the design of automated telephone services', Voice95 (1995).

7.   Dialogue Engineering at the Centre for Communiction Interface Research, Edinburgh University: http://www.ccir.ed.ac.uk/dialogue/diaeng1a.html

8.   Bruce I P C: 'Engineering an intelligent voice dialogue controller', BT Technol J, 5, No 4, pp 61-69 (October 1987).

9.   Hanes R B *et al*: 'Service creation tools for creating speech intractive services', EuroSpeech 1993, Berlin (1993).

10.   Chester A *et al*: 'Service creation tools for speech interactive services', BT Technol Journal, 14, No 2, pp 43-51 (April 1996).

11.   Sidhu C K and Coyle G: 'Usability engineering of speech based services', British Telecommunications Eng J, 14, Pt 4 (January 1996).

12.   Whittaker S J *et al*: 'Improving customer service through speech technology', BT Technol J, 14, No 2, pp 75-83 (April 1996).

13.   Attwater D J and Whittaker S J: 'Large vocabulary access to corporate directories', Institute of Acoustics Autumn Conference on Speech and Hearing, Windermere (1996).

14.   Bonner M: 'Call centres — doing better business by telephone', British Telecommunications Eng J, 13, Pt 2 (July 1994).

15.   Catchpole A *et al*: 'Introduction to computer telephony integration', British Telecommunications Eng J, 14, Pt 2 (July 1995).

16.   Beacham K and Barrington S: 'CallMinder — the development of BT's new answering service', BT Technol J, 14, No 2, pp 52-59 (April 1996) .

17.   Orrey D A and Adams L R: 'Operator assistance voice automation trials' , BT Technol J, 14, No 2, pp 69-74 (April 1996).

18.   Whittaker S J and Attwater D J: 'The design of complex telephony applications using large vocabulary speech technology', International Conference on Spoken Language Processing (ICSLP), 2, pp 705-708 (1996).

16

ISSUES IN LARGE-VOCABULARY INTERACTIVE SPEECH SYSTEMS

D J Attwater and S J Whittaker

16.1　INTRODUCTION

Telephone-based interactive voice response (IVR) systems have been available for some time in such areas as call control, voice mail, and simple information retrieval tasks. In many cases they have been hugely successful in reducing the cost of providing telephone services and have consequently enabled a number of new services. Such speech systems are usually based upon dual tone multifrequency (DTMF) telephone keypad entry. This provides the caller with the ability to either perform a menu-style navigation through a structured speech dialogue, or to enter numeric information, such as telephone numbers or account numbers. Recorded speech phrases are usually used as the method of prompting the caller or for presenting information to them.

　　Recent developments in speech recognition technology have facilitated the introduction of small vocabulary (tens of words) speech recognition into network-based services. The recently launched BT CallMinder™ product, a network-based answering machine, is an example of the effective use of this. It incorporates 'yes/no', isolated digit and keyword recognition into a highly usable service. This replaces the need for callers to remember DTMF key sequences, which can be daunting for some customers. Many potential applications, however, require even more versatility in their caller interface. The entry of names and addresses, for example, is difficult with DTMF requiring that telephone keypads are labelled with the letters of the alphabet — each key representing a group of letters. This is inherently an ambiguous form of entry, and many customers in the United Kingdom do not have alphabetic labels on their telephones.

Recognition technology is maturing quickly bringing improvements in the accuracy and flexibility of large-vocabulary speech recognition. This, coupled with the continued trend towards faster, cheaper computing platforms, has made the use of large-vocabulary speech recognition in interactive network services a real possibility.

By developing an automatic corporate directory enquiry system giving access to thousands of telephone numbers, BT Laboratories have demonstrated that the access of databases over the telephone using large-vocabulary speech recognition is now viable.

This chapter discusses the issues that surround the development of such systems, and describes techniques which address these issues.

16.2   LARGE-VOCABULARY IVR SYSTEM ARCHITECTURE

Figure 16.1a shows a system architecture for a traditional IVR system which needs to access an application database.

The 'dialogue manager' component defines the interface as seen by the caller. It is responsible for the sequencing of dialogue events, such as the playing out of prompts, and initiating speech recognition. In simple IVR applications the dialogue manager makes database searches directly using the results from DTMF entry or simple speech recognition.

The 'speech platform' component provides the speech functions required by the dialogue manager. It contains all of the speech recognition algorithms, DTMF detection, speech recording, dialogue prompt playback, and a telephony interface.

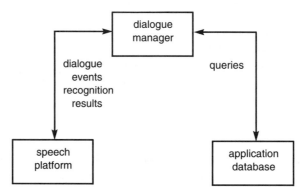

Fig. 16.1a   Traditional IVR architecture.

The 'application database' component defines all of the relationships between pieces of information used in the application. Commercially available database engines may be used for the application database component.

As this chapter will show, it is desirable for systems requiring large-vocabulary speech access to databases to shield the dialogue manager from the need for direct application database interaction. This is because tight coupling is often required between the database and the speech recognizers in order to constrain the recognition vocabularies to the most likely anticipated responses.

In order to abstract this function from the dialogue manager, a new 'information manager' component may be added to the architecture as shown in Fig. 16.1b. This component is responsible for mediating high-level strategic requests from the dialogue manager and converting them into tactical actions. This will involve maintaining alternative hypotheses, querying the database for speech recognition constraints, and preparing speech recognizers with these constraints. The information manager will prepare information such as confirmation items for the dialogue manager to offer to the caller. It will also ensure that vocabulary aliases are correctly managed throughout all of these actions.

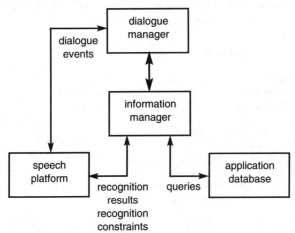

Fig. 16.1b   Large-vocabulary IVR architecture.

In large-vocabulary IVR systems additional speech platform features are required over and above those provided by traditional IVR platforms. These include:

• large-vocabulary, speaker-independent, sub-word-based speech recognition;

• dynamic recognition vocabulary modification and/or dynamic vocabulary generation in real time;

- text to speech.

Current large-vocabulary IVR development at BT Laboratories uses the ISAP speech platform [1] which meets all of these requirements.

16.3   CORPORATE DIRECTORY ENQUIRY SYSTEM

BT Laboratories has produced a demonstration system providing access to a corporate telephone directory covering several thousand BT people from the Laboratories and the surrounding area.

The system was built to demonstrate the feasibility of large-vocabulary IVR applications, and to provide a test bed on which to explore the issues involved. Emphasis was placed on producing a robust, usable service. Application independence was also sought for as much of the system design as possible.

This system provides prompt and accurate telephone number information for people, given only the first name and surname. It ensures that synonyms such as 'David' and 'Dave', or homophones such as 'Bayley' and 'Bailey', are considered to be equivalent during database access and in the subsequent confirmation and offering stages.

An example dialogue from this system is presented in Fig. 16.2.

BT Labs directory. Which surname please?	Bailey
How do you spell that?	B-A-I-L-E-Y
Please say the first name	David
Is the name Dave Bailey?	Yes
I have three entries matching that information,	
The number for Dave Bailey of	
Networks and Systems is 01473 64xxxx	
The number for David Bayley of Facilities	
and Technical Management is 01473 64xxxx	
The number for David Bailey of Business	
Operations is 01473 64xxxx	
I'll repeat that ..	

Fig. 16.2   Example corporate directory system dialogue.

The issues that surround the development of such a system will be discussed in the following sections.

16.4   DIALOGUE DESIGN AND HUMAN FACTORS

The dialogue design of a large-vocabulary IVR application can have a dramatic impact on the performance of the system, given a particular underlying recognition performance [2], and consequently its acceptability to callers.

16.4.1   Choosing vocabularies

In order to keep a dialogue brief and reliable, it should start by asking for information that is accurately known by most callers. Callers must also know how to express this information when asked for it. Once all of the well-known information has been presented, the dialogue can proceed to ask for less well-known information, if this is needed to satisfy the enquiry. In these instances, a 'yes/no' question may be required to check whether a particular piece of information is known prior to asking for it.

The suitability of a particular type of information for input should also consider how many ways a given item or concept can be expressed. Vocabularies with a single form for each identifiable concept are highly predictable. An example of such vocabularies is postcodes. Some vocabularies are very difficult to define, e.g. business areas in a corporate enquiry system, as separate callers may use a wide range of different vocabularies to describe the same item. It is very difficult to build accurate speech recognizers for these unpredictable vocabularies.

Spelt or numeric vocabularies are generally highly predictable. If the information for that vocabulary is also well known for the given application, then it will probably provide a sound starting point on which to base an enquiry. Examples of such systems include financial systems where knowing a personal account number is a reasonable prerequisite for using the system, or postcodes and spellings in applications requiring self-identification of the caller.

16.4.2   Message design for predictability

The design of the messages prompting the caller for desired information will itself affect the predictability of the response received. The use of clear, closed questions is sufficient to constrain many callers to present concise answers without extraneous speech.

Early prototyping and small-scale trials of systems are essential in eliminating weaknesses in the dialogue design.

16.4.3   Mixed modality

There are several different modes of data entry available to the designer of a large-vocabulary IVR system. Among these are DTMF, large-vocabulary spoken recognition, large-vocabulary spelt recognition, and the use of 'yes/no' or DTMF response for menu selection. All of these may be mixed together in a single application, but it is important to remain consistent in approach throughout the dialogue.

For example, in systems which mix spoken and spelt responses, experiments have shown that callers may be confused with dialogues, such as the one shown in Fig. 16.3.

| Please spell the surname | F-O-S-T-E-R |
| Now, please say the first name | David |

Fig. 16.3   Example from system using mix of spoken and spelt responses.

With this dialogue, callers may have a tendency not to listen closely to the prompts, and use the modality of a previous question to answer the current one. So, in the example in Fig. 16.3, the first name may well be spelt rather than spoken.

If spelling is to be mixed with spoken responses, therefore, then it is advisable to always ask for the spoken word before asking for its spelling, even if the spoken response is never used by the system. Thus, the example in Fig. 16.3 would become as shown in Fig. 16.4.

Please say the surname	Foster
And now, please SPELL the surname	F-O-S-T-E-R
Now, please say the first name	David

Fig. 16.4   Improved configuration for example of Fig. 16.3.

Similar issues surround the mixing of DTMF with other modalities.

DTMF entry may be successfully used as a simultaneous alternative to speech input. In these instances, hybrid dialogue prompts, such as 'Please say or use the keypad to enter the account number', should be avoided. The prompt should only ask for a spoken response leaving DTMF as a fast-track option for experienced users who have found out about it from another source. BT's Call-Minder is a successful example of this approach permitting the digits '1' and '2' to be used instead of 'yes' and 'no' if desired.

16.4.4   Confirmation and offering

It is often helpful to confirm the caller's request with them before offering the desired information, or service, related to the request. Such confirmation ensures that recognition errors are detected by the dialogue, and it also helps to orient the caller within the dialogue in preparation for the next phase.

Confirmation enables the caller and dialogue to proceed with confidence. It may not be necessary to confirm all of the information entered by the user, and it is acceptable to ask for further clarifying information after confirmation has occurred. Consider the corporate directory dialogue shown in Fig. 16.5.

Please say the surname	*Foster*
And now, please SPELL the surname	*F-O-S-T-E-R*
Now, please say the first name	*David*
Please say yes or no,	
is the surname Foster?	*Yes*
I have a number of entries	
matching that information.	
Please answer yes or no,	
do you know the town name	*Yes*

Fig. 16.5   Example of mid-query confirmation.

In the example of Fig. 16.5, the caller is given confidence that the enquiry is proceeding well, and the system is now sure that the surname is 'Foster'. If the wrong surname is offered at this point, the caller is able to realize that the enquiry is in trouble prior to engaging in further lengthy dialogue to gather extra information.

Experience has shown that it can be confusing to expect callers to confirm information which they have not offered. For example, the dialogue shown in Fig. 16.6 may be confusing to the caller.

Please say the surname	*Foster*
Please say yes or no,	
is the first name David?	*Yes*

Fig. 16.6   Example which may cause confusion to the caller.

The exact words used for confirmation can be important. The choice of these words can come from two origins. Firstly, the spoken vocabulary, which was originally recognized, could be used directly. This has the advantage of confirming with the user the exact word which was spoken. Alternatively it may be necessary to confirm using portions of database entries which have matched the spoken information. This method has the disadvantage that if there are synonyms for the given vocabulary, the caller may be asked to confirm information using a synonym of the word they originally spoke rather than the word itself.

In the final stage of a dialogue, information is often offered to the caller in response to their enquiry, e.g. in the corporate directory example a telephone number must be offered. The issues surrounding offering are similar to those of confirmation.

16.4.5   Distinguishability

It is sometimes desirable to confirm or offer items from a list of potential responses. In these instances, it is not wise to confirm lists of more than three items. The system must ensure that it does not try to draw distinctions between things which are indistinguishable to the user.

For example, consider the confirmation sub-dialogue in Fig. 16.7.

Please say yes or no,
    is the name David Bailey?   *No*
Then is the name Dave Bayley?   ....

Fig. 16.7 Example of an inappropriate confirmation sub-dialogue.

In the instance shown in Fig. 16.7, it is inappropriate to expect the caller to draw a distinction between the two surname homophones 'Bayley' and 'Bailey'. It might also be inappropriate to expect the caller to distinguish between the synonyms 'David' and 'Dave'.

In order to avoid this, systems should test whether spoken vocabulary items share common pronunciations or have potential synonym confusions.

If an item is positively confirmed by the caller, all other items considered indistinguishable from that item should also be considered confirmed. This will mean the addition of homophones, and possibly synonyms also, prior to any re-searching of the database using the confirmed information.

16.4.6 Speech generation

Large-vocabulary speech output may be handled in a few different ways. One simple way to achieve it is to make recordings of each vocabulary item that may need to be spoken out by the dialogue. This is time consuming, and difficult to maintain, but it can produce a very high quality interface. This approach is suited to applications whose vocabularies are small to medium sized, and change slowly, if at all. One example of this may be a railway timetable enquiry system where the vocabulary of towns with stations changes very slowly. These towns, and the component words of dates and times may be recorded. Careful concatenation of these recordings in different contexts is able to provide a high-quality interface.

For larger vocabularies, text to speech (TTS) is the only viable alternative. The designer must then choose whether all of the dialogue prompts are to be TTS generated, or simply the variable fields. Many recent TTS systems, including BT's Laureate, can be based on a particular individual's voice. This enables TTS-generated variable fields to be spliced into recorded sentences from the same speaker. Care must then be taken with the prosody and intonation of the resulting sentences.

16.5   VOCABULARY DESIGN AND DATA MODELLING

16.5.1   The vocabulary model

As has been seen already, a database entry may be referred to using many different forms. Often, the caller cannot be sure that the word or spelling that they know for a certain piece of information is the form which found its way into the database.

Figure 16.8 shows a vocabulary model, a generalized entity relationship (ERE) diagram relating together the different representations used for a given vocabulary in large-vocabulary systems. These representations are shown as 'spoken vocabulary', 'spelling vocabulary', 'pronunciations', and 'database representation'.

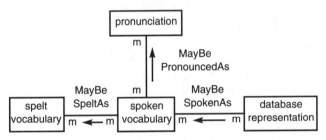

Fig. 16.8   ERE diagram relating different representations of vocabulary items.

The database representation is the form in which an item is stored in the database. It is the relationship between various concepts that the applications database is modelling. These database representations may be referred to using many different spoken words for a given item. Each spoken item has a spelling, but there may be more than one 'valid' spelling of an item in an application if callers are permitted to spell homophones or make common spelling errors. For this reason, spelt vocabulary representations are considered to be separate from spoken representations.

Each word will have a pronunciation, but there may be more than one pronunciation for a given word (homonyms), and different words that share pronunciations (homophones).

16.5.2   Data management

Given the rich set of representations and relationships required by large-vocabulary applications, data preparation and management are important issues.

Figure 16.9 shows a typical data preparation process required to get from the source database text to the spoken, spelt and pronunciation forms as defined in Fig. 16.8. Figure 16.9 only shows the preparation of the sets of valid tags for the entities of Fig. 16.8. The process also implies the definition of the relations shown in Fig. 16.8.

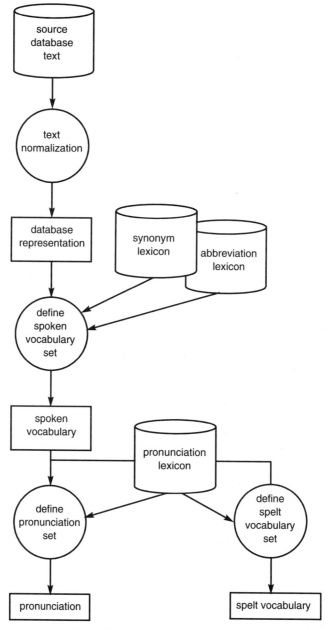

Fig. 16.9   A typical data preparation process.

In text-based information systems, anomalies in the text stored in the database may often be compensated for by the re-keying of alternatives and visual browsing by the user. This is especially true of name and address information where the original data entry may have introduced spelling errors, or homophones of the correct words.

In large-vocabulary IVR applications, the information in the database must be associated with the vocabulary with which to describe it, and the valid pronunciations of this vocabulary. This requires the use of standardized synonym, abbreviation and pronunciation lexicons. Thus, text normalization is often required on the raw database information in order to establish a regular set of database representations against which vocabularies and their relationships to the database may be prepared.

In the building of large-vocabulary recognizers from sub-word units, the quality of the pronunciation lexicons is often a significant factor affecting the recognition performance. As large-vocabulary IVR systems become more widespread, the management of corporate lexical information such as pronunciations and synonyms will become an increasingly important issue. The EU Onomastica project, which derived accurate pronunciations for many names, roads, towns and businesses in Europe attempted to anticipate this [3].

It is possible to set a balance between how much data preparation is performed off-line from the live application, and how much is performed dynamically. This is a design decision driven by the characteristics of the target application platform and the data processing requirements.

16.6   RECOGNITION TECHNOLOGY

Accurate speech recognition is essential if reliable services are to be offered. Large-vocabulary speech recognition performance is affected by several factors, including:

- vocabulary — the number and similarity of words or phrases in the target vocabulary;

- training — whether the speaker has trained the recognizer prior to use;

- channel — effects of the channel (bandwidth, etc) and background noise;

- predictability — the predictability of response from the caller;

- pronunciations — the accuracy of the pronunciations upon which sub-word recognizers are based.

In network-based services with large vocabularies and an open caller group, all of these factors mitigate against accurate speech recognition. By way of example, dictation systems in which the user pauses between words can achieve a word accuracy in excess of 90% for vocabularies of tens of thousands of words. In order to achieve this, they must be trained by the particular user, use a high-quality microphone and use knowledge of the language being spoken to assist the recognition.

In contrast, speaker-independent, isolated-word recognition over the telephone network for some vocabularies will give comparable performance for vocabularies comprising only tens or hundreds of words. Fluent spelt and alpha-numeric recognizers can work considerably more accurately over the network under the same conditions, giving comparable performance to above for vocabularies of thousands of words.

The need for careful engineering of systems with such performance is evident. It has been the approach of the authors to design for high application performance, attempting to prevent, hide, or minimize the impact of speech recognition errors. In order to achieve this, it is often helpful to use contextual constraints on the recognizers throughout all stages of the enquiry, and to maintain many alternative speech recognition hypotheses in the enquiry until corroborating information can be used to choose between them. The next section discusses these techniques in more detail.

16.7   MAINTAINING UNCERTAINTY AND IMPOSING CONSTRAINTS

16.7.1   The benefits of uncertainty

Many current speech recognizers do not simply return a 'best guess' candidate from a recognition, they can return several. This is termed the 'N-best' list of candidates. Generally, each candidate word in this N-best list will have associated with it a score to indicate a confidence relative to the other words recognized.

The maximum number of candidates which may be returned will depend on the computational cost of keeping multiple word hypotheses active to the end of the recognition. It may also depend on certain rejection criteria in an attempt to return only credibly similar words from the recognition.

Where there are a number of words which are acoustically similar, it is likely that the correct word may be present in the N-best candidate words, but incorrectly ranked. Figure 16.10 shows the typical performance trend of a best-1 recognition, and a best-5 recognition for a spoken large-vocabulary network-based speech recognition task.

In some interactive speech systems it is not necessary to know the exact vocabulary item entered at the moment it is presented. This means that a number of uncertain candidates may be retained, and corroborative information may be used later to re-rank the candidates giving an opportunity for error recovery.

As a simple example, consider a directory access system. The caller is asked to state the first name, and then asked to state the surname. Two independent recognitions of these words may be performed, keeping the N-best candidate names from each recognition. A database search may then be performed finding all of the entries which match one of the first names in the list and one of the surnames in the list. It is probable that, due to redundancy in the database, only certain combinations of first names and surnames will be present in the result. If any matching pairs are located, then they may be rank ordered according to the scores of both of the candidates which matched, and presented to the caller for confirmation. In this simple example, a considerable application performance gain may be achieved over using the best-1 candidate alone for the search.

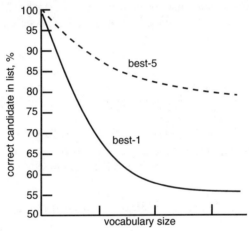

Fig. 16.10   Best-1 and best-5 recognition performance versus vocabulary size for a typical large-vocabulary speech recognition task.

16.7.2   Constraining speech recognizers

The computational cost of large-vocabulary speech recognition often demands that the pruning of unlikely paths in the recognition network is performed throughout the matching of the word [4]. This can sometimes result in the correct candidate being pruned out early on in the recognition due to a poor match at the start of the utterance. It cannot then be recovered, even if the final recognition score would have been high.

In the example system described in the section above, this kind of error cannot be guarded against, and pruning errors will occur. This kind of effect is also noticeable in systems where acoustic fast-matches are used to reduce the search space of a recognition prior to a detailed acoustical match.

As already mentioned in section 16.6, the recognition performance is directly related to the degree to which one may predict, *a priori,* a caller's response to a question. If it is known that certain words are more likely to be said than others prior to an attempt to recognize them, then the recognizer can be weighted towards these words making it less likely that these words will be pruned out during the pattern match, and making them more likely to be recognized.

In some cases, as will be discussed in the next section, it is sensible to 'turn off' certain vocabulary items altogether. That is to say, inform the recognizer that there is no possibility that the particular word will be spoken. This mechanism is termed 'subsetting' throughout this chapter. The benefit of vocabulary subsetting is that pattern matching effort is not wasted on improbable vocabulary items, and hence for a given computational resource, more effort can be channelled into recognizing the word against a smaller set of reference models, more quickly, or with a higher accuracy.

With this technique, different weightings may still be applied to the remaining vocabulary items, termed 'soft subsetting', but this may not be desirable or possible with some speech recognizers, and a simple on/off constraint may be used, the surviving vocabulary items being treated as equally probable. This is termed 'hard subsetting'.

The use of *a priori* constraints on a pruning recognizer is a common approach in speech recognition, known generally as language modelling. The next section discusses the use of the database itself as a language model.

16.7.3   Database constraints

In large-vocabulary IVR systems, a language model may be formed dynamically from the contents of the database itself and applied as *a priori* constraints on the speech recognizers. These constraints are derived both from previous recognition results and the known database contents, and are termed database constraints. This can provide a powerful method for reducing recognition pruning errors, reducing vocabulary sizes, and speeding up application performance.

The use of database constraints depends on the assumption that the caller is attempting to obtain an entry which is represented in the database. Figure 16.11 illustrates the use of database constraints by reconsidering the corporate directory example discussed in the previous sections.

Consider the enquiry as a process of gradually closing down the number of possible database entries that the caller may be trying to access until a single entry may be selected with a certain degree of confidence. Initially, any one of the entries may be desired by the caller with equal probability. This hypothesis is labelled 'I' in Fig. 16.11.

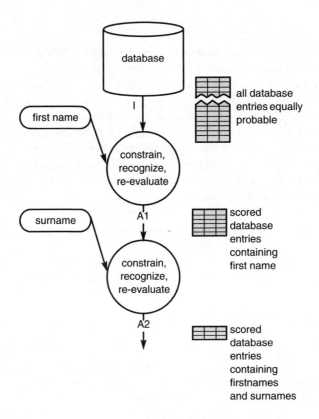

Fig. 16.11   Using database constraints to improve application performance.

The first name recognizer is then constrained by this hypothesis, which in this case means that the recognizer uses all of the valid vocabulary represented in the database. So far, this is no different from the previous example. The first name is then recognized, and the database entry hypothesis is re-formulated to keep all entries which match one of the N-best first names, and discard those that do not. Of those that match, they are assigned relative scores depending on the recognition scores. This new hypothesis is labelled A1 in Fig. 16.11.

Now, under the assumption that this recognition contained the correct result, the surname recognizer is then constrained to only recognize surnames which are still in A1. The surname is then recognized, and the database entries are re-scored again if they match one of the N-best surnames, or further entries are discarded if they do not. The resulting entry hypothesis is labelled A2.

A test may then be made on A2 to see whether the potential entries are now so few in number that the most probable of them may be confirmed with the caller. If there are too many still, then further information may be sought, for example, a town or building name.

This system cannot have worse performance than the previous example, and may well have better performance in terms of the proportion of correct entries that are offered. In some cases, this method can have dramatic performance gains. However, because of the assumptions on which it is based, this approach is sensitive to recognition errors in previous vocabularies, and it is also sensitive to callers presenting information which is not present in the database (e.g. the name of a colleague who has left the company). In either case, it may on occasion return database entries to the caller which bear no similarity to the original information. The following section discusses these weaknesses in more detail, and proposes mechanisms for avoiding them.

16.7.4   Alternative hypotheses and error recovery

The example presented in the previous section represented a single track approach to accessing the database. In that system it was assumed at all stages that:

- the previous recognition was correct,

- the caller presented information that was represented in the database.

In real systems, both of these assumptions have limited validity, and the application can be made much more robust to errors by attempting to detect when mistakes have been made and correct for them.

Consider the corporate directory system of Fig. 16.12. The derivation of A2 is the same as described above. Assume, however, in this instance that the surname recognizer used to derive A2 indicated that an out-of-vocabulary item was detected. This may suggest that the caller presented a first name/surname pair that was not in the database, or it may indicate that an error was made recognizing the first name. By re-recognizing the surname without the first name constraint, a new estimate of the entries which are desired may be derived. This is shown as B1 in the diagram. There is now an alternative, possibly lower confidence, hypothesis track which the application may pursue. This may be explored in a similar manner to the previous track until it is either found to lead to the correct solution, or it too is found to be lacking.

This is an example of the use of an out-of-vocabulary rejection and an alternative lower confidence back-off strategy in an attempt to recover from an error. Alternatively multiple entry hypotheses may be compared with one another to cross-check that the enquiry is proceeding well.

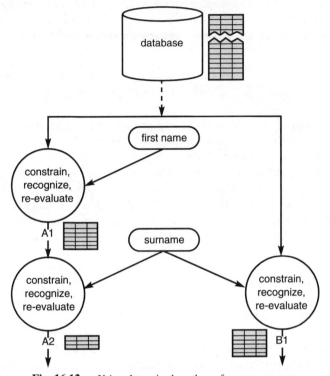

Fig. 16.12   Using alternative hypotheses for error recovery.

Often, all of these approaches can be used together, with great effect, to provide error recovery and give indications of confidence.

16.7.5   Spelling and alphanumerics

As previously mentioned in section 16.6, the recognition of fluently spoken alphanumerics is generally more reliable for a given vocabulary size than spoken recognition. This fact, coupled with issues of predictability discussed in section 16.4, mean that alphanumeric entry can provide a very reliable first-level constraint in some large-vocabulary IVR systems.

Examples of the effective use of alphanumerics include the customer account numbers, or the spelling of a surname. If the item is entirely numeric, e.g. a telephone number, then DTMF may provide a totally reliable form of entry. Subsequent recognitions may be constrained by this information using a database constraint.

Of specific interest is the speaking and spelling of the same vocabulary item, for example a surname. Where this is acceptable to callers, using the techniques already discussed, combined spelt and spoken recognition may be made very reliable.

16.7.6   Application usage constraints

The application of constraints on the recognition need not be based on previous recognition results. They may be based on any statistical knowledge that is available to the application designers.

For example, an application where this may prove useful is catalogue shopping where customers are identifying themselves by their own name and address. Of the tens of thousands of towns in the UK, the majority of callers may be concentrated in the top few cities and towns, the remaining towns being represented with decreasing frequency. Figure 16.13 shows a typical town vocabulary distribution that such an application may display.

This knowledge may be used in a similar way to other constraints. One approach may be to use a full vocabulary recognizer with weightings reflecting the sorts of distributions shown in Fig. 16.13. If carefully balanced and tuned, such a system could provide improved average application performance while still enabling enquiries in the statistical tail of the graph to succeed.

An alternative approach is to deliberately exclude the unlikely vocabulary entries (i.e. use a soft or hard vocabulary subset), for example, keeping only the towns to the left of line A on Fig. 16.13. This is termed 'partial vocabulary recognition'. By excluding the tail of the distribution, the recognition performance on the top few likely words will be very reliable. Out-of-vocabulary rejection on such a small vocabulary may be reliable enough to detect when callers are presenting information in the tail of the distribution and react accordingly, for example asking for a corroborating spelling of the vocabulary item.

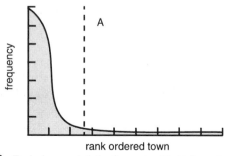

Fig. 16.13   Typical usage statistics for geographical information systems.

16.8   ENGINEERING LARGE-VOCABULARY DATABASE ACCESS SYSTEMS

16.8.1   Information manager design

Section 16.2 discussed the role of an information manager in the system architecture of large vocabulary IVR systems. Figure 16.14 shows a general-purpose design for an information manager. As stated above, the purpose of the information management component is to translate high-level strategic information requests into tactical operations. As an enquiry progresses, the information manager should have an ever-narrowing idea, with associated confidence levels, of which specific entries are being requested.

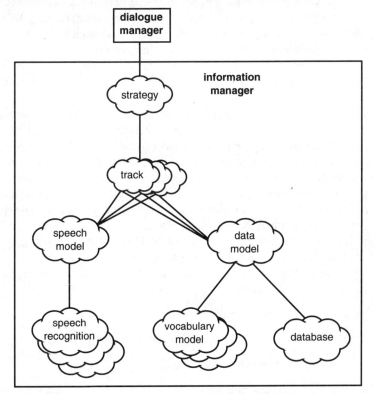

Fig. 16.14   Information manager design.

Typical services provided by the information manager to the dialogue manager are:

- note the *a posteriori* recognition results in order to narrow down the search space for all of the following tasks;

- prepare the recognizers with the most appropriate *a priori* constraints, given selected recognition results to date, and the database contents [5];

- recommend whether enough information has been gathered and the system is confident enough that confirmation may be attempted or that one or more entries can be offered;

- present these entries or partial entries for confirmation or offering.

In order to achieve this, the information manager requires a 'strategy'. This will be specific to the task which is being attempted and will manage a number of competing database hypotheses, named 'tracks'. Probabilistic constraints, such as speech recognition results, or other application information, may be selectively added by the strategy on to specific tracks. Thus each track represents a view into the application database, with each entry or part entry having a confidence score. Single tracks or combinations of them may then be selected by the strategy in order to satisfy requests from the dialogue manager as described above.

Tracks can be manipulated straightforwardly and combined to perform complex processes. In general, an application will maintain a primary track hypothesis, spawning off competing tracks at points of uncertainty in the enquiry and subsequently choosing one track or combining more than one track together when a more accurate decision may be made as to which track is maintaining the more probable constraints. For example, in the corporate enquiry application, the spelt and spoken surname recognition results may be maintained in two separate tracks until a decision may be made as to which track (it may be both or neither of course) is likely to contain the correct result. This process can be largely independent of the ordering of dialogue prompts.

In order to perform this task satisfactorily, tracks require knowledge of the following:

- the identity of the speech recognizers;

- the relationships within vocabularies (such as synonyms and homophones);

- the core database that is being accessed.

The first of these functions is provided by the speech model. This model simply relates the fields in the database to the specific speech recognition technology available in the application.

The second and third functions are provided by the data model. This contains a number of vocabulary models, each associated with one vocabulary within the application, and the core database. As already discussed, the vocabulary models distinguish between items of vocabulary as represented in the database and the alternative ways in which they may be spoken and spelt by the caller. The data model also manages the link between the vocabulary models and the core application database.

16.8.2   Database integration

The techniques of using database constraints, as discussed in section 16.7, may require high-performance database engines to provide real-time response. This is one limitation on the use of the technique, and presents serious engineering problems for multichannel systems which are to interface into existing corporate databases [6].

It may be necessary to cache portions of the database, or to use more loosely derived statistical models of the database in order to get real-time constraint performance.

Local data caching is realistic for read-only databases with slow update cycles, but more sophisticated approaches will be required for multichannel access to fast-changing information sources, such as airline information systems.

16.8.3   Dynamic vocabulary modification

Section 16.7 discussed the benefits of vocabulary subsetting in speech recognizers. There are several different ways in which this can be implemented in practice, some of which are listed below:

- multiple recognition vocabularies, which have been pre-built for each expected context, are used — this approach quickly becomes unmanageable for even small systems;

- dynamic *a priori* probability modification of an existing large-vocabulary recognition vocabulary can be achieved in real time, and is a very effective approach for large vocabularies — Chapter 11 describes how BT speech recognizers implement this technique for finite state grammar recognizers (this can provide soft subsetting);

- dynamic vocabulary generation — using this method, recognition vocabularies are only built when they are required; this is currently a very effective technique for small vocabularies, and it is especially attractive for rapidly changing databases. This can provide hard subsetting, or soft subsetting when used in conjunction with dynamic *a priori* probability modification.

16.9   CONCLUSIONS

This chapter has presented a number of techniques which enable some large-vocabulary IVR applications to become practical propositions with the current technology that is available.

These techniques are based on the philosophy that the speech recognition algorithms must have the maximum possible application information available to them prior to attempting recognition. The predictability of response from the caller is also a major influence on performance. Dialogue designs must be well considered and tested thoroughly.

Such approaches will maximize the performance of an application for a given speech recognition accuracy. However, some classes of application are still too large or undefined to be made sufficiently robust for use with current recognition algorithms. These will require fundamental advances in accuracy before they become viable.

The approaches presented have been tested by the development of a corporate directory system for BT Laboratories, providing reliable number information based on first name and surname alone for several thousand people. This system represents a major step forward, demonstrating that a whole new class of applications are now suitable for automation in the immediate future.

REFERENCES

1. Rose K R and Hughes P M: 'The BT speech applications platform'. British Telecommunications Eng J, 13, pp 304-31 (January 1995).

2. 'Now you're talking ... Voice Services', BT publication (1994).

3. Schmidt M and Jack M: 'A multilingual pronunciation dictionary of proper nouns and placenames: Onomastica', Proc of the ELSNET Language Engineering Convention, pp 125-128 (1994).

4. Steinbiss G, Bach-Hiep I and Ney D: 'Improvements in beam search', ICSLP, 4, pp 2143-2146 (1994).

5. Seide F, Rueber B and Kellner A: 'Improving speech understanding by incorporating database constraints and dialogue history', ICSLP, 2, pp 1017-1020 (1996).

6. Attwater D J and Whittaker S J: 'Large vocabulary access to corporate directories', IOA Speech and Hearing, 18, Pt 9, pp 409-416 (1996).

17

SPOKEN LANGUAGE SYSTEMS — BEYOND PROMPT AND RESPONSE

P J Wyard, A D Simons, S Appleby, E Kaneen, S H Williams and K R Preston

17.1 INTRODUCTION

No science fiction image of the future is complete without the ever-present personable computer which can understand every word said to them. In spite of these popular media images, the goal of completely natural interaction between humans and machines is still some way off.

Interactive voice response (IVR) systems, which provide services over the telephone network, have been available since the mid-1980s. Initially they were restricted to interactive TouchTone® input with voice providing the response to the user. The use of such services was therefore limited to the population with TouchTone keypads. More recently applications using automatic speech recognition (ASR) have been developed. These often simply allow the option of spoken digit recognition as an alternative to keypad entry, thus allowing the service to be launched even in areas where TouchTone penetration is poor. Moving on from such systems the words which are spoken can be matched to the service. This allows these ASR-based services to be more user-friendly than their TouchTone counterparts because the user can directly answer the question: 'Which service do you require?' with 'weather' or 'sport' rather than 'for weather press 1 for sport press 2', etc. However, they still rely on selection from a predetermined menu of items at any point in the dialogue.

More sophisticated services are now becoming possible using emerging larger vocabulary speech recognition technology. However, it is not sensible to simply extend the menu-based approach to accommodate larger vocabularies.

Although well-engineered simple applications may be easy to use, more advanced services are likely to have complicated menu structures. If information can only be provided one item at a time, using a 'prompt and response' dialogue, rigid interaction styles may steer the user through a complex dialogue. This can result in the user becoming lost, or ending up with the wrong information. These problems are particularly significant for inexperienced users. On the other hand, experienced users may become bored by the large number of responses needed when they know exactly what they want. The menu-based structure required by systems which rely on isolated word input is often the limiting factor for new services. This limitation of the user interface is one of the greatest barriers to the usability of many IVR services.

Moving beyond the menu-style interaction towards conversational spoken language will allow users to express their requirements more directly and avoid tedious navigation through menus. This approach will also allow the user to take control of the interaction rather than using the more common 'prompt and response' dialogue.

BT is interested in the development of spoken language systems (SLS) to provide a key competitive advantage. SLSs allow users to interact with computers using conversational language rather than simply responding to system prompts with short or one word utterances. With the rapid increase in competition, service differentiation becomes a key factor in gaining market share. Systems which allow users 24-hour remote access to information provide a very useful service for people who are in different time zones, or away from their office, or who need information immediately during unsocial hours. SLSs can be used to automate such services and also those which currently require human operators, thus freeing their time to deal with difficult situations where more complex, or more personalized advice is needed.

Current trends in information networking and the phenomenal growth of the Internet bring their attendant problems for our customers in keeping up with technology, finding what they need, and using information to their best advantage. Spoken language system technology can greatly enhance our customers' ease of access to information, thus increasing network revenue through new and increased usage. Systems which combine several modes of input and output, such as speech, graphics, text, video, mouse-control, touch and virtual reality, are known as multimodal spoken language systems. These allow far greater freedom of expression for users who, as a result, should feel more comfortable and less as though they are 'talking to a computer'. They are able to point, use gestures, speak, type; whatever comes most naturally to them. Spoken language systems will become increasingly important as progress in technology becomes more widely available.

The goal is to be able to build systems which are not restricted only to those motivated users who are prepared to spend time learning the language the machine understands. These new systems can be used by anyone who wants occasional access to a particular service. They will also help the user successfully gain the information or service they require by simply calling a number and asking for what they want. In fact, the aim is to put back some of the intelligence which existed in the network 50 years ago when a user simply lifted the handset and asked to be connected to the service or number required.

This chapter discusses the design and implementation of spoken language systems and is organized as follows. Section 17.2 gives an outline of the architecture of an SLS. Section 17.4 describes the components of an SLS in some detail, giving concrete examples from current systems. Section 17.3 discusses some of the systems currently under development at BTL. These include a multi-modal system for access to the BT Business Catalogue, a speech-in/ speech-out system for remote e-mail access and a system for accessing information about films. Section 17.5 discusses future work which needs to be carried out to improve the quality and usability of SLSs, and section 17.6 draws some conclusions.

17.2   SYSTEM OVERVIEW

This section outlines a typical spoken language system architecture, from the information processing point of view (platform and inter-process communication issues are not dealt with to any great extent in this chapter). The architecture and the key processing components are outlined.

The most basic form of SLS, a speech-in/speech-out (rather than multimodal) system, requires at least the following major components (described briefly below and in more detail in section 17.4).

- Speech recognition — to convert an input speech utterance to a string of words.

- Meaning extraction — to extract as much of the meaning as is necessary for the application from the recognizer output and encode it into a suitable meaning representation.

- Database query — to retrieve the information specified by the output of the meaning extraction component. Some applications (e.g. home banking) may require a specific transaction to occur. Many applications may be a mixture of database query and transaction processing.

- Dialogue manager — this controls the interaction or 'dialogue' between the system and the user, and co-ordinates the operation of all the other system components. It uses a dialogue model (generic information about how conversations progress) to aid the final interpretation of an utterance. This may not have been achieved by the 'meaning extraction' component, because the interpretation relies on an understanding of the conversation as a whole.

- Response generation — to generate the text to be output in spoken form. Information retrieved by the database query component will be passed to the response generation component, together with instructions from the dialogue manager about how to generate the text (e.g. terse/verbose, polite/curt).

- Speech output module (text-to-speech synthesis or recorded speech).

At its simplest, processing consists of a linear sequence of calls to each component, as shown in Fig. 17.1. A typical output of each stage from an application which accesses the BT Business Catalogue is shown. It is not necessary to understand the output of the 'meaning extraction' component in detail to realize that meaning extraction can be a non-trivial exercise. The simple linear sequence

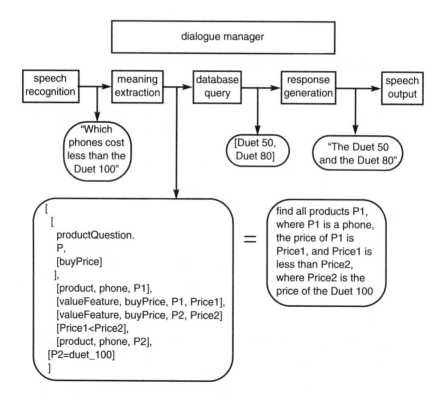

Fig. 17.1   Example of a linear process flow in a spoken language system.

shown in Fig. 17.1 is, in general, too inflexible. It is better if the dialogue manager is given greater control, to call the other components in a flexible order, according to the results at each stage. This leads to an architecture of the type shown in Fig. 17.2.

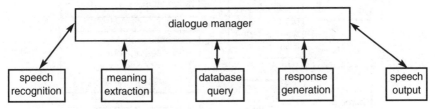

Fig. 17.2 Role of a dialogue manager in a spoken language system.

The need for this more flexible architecture is illustrated by the processing sequence in Fig. 17.3 which shows the dialogue manager as control centre, calling each component in an order determined by the results of processing at each stage. Although every processing stage is passed through the dialogue manager, this is not included in the sequence unless some non-trivial decision or action is taken. The example given in Fig. 17.3 is largely driven by limitations of the recognizer, but the need for this sort of flexible architecture goes far beyond this. It will eventually enable the dialogue manager to act in an intelligent manner, coordinating the components and combining their outputs in a nonlinear manner.

So far in this section, the discussion has covered speech in/speech out systems. However, systems such as the BT Business Catalogue access system (see section 17.3.1) are multimodal and require a screen and a means of inputting text and mouse clicks and outputting text and graphics. These components must be added to the architecture shown in Fig. 17.2 and the dialogue manager and response generator must be upgraded to deal with the extra modalities. However, most of the discussion of this section applies equally to multimodal systems.

17.3 EXAMPLE SYSTEMS

In this section three spoken language systems under development at BT Laboratories are described:

- access to the BT Business Catalogue, known as BusCat — this was the first multimodal continuous speech input spoken language system;

- an e-mail access system, which is speech in/speech out only, but has the conversational features described in this chapter — it is also a dial-up service over the telephone network;

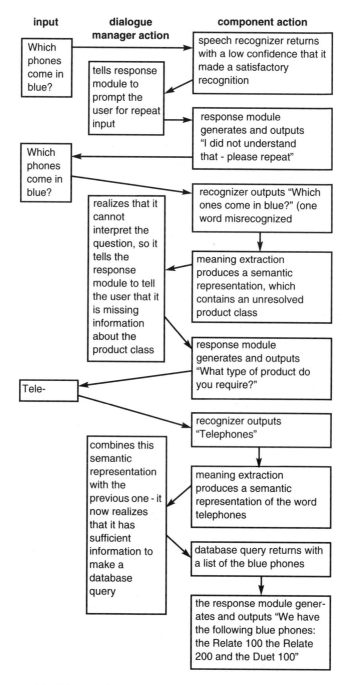

input | dialogue manager action | component action

Which phones come in blue?

speech recognizer returns with a low confidence that it made a satisfactory recognition

tells response module to prompt the user for repeat input

response module generates and outputs "I did not understand that - please repeat"

Which phones come in blue?

realizes that it cannot interpret the question, so it tells the response module to tell the user that it is missing information about the product class

recognizer outputs "Which ones come in blue?" (one word misrecognized

meaning extraction produces a semantic representation, which contains an unresolved product class

response module generates and outputs "What type of product do you require?"

Tele-

recognizer outputs "Telephones"

combines this semantic representation with the previous one - it now realizes that it has sufficient information to make a database query

meaning extraction produces a semantic representation of the word telephones

database query returns with a list of the blue phones

the response module generates and outputs "We have the following blue phones: the Relate 100 the Relate 200 and the Duet 100"

Fig. 17.3   Nonlinear process flow in spoken language systems.

- a film access system, in which users will be able to select films and videos using continuous speech and button pushes on a remote control handset — this system is targeted at the interactive TV environment.

17.3.1   BusCat

BusCat is an SLS which provides direct access to a subset of the BT Business Catalogue, which covers a range of products such as telephones, answering machines and phone systems. The user has a screen displaying a Netscape WWW browser and speech input/output facilities. All the normal WWW browser features are present, such as the ability to click on links to other pages, and a display consisting of mixed text and graphics (see Fig. 17.4). Additionally, in this system users may use continuous speech input, type questions into a free-text window, and listen to speech output generated by a text-to-speech (TTS) system. This multimodal interface enables users to request specific information about the products in the catalogue, or to browse through the catalogue.

Fig. 17.4   The SLS BusCat system in use.

The overall structure of the system is shown in Fig. 17.5. The system can cope with multiple simultaneous users.

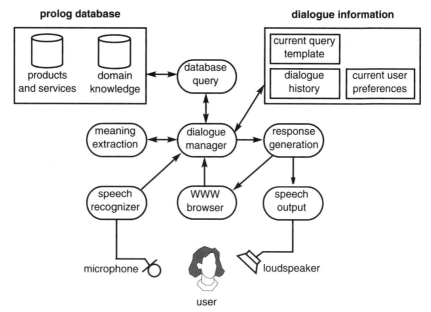

Fig. 17.5   Architecture of Bus Cat.

In addition to its internal knowledge bases, the system has the capability to access external databases across a network. One application for this might be to provide a multimodal interface for such databases. Another is to allow the internal knowledge bases to be periodically updated from an external database.

The speech recognizer used is BT's Stap recognizer (see Chapter 11), and the text-to-speech system is BT's Laureate system (see Chapters 5, 6 and 7).

The example in Table 17.1 gives a flavour of what it feels like to interact with the system. Here the user is already logged on to the system. From each WWW page there is a choice of:

- speaking to the system;
- clicking on a link;
- typing into the free-text field.

In the interaction the user wants to know what on-hook dialling is. Having received an explanation of this feature, he decides he wants a phone with on-hook dialling which costs less than £60. Then he remembers he also wants it in grey to match his living room. He finally selects the Relate 200 telephone.

Table 17.1   An example session with BusCat.

User input	System response
'What is on-hook dialling?'	Textual (and optionally spoken) explanation of on-hook dialling: 'Time spent waiting for someone to answer the phone can often be lost time. But with this feature, you can dial without picking up the phone handset, leaving you free to carry on with something else until the second your call connects,' and a list of five phones which have this feature: Vanguard 10e, Relate 200, Relate 300, Relate 400, Converse 300.
'Which phones have on-hook dialling and cost less than 60 pounds?'	Text: 'The following products meet your requirements,' and a list of four phones, each with a small picture, a short description and a price (Vanguard 10e, Relate 200, Relate 300, Converse 300).
'Which ones come in grey?'	Text: 'The following products meet your requirements,' and a list of three phones, each with a small picture, a short description and a price (Vanguard 10e, Relate 200, Relate 300).
The user clicks on the link next to the picture of the Relate 200.	The system responds with a large picture of the Relate 200, a full description including all its features and a price.

17.3.2   E-mail access

BT is very interested in the mobile telephony market. Speech-only natural language systems are very attractive to this market because people want to be able to keep in touch while on the move. They are likely to only have a mobile phone with which to do so. Speech access to information can have an added safety advantage over TouchTone interfaces as well as helping to improve the usability of the service.

E-mail has been chosen as the vehicle for producing a speech in/speech out natural language system because the information source (the user's own e-mail) is readily available. An existing system allows selected users to access their e-mail using TouchTone commands, thus providing information on whether natural language access confers real benefits to the users. It will enable queries which are not possible currently, such as: 'Do I have any messages from John?'

The main differences between the e-mail service and the BusCat service are as follows. The e-mail service uses telephony speech only, rather than a multimodal system. The dialogue is more robust to errors and will allow the user or the system to take the inititive whereas the Business Catalogue system is based on the user taking the initiative. The e-mail system architecture is being designed to decouple domain-independent and domain-dependent parts of the system, and the same approach is being used for the film access system (see section 17.3.3). This means that in the future it will be possible to produce spoken language systems for new domains more quickly than at present.

An example of a dialogue the system might have with the user is shown in Table 17.2.

Table. 17.2   An example session with an e-mail service.

User	System	Note
(dials up system)	Good morning Alison. You have ten new messages and 20 old messages. Your first new message is from Mike Harper, regarding, Thanks for demo.	The default startup is to tell the user how many old and new messages there are and to start listing the new ones.
Read it.	Message is: 'Thank you for your demo yesterday. The customers were most impressed.'	
Forward it to Peter Smith and Sandra	Sandra Jones is already on the distribution list	Resolution of partially specified name. Check on preconditions for forwarding
	Forwarding message to Peter Smith	
	Next message is from Colin North, regarding, Team Meeting.	System assumes you want to carry on listing new messages unless you tell it otherwise.
Forward to Language Group	Forwarding message to all the people on your distribution list called 'Language Group'.	Does not confirm on forwarding a message.
	Next message is from Laurence Cooper: Plans for Workshop	
Let's hear it.	Message is: 'Hi Alison, Have you completed the plans yet?'	
Reply	I have four possible responses, yes, OK, no, or I'll call you.	Choice of several stock replies
Tell him yes	Reply to Laurence Cooper: Text is 'Yes'. Please confirm.	Confirms before sending a reply
OK	Reply sent. Next message is from...	
Stop.	... OK. What would you like?	Detection of user speaking terminates prompt. Listing new messages ended by user.
Have I any recent messages from Steve Brown or Steve Clark	In the last week you have received 5 messages from Steve Brown and 4 from Steve Clark	Slightly more complex user query. System makes explicit what it thinks of as 'recent'.
List the ones from Clark.	Message 1 is about Meeting on the 31st Message 2 is about Agenda for team meeting	Resolution of partial specified name.

17.3.3   Film access for interactive multimedia services

Interactive multimedia services (IMS) are currently being implemented by BT in Colchester and Ipswich [1]. The usability of the interface is still constrained to navigation using a TV remote control. This means that multiple layers of menus need to be traversed to get to the information required. Spoken language access would allow users to go straight to the information they are searching for, without requiring them to learn complex navigation procedures.

The video-on-demand subset of the IMS, which consists of over 4000 hours of material, including films, educational programmes, children's programmes, etc, was chosen. The SLS will allow users to give instructions such as: 'I want a comedy film starring Harrison Ford'. Part of the benefit of developing such a system, is to ensure that the generic SLS framework is truly domain independent.

There is currently a text-based interface to the Internet movie database [2]. This allows users to enter queries such as: 'Tell me the ratings of comedy movies starring Harrison Ford'.

The system performs the meaning extraction using a caseframe parser (section 17.4.2). This allows it to pick out the salient infor-mation from among extraneous words.

It seems likely, from human analysis of typical queries about films, that this method is suitable.

An issue yet to be addressed is how to best reconcile the advantages of using speech, with the limitations of current recognition technology. This is clearly illustrated in the present example, since the text-based interface can query the database of over 50 000 films and 100 000 cast names. No speech recognizer yet built can cope with this range of vocabulary. The obvious solution is to restrict the size of the database.

A possible step in the right direction would be to couple the 'meaning extrac-tion' component and recognizer much more closely, so that meaning extraction and recognition happen simultaneously. This might enable the recognizer to cut down the vocabulary size 'on the fly'. For example, given the input sentence: 'Which comedy movies star Burt Lancaster,' it could be established straightaway that the user was talking about comedies, then only about cinema films, and finally that the user was only interested in an actor. Therefore, by the time the recognizer gets to the name 'Burt Lancaster,' the number of possible words has reduced considerably.

This is the subject of further research and is discussed in more detail in the next section.

17.4   COMPONENTS OF A SPOKEN LANGUAGE SYSTEM

17.4.1   Speech recognition

The job of a speech recognizer is typically thought of as converting a speech utterance into a string of text. The internal workings of speech recognizers are explained in some depth in Chapter 9. This section looks at the recognizer's place within an SLS, and, in particular, at the language model (LM) the recognizer uses and the form of output that it provides.

A language model embodies information about which words or phrases are more likely than others at a given point in a dialogue.

One might imagine that in a system that accepts fluent language, for example an automated travel agent, the speech recognizer might need only one language model, that of the entire English language. It could then recognize anything that anyone said to it (assuming they are speaking English) and could inform the dialogue manager accordingly. Speech recognition is not yet accurate enough and a model of the entire English language does not exist. Instead, to get a working system, the recognizer must be given as much help as possible. It must be given hints about what the user is likely to say next to improve the chances of correctly recognizing what has been said. If the dialogue manager knows that the customer wants to go on a cruise and has just asked them where they would like to go, it should prime the recognizer to be expecting a response that may well concern one of a number of specified cruise ports and, by the same token, is unlikely to have anything to do with backpacking in Nepal.

Recognizers use a language model to hold this information. There are a number of ways that the dialogue manager can update the information that the recognizer is using. The simplest option is just to tell the recognizer which of a predefined set of language models to use. Then there are a range of possibilities for updating or modifying predefined language models. Certain portions of a grammar can be made more likely than others, e.g. phrases to do with the time of day might be expected at one stage in the dialogue, while requests concerning holiday destinations might be more likely at another, and the language model can be adjusted accordingly. Alternatively, the recognizer might have a grammar, a portion of which allows the sequence 'from <airport> to <city>' where the range of possible airports and cities is specified by the dialogue manager only immediately prior to recognition. This could be dependent, say, on which country is under discussion.

The following subsections discuss in more detail;

- language models;

- perplexity of a language model;

- advantages and disadvantages of language models;

- loading language models into the recognizer;

- output from the recognizer.

17.4.1.1   Language models for the recognizer

The primary knowledge source for the speech recognition component is a set of statistical models, known as hidden Markov models (HMMs), which encode how likely a given acoustic utterance is, given a string of spoken words. A recognizer can decode a speech utterance purely on the basis of this acoustic-phonetic knowledge, and this is basically what happens in the case of single isolated-word recognition. However, in the case of recognizing a string of words (which form part of a spoken language), the recognizer can use a second knowledge source, namely the intrinsic probability of the given string. This second knowledge source is known as the language model.

To take a classic example, a given utterance may have almost equal acoustic-phonetic probabilities of being 'recognize speech' or 'wreck a nice beach'. However, the intrinsic probability of the first string is likely to be higher than that of the second, particularly if this utterance came from the domain of a technical journal on speech technology.

This can be expressed mathematically as follows. Let X be the acoustic utterance and let S be the sentence to be recognized. The task is to find the sentence S for which the posterior probability $p(S|X)$ is a maximum. Using Bayes' rule this can be rewritten as a requirement to find:

$$\text{argmax } \{p(S) * p(X|S)\}$$
$$S$$

In this formula, $p(S)$ is the prior probability of the sentence according to the language model, and $p(X|S)$ is the conditional probability of observing the acoustic utterance given the sentence (encoded in the HMMs of the recognizer).

In general, the language model in a recognizer consists of all the language information to which it has access in order to help constrain the recognition, by making certain word strings less likely than others, or indeed impossible. This language information may be the same information as used in the 'meaning extraction' component of the system (see section 17.4.2), for example, when the grammar in the meaning extraction component is compiled down to a finite state network (FSN) for use in the recognizer. Figure 17.6 gives an example of a recognizer FSN. More commonly, the language information in the recognizer is represented by a statistical model, possibly derived from the same corpus data as the language information in the 'meaning extraction' component, but generally only employing recent context (one or two words). This type of model is known as an *n*-gram model. Such models are easy to integrate with the standard acoustic decoding architecture in speech recognizers, but ignore some of the available language information which might improve accuracy.

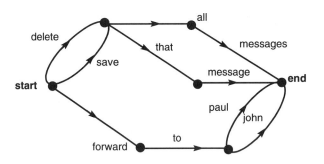

Fig. 17.6   A recognizer finite state network.

N-gram models can be word *n*-grams, class *n*-grams or a hybrid of the two. *N* is typically 2 or 3, and these models are referred to as bigrams and trigrams respectively. A word bigram model gives the probability of all possible next words on the basis of the current word only, i.e. it is of the form $p(w_2|w_1)$. A trigram model is based on two words of context, $p(w_3|w_1w_2)$. The probability of a sentence such as 'delete that one' is obtained by multiplying the probabilities for each word, given its predecessor(s), e.g. for a bigram model:

$$p(\text{Sent}) = p(\text{delete}|\text{start}) * p(\text{that}|\text{delete}) * p(\text{one}|\text{that}) * p(\text{end}|\text{one})$$

A class *n*-gram model is one in which the words are grouped into classes before computing the *n*-gram statistics. These classes may be syntactic categories such as nouns or verbs, hand-crafted semantic categories (e.g. all days of the week might be grouped together, all names of people, etc), or automatically learnt classes. The probability of a word w_3, given two previous classes, c_1, c_2, is given by:

$$p(w_3|c_1 \; c_2) = p(c_3|c_1 \; c_2) * p(w_3|c_3)$$

The advantage of a class *n*-gram model is that there are fewer parameters to estimate. The disadvantage is that, as a word is predicted only on the basis of the class history, some distinctions which may be present in the training data are lost; for example, 'Monday morning' may be more likely than 'Monday afternoon', whereas for all the other days of the week, 'DAY morning' and 'DAY afternoon' may be equally likely.

An FSN gives all possible word sequences in the language model, and it may have probabilities attached to the transitions if reliable statistics for these can be obtained. For example, a fragment of an FSN LM is shown in Fig. 17.6.

Both n-gram and FSNs give a probability for the next word w_n given a word history $w_i......w_{n-1}$. In the case of n-grams, the history is only one or two words, as stated above (e.g. $i = n - 2$ for a trigram model). In the case of FSNs, the history extends back to the start of the sentence (i.e. $i = 1$).

17.4.1.2   Perplexity

In some cases, the constraints imposed by the language model are so great that the job of acoustic decoding becomes much easier than one would have imagined from the size of the vocabulary. This is because the LM is effectively ruling out many possible strings from the search space in which the acoustic decoder is operating. It is possible to calculate a figure called the perplexity associated with a given LM, which is a measure of how difficult the acoustic matching task for the recognizer is. Usually, the lower the perplexity, the higher the recognition accuracy. Perplexity can be roughly defined as the average branching factor, i.e. the average number of words which are allowed to follow a given word. Perplexity may either be calculated intrinsic to the LM (for an FSN, perplexity is literally the average branching factor of the FSN), or with reference to a test corpus, which is the way one usually calculates the perplexity of an n-gram LM. In the latter case, perplexity is defined as the reciprocal of the geometric mean of the probability of the next word, for all words in the test corpus. Intuitively, perplexity is low if the language model tends to know what is coming next, which means that the next word probabilities are relatively high. The formula for test-set perplexity is:

$$PP = \left[\prod_i P(w_i) \right]^{-1/N}$$

where $P(w_i)$ is the probability of the ith word,
    N is the number of words in the test set,
    and the index i runs over each word in the test set.

Typical values of perplexity may range from about 40 to several hundred in current research systems.

17.4.1.3   Advantages and disadvantages of FSNs and n-gram LMs

An advantage of FSNs is that they have a fairly low perplexity, i.e. the number of legal following words is usually a small subset of the entire vocabulary. They can also be constructed manually, before a proper training corpus for the domain exists.

One disadvantage of FSN language models is that they can easily be over-restrictive, ruling out many perfectly valid input strings, simply because of the impossibility of anticipating all the different things users will want to say. This disadvantage can be mitigated by making the FSN more flexible, with liberal use of optional phrases (as in BusCat), or going some way to turning it into an unconstrained phrase network, where any phrase (a few words) can follow any phrase. This of course has the effect of increasing the perplexity, and at a certain point it becomes preferable to move to an n-gram LM. A second disadvantage of FSN language models is that as one moves towards a broader, more habitable user language, they very quickly become large, making the recognition time unacceptably long. The FSNs themselves become cumbersome and difficult to maintain.

N-gram language models, on the other hand, usually have a higher perplexity than FSNs, because all words may be legal followers of the current word, albeit many of them having low probabilities.

The disadvantage of n-gram LMs is that they are not as constraining as FSN LMs, which means that if the acoustic match accuracy is not good enough, the output of the recognizer may be so poor that the meaning extraction component has little chance of interpreting it, however robust its algorithm.

17.4.1.4   Loading the LM into the recognizer

The LM(s) may be loaded into the recognizer:

- once at the start of an application;

- repeatedly, according to where the user is in the dialogue immediately prior to recognition;

- during a recognition by the recognizer.

These three possibilities represent increasing degrees of sophistication. In the first case, there is just one LM for the whole application. In the second case, use is made of the fact that at different points in the dialogue the relative probabilities of different words will change greatly. For example, if the system asks: 'Which day do you wish to travel on?', there is higher probability than normal that the user's answer will contain a day of the week. Similarly, if it has already been established that the user wishes to travel from Ipswich, there is a higher probability than normal that the destination will be somewhere in the Anglia region.

In the third case, the LM is changing 'on the fly', during the decoding of the speech utterance. For example, if the first words in the sentence are 'I want to go from Ipswich to..', then the LM can increase the probabilities of stations which have direct trains from Ipswich before recognizing the rest of the sentence.

The LM(s) passed to the recognizer may be complete models or modifications to the current model. A modification to a current model might consist of:

- a list of words to be added or deleted,

- modified probabilities for words which are already within the language model.

17.4.1.5   Output from the recognizer

If the job of a speech recognizer is to convert a speech utterance into a string of text, then, when someone says: 'What time does the flight leave?' it is hoped that the recognizer might come up with the string: 'What time does the flight leave'. In practice the recognizer's best guess may not be correct, but it may be able to give a number of scored alternative possibilities which the analysis module might use if the top choice does not make sense, for example:

1   'what time does the white leaf' 1245.6

2   'what time does the flight leave' 1250.1

3   'what time does a flight leave' 1252.3

4   'what time did the flight leave' 1270.1

5   'what time did a flight leave' 1272.3

The analysis module may then use its own rules to re-rank the list of possibilities or extract those portions that carry useful information (perhaps taking account of the recognizer scores or perhaps ignoring them).

There are, however, significant drawbacks in representing the recognizer output as a ranked list of word strings. The list can become very long indeed, containing large numbers of sentences that only differ by one or two words.

A much more compact representation of the same information can be achieved through the use of directed graphs (see Fig. 17.7).

A third option that is sometimes used is the word lattice (see Fig. 17.8). This consists of a lattice with entries, each of which specifies a possible word, its start time and its end time (but not which other words it follows or precedes). The benefit of the word lattice is also its weakness. It can offer in a compact form for storing a very large number of candidate sentences — sentences can be generated by connecting any sequence of words such that one word starts where the previous one finishes. Unfortunately this allows one to generate sentences that the recognizer did not consider very likely, e.g. 'what time did a flight leaf'.

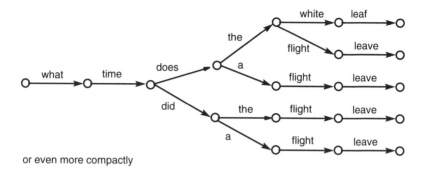

or even more compactly

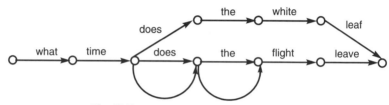

Fig. 17.7   Directed graph for recognizer output.

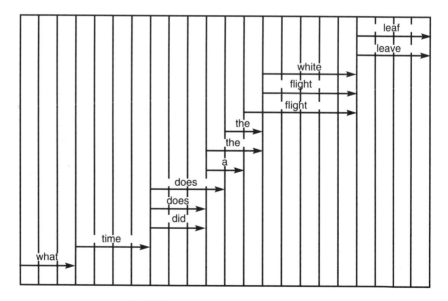

Fig. 17.8   Word lattice for recognizer output.

17.4.2   Meaning extraction

17.4.2.1   Purpose of the meaning extraction component

The input to this component is a representation of the user's utterance that is the output from the speech recognizer. This representation may take one of several forms as described in the previous section. The meaning extraction component of the SLS converts the input to a representation of the meaning of the utterance, which is used to determine the next step in the dialogue.

The user's utterance will contain several pieces of information, some of which will need to be represented in the output from the 'meaning analysis' component. One piece is the type of utterance that the user made, e.g. an instruction, a statement or a particular kind of question. Another concerns the expectations present in the user's utterance. These are often expectations about the result of a query and are not normally intended to be used as constraints on the query itself (such as the use of a plural form to indicate that the user is expecting more than one item to satisfy a request). A third piece of information consists of the entities referred to in the user's utterance, and the relationship between them. These will form the basis of the constraints on the database query. The purpose of the 'meaning extraction' component is to make some of this information explicit so that it may be more easily processed by subsequent components.

Although the role of this component can be described simply enough, the task is daunting. There are at least two main difficulties that need to be overcome during meaning extraction in SLS:

- ambiguity;

- ill-formed input.

Ambiguity [3] may be present in words as lexical ambiguity (e.g. 'saw' can either be a noun or a verb) or sense ambiguity (e.g. 'bank' may be a place in which to store money, or from which to fish), or it can be present in the relationship between phrases as structural ambiguity. To take the ubiquitous example:

'The man saw the boy with the telescope in the park.'

This sentence is structurally ambiguous in what was seen, the instrument of seeing, and the location of the person seeing. Often, this type of ambiguity cannot be resolved by the meaning extraction component, and has to be maintained in the meaning representation, to be deciphered by later components.

Ambiguity is inherent in all natural language, while ill-formedness can be caused by misrecognition in the speech recognizer, or by the user making an ungrammatical utterance (as defined by the system). Speech recognizers are becoming increasingly competent, but they still mishear quite often. This is particularly a problem with recognizers which use an *n*-gram language model which are less restricted in the ordering of the words they can recognize. For example, in a 'hotel enquiry' domain [4], the parser was faced with input such as:

'do the noise outsiders use your lake'

when the user actually said:

'there is a noise outside that keeps me awake'

It is doubtful that any meaning extraction component would be able to cope with such a discrepancy between the actual utterance and the recognized one, but there are mechanisms by which simpler ill-formed utterances can be analysed [5]. These will be explored in more detail below, but essentially they attempt to pick out those bits which can be analysed, and put them together in the best way possible.

If meaning extraction is viewed as performing a translation between an input text and its meaning, then, quite apart from the two particular problems described above, the difficulty of defining the translation is faced. This is a many-to-many mapping, since many English sentences can have essentially the same meaning (in the sense that they are true in the same situations) and some sentences can have more than one meaning. For instance, the following two sentences, to all intents and purposes, mean the same, but have a different structure:

'Romeo loves Juliet' and 'Juliet is loved by Romeo'

Assuming that all of the above problems can be overcome, a suitable representation for the output meaning must be chosen. The issues to be considered are:

- suitability for successive stages of processing — whether the representation encodes all the information necessary for further processing;

- explicitness versus conciseness — whether one representation can encode all unresolved ambiguity, while maintaining a clear meaning;

- extendibility — whether the representation will only cope with the particular sentences chosen for development.

The representation generally depends on the technology used, but various logics [6] are popular forms of representation. This is because they are well-understood, and fully-specified. Perhaps more importantly, a meaning represented in logic can be reasoned with, using the proof calculus of that logic. This is useful in database query systems [7]. The other main representation is to use a frame which encodes predefined meanings (such as the act of 'seeing') [8]. These frames will take arguments to fill in the details ('who', 'what', 'where', etc).

The rest of this section is devoted to a particular representation based on a logical representation. This representation, which is referred to as an extended logical form (ELF), has three fields corresponding to three types of information present in the user's utterance. An example ELF, representing the utterance 'read the message from John' is given in Table 17.3.

Table 17.3   Fields in an extended logical form representation.

Field name	Field content
Type	read(A),
Expectations	salient(A), singular(A)
Entities and relationships	message(A), named(B john), from (A,B)

The type field is used to indicate the type of the utterance, such as imperative, indicative, question, among others.

The expectations field contains a list of constraints that are the user's assumptions. In the example above, the user is presupposing that there is only one message from someone called John, therefore 'singular(A)' will be an item on the list of expectations. The constraint, 'salient(A)', is used to indicate that the user has assumed that there is some specific (usually recently mentioned) message that is the object of the instruction. This contrasts with 'read a message from John' where there would not be a 'salient(A)' constraint on the list.

The entities and relationships field of the ELF contains a list of conjunctive constraints which express the entities referred to in the user's utterance and the relationships between them.

The frame representation and the logical representation are indicative of the separation between different meaning extraction technologies. The logical form of representation tends to be used by grammar-based parsers, while the frame form tends to be used by caseframe parsers. These different types of parser will be described in the next two sections.

17.4.2.2   Grammar-based parsers

Traditionally, schoolchildren who are given the exercise of analysing English sentences according to a prescribed grammar of English have been doing parsing [9]. Parsing has the purpose of assigning a structure to an input utterance. This can be done by hand, but is now more often done using an automatic parser. More precisely, what has just been described would be a syntactic parser, performing a syntactic analysis. Here, the sentence is being analysed according to a syntactic grammar [10], where the relationship between nouns and verbs, for example, is made explicit. Syntactic analysis gives no representation of the meaning of the sentence. Indeed, the following two sentences would be assigned the same syntactic structure [11]:

'Bright red buses drive quickly' and 'Colourless green ideas sleep furiously'.

Of course, they clearly have a different meaning, and semantic analysis [12] is concerned with providing such a meaning, solely from the combination of the meaning of the individual words.

Grammar-based parsers take a linguistic grammar of, say, English, and assign a structure to the sentence based upon that grammar. Two stages are necessary — firstly, assign each word a part of speech (POS), such as noun, and, secondly, from these POS, apply the grammar rules.

So, given the phrase:

'the man'

the lexicon which contains the following rewrite rules:

Det   →   'the'

Noun   →   'man'

and the grammar rule:

NounPhrase   →   Det   Noun

the parser would build the syntactic structure:

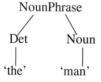

Syntactic analysis is not sufficient for meaning extraction as mentioned previously, and a semantic analysis stage is needed. This can either take place after the syntactic analysis or concurrently with the syntactic analysis. Indeed, semantics is often built up in parallel with the syntax, using the rule-rule hypothesis [12]. This basically states that syntactic and semantic rules should only exist in pairs, the semantic analysis being built by 'piggy-backing' on the syntactic analysis. This means that every part of the sentence contributes to its overall meaning, and that fine-grained syntax produces fine-grained semantics. Extremely complex grammars can be written [13], encapsulating fine points of language. These details may be important in some systems, and are thus represented in the final meaning representation.

The primary problem with grammar-based parsers, from the point of view of SLSs, is that they need to analyse the whole utterance before they can produce a representation. As has already been stated, there is quite a high likelihood of misrecognition, thereby producing input which is outside the scope of the grammar (ungrammatical input). One option is to constrain the speech recognizer using a finite-state network which mirrors the grammar of the meaning analysis component (section 17.4.1), thereby forcing the 'speech recognition' component to produce only those sentences which can be analysed using the grammar. An alternative is to dispense with the grammar-based approach altogether and use a caseframe parser.

17.4.2.3   Caseframe parsers

An alternative view of parsing has emerged as researchers have begun to use natural-language processing in practical applications. Caseframe parsers [14] do not use an intermediate syntactic stage of processing, but attempt to go straight from the input representation to a representation of its meaning. This is done by assuming that phrases in the sentence have meaning, but only in the context of a predefined frame, or meaning template. An example of a frame is shown in Table 17.4. As many as possible of the slots of the frame are filled in, using simplistic syntactic analysis, from the phrases of the input sentence. This means that the phrases can appear in the middle of any amount of extraneous information, because only the particular phrases which are relevant to the frame are analysed. This approach is more robust than the grammar-based technique. For example, the following three sentences would be mapped on to the 'spilling' frame as shown.

'the waiter spilled the wine on the couple'

'It is the case, I believe, that earlier last week the waiter, what an appalling employee, spilled some of the wine which we use for only our best customers, on the couple visiting for the first time.'

'when waiter spilled the wine on the couple food'

Table 17.4   An example of a frame.

Spilling frame	
Actor	waiter
Patient	wine
Location	couple

However, there is a step in the above description that is obviously missing — how to pick the correct frame for the given utterance. This is usually done on the basis of the information contained in the sentence. One method of choosing the correct frame is to allow each piece of information within a sentence to vote for a particular frame. The frame used is the one which has the most votes for the given sentence.

A circular process can develop, however, where only the required information to fill the frame is analysed, but all the information needs to be analysed to pick the right frame. For this reason, caseframe parsers are generally only used in limited domains, where only a small number of frames are considered.

Caseframe parsers are currently not capable of the fine detail which a good grammar-parser can produce. This may be unnecessary in most applications, but could lead to confusion if the kind of input encountered really can only be distinguished using detailed analysis.

17.4.2.4   A hybrid approach

If grammar-based parsers are good at producing detailed analyses but poor at handling ill-formed input, while caseframe parsers are good with ill-formed input but unable to produce highly detailed analyses, then a combination of the two would seem ideal. Stallard and Bobrow [15] propose a two-stage process, combining a grammar-based parser followed by a caseframe parser, using a chart [16] as the intermediate representation.

Charts are used by a particular kind of grammar-based parser called a chart parser (this is something of a misnomer, since it is parsing words, not charts). This kind of parser produces all possible analyses of the text, storing intermediate results in a chart. This means that if there is no known analysis for the whole of the input, then there is the potential to look in the chart for incomplete analyses, which will represent those parts of the sentence which can be analysed by the grammar-based parser.

A good way of combining these fragments is to build them into a frame. A caseframe parser can then be used to extract the meaning of the input utterance, not from the words, but from the chart produced in the first stage. This frame will then provide a meaning representation with as much detail as the grammar will allow, thus improving on the standard performance of caseframe parsers.

17.4.3   Dialogue manager

The dialogue manager (DM) is the central controller for an SLS. On receiving input from the speech recognizer, the DM manages and co-ordinates calls to the rest of the components in the system (see Fig. 17.2).

The DM knows about the generic structure of conversations. It uses a conversational modeller to build up a dynamic model of the conversation between a user and the system as it progresses. It models conversational turns from both participants, matching questions with their answers, instructions with acknowledgements, and so on.

The syntactic and semantic analysis of individual utterances (or sentences) produced by the 'meaning extraction' component is used as a base and the DM goes beyond these single-sentence analyses to produce a dialogue model, which could be thought of as a grammar for the dialogue.

17.4.3.1   System-driven, user-driven, and mixed-initiative dialogues

Conversations with existing IVR dialogue systems, such as telephone banking systems, can be frustrating for callers because the conversation must always follow a strict pattern. A caller cannot say anything he or she wants, but is forced to give the correct reply at the appropriate time, otherwise the interaction cannot proceed. In this kind of dialogue, the system takes all the initiative and callers are not allowed to ask questions, or say anything which is not specifically asked for. In other words, the dialogue is entirely system-driven.

At the other end of the scale are user-driven systems where the user takes all the initiative. The interface to the BT Business Catalogue (see section 17.3.1), operates in this way. The system accesses the World Wide Web (WWW) and waits until the user inputs a query.

Dialogue managers are being developed that allow more natural conversations to take place between humans and machines. The intention is to produce a two-way flow of communication where the initiative can be taken either by a user, or by the system, and information may be given, asked for, and received by either party. Here is an example mixed-initiative conversation with an imaginary movie database query system of the future:

User: I'd like to see a film starring Meryl Streep.
System: There are quite a few of them. Are you interested in thrillers, comedies or historical dramas?
User: What about that one where she plays a Danish farmer.
System: Do you mean 'Out of Africa'?
User: Yes, that's the one.

The user instructs the system to find it a film starring Meryl Streep (user-initiative), but the system finds there is more than one and decides to ask the user what category of film (system-initiative). The user chooses not to answer this question directly, e.g. 'comedies', but gives instead some more information about a particular film (user-initiative). Now the system has enough information to pinpoint the film and the user confirms the system is correct.

17.4.3.2   Ellipsis and anaphoric reference resolution

Speakers use words or phrases to refer to things in the real world. Anaphoric references are a specific type of referring expression and are used when a speaker refers back to something mentioned earlier, e.g. 'he', 'him', 'her', 'that one', 'that way', 'the house' can all be anaphoric reference expressions:

User: Are there any messages from Peter?
System: You have two messages from Peter Wyard.
User: Read his second one.

In the above exchange, the user's second utterance contains two anaphoric references: 'his' and 'second one'. Assuming a grammar-based parser, it will represent 'Read his second one' as an ELF (see section 17.4.2). The ELF below shows that there is something salient and singular (A) which is to be read, it is the second one, and it is from a masculine person of unknown name:

ELF:

       read(A),
       [salient(A),singular(A)],
       [ord(A,2),named(B,C),gender(B,masculine),from(A,B)]

It is the reference resolution process which fills in missing information in the ELF. It does this by searching back through the preceding conversation to find something which can be read, and a masculine person. It is assumed that the last masculine person mentioned will be referred to as 'his', although this might not always be the case. The ELF output by the meaning extraction component is thus converted into a resolved extended logical form (RELF) where the missing information (message(A) and name (B,peter)) is filled in:

RELF:

> read(A),
> [salient(A),singular(A)]
> [message(A),from(A,B),ord(A,2),name(B,peter),
>   gender(B,masculine)]

Ellipses occur when something is left out of an utterance which can be determined from what has gone before:

User: Do I have any messages from Peter?
System: You have two messages from Peter Wyard.
User: And David?

Here the user has left out: 'Do I have any messages from..' and has simply said: 'And David?' The 'meaning extraction' component will produce an extended logical form:

ELF:

> A,
> [],
> [name(B,david)]

and the RELF will include the missing information which was found in the previous logical form:

RELF:

> list(A),
> [],
> [message(A),from(A,B),name(B,david),
> gender(B,masculine)]

17.4.3.3   Co-operative responses

Much of the ground work on the nature of conversation has been carried out by philosophers of language such as Grice. Grice devised a set of maxims, or co-operative principles, to which he considered people generally conform when participating in a conversation [17]. Four of the most important maxims are of quality (be truthful and avoid statements for which you have too little evidence), quantity (be as informative as necessary, but do not give too much information), relation (do not give irrelevant information), and manner (do not present information in a way that is obscure, ambiguous or too lengthy). People tend to be co-operative in conversation for many reasons, for instance:

- to make the conversation flow more easily;

- to assist understanding;

- to pass information efficiently avoiding irrelevant information, false information, or too much detail;

- to promote good social relations.

If the maxims were ignored, then conversation would become very difficult. Of course there are many examples where the maxims are deliberately broken for the sake of humour, pathos or sarcasm. At present, these are outside the scope of this work and, unfortunately, the DMs do not have a sense of humour!

The DM tries to be as co-operative as it can in responding to the user — it tries to obey at least the maxims of quality, quantity and manner. It does this by being as brief in its responses as it can, by ensuring the database is accurate and up-to-date, and by supplying helpful information. The maxim of relation can be harder to satisfy in that the DM cannot always be sure that the information it is giving is maximally relevant to the user. For instance, if the database query finds nothing, then rather than just saying 'none found', the DM attempts to give some information which the user might find helpful. For instance:

User: Do I have any e-mails from Anna about aardvarks?
System: Unfortunately you have no matching message,
         however, you have one message about aardvarks.
User: OK, but do I have any e-mails at all from Anna?

Here there are no e-mails from Anna about aardvarks, but the system finds a message which is not from Anna but is about aardvarks. The DM does this by generalizing the query it is making to the database. In order to do so it must decide which parts of the user's specification are the most important. Is it most important that the e-mail is from Anna or is it most important that the message is about aardvarks? Or are the two equally important? Here it wrongly assumes that the user is primarily interested in aardvarks. Our DM at present works from a priority list compiled from intuition about what might be most important to a user. The whole area of co-operation is one intended for future investigation.

When the database query is generalized with a successful result, the DM produces another logical form, the co-operative extended logical form (CoopELF). For the above example, the RELF and CoopELF are as follows:

RELF:

> list(A),
> [plural(A)]
> [message(A),from(A,B),name(B,anna),
> gender(B,feminine),
> about(A,aardvarks)]

CoopELF:

> list(A),
> [plural(A)] [message(A),about(A,aardvarks)]

Both logical forms are then sent to the response module (see section 17.4.5).

17.4.3.4   Error recovery

There are several reasons why a system may not understand a user's request — speech recognition errors, meaning extraction errors, and conversational modelling errors.

Unfortunately speech recognizers are not 100% accurate and a user does not always use phrases the computer can understand. It can be frustrating for a user if the system simply keeps repeating 'Sorry I didn't understand that'. Sometimes asking a user to repeat an utterance will result in a successful recognition, if, for example, a quiet utterance or background noise resulted in a recognition error. However, sometimes the user will need to rephrase the question, or speak later, or earlier, etc. It is necessary to have an error recovery dialogue which helps the user to try different strategies when difficulties occur, and which seems helpful rather than repetitive.

Sometimes conversational modelling errors occur; for instance, when a user says 'Read her e-mail' and the DM has no record of a female person being mentioned previously, the DM must then ask the user an appropriate question, e.g. 'Who do you mean?'

17.4.4   Database query

When the DM has prepared the query, it will be passed to the database query component. The database query component's purpose is to convert the query from the DM into one or more queries which can be used to find the required information from within the database. Having established the queries, the database query component then extracts the actual information from the database.

This is not as straightforward as it sounds and may involve a number of steps.

- Using general and domain-specific knowledge to attempt to satisfy the query — an example of this is the use of class taxonomies. If a user asked about 'reptiles', but the database knew about 'snakes', then the database query component could look down a reptiles taxonomy to realize that there is a link between query and data. Similarly, a user asking about 'turtles' from a database of 'terrapins' could have the query satisfied by the database query component if it looked sideways in a reptiles taxonomy.

- Optimizing the query — by carefully arranging the order of execution of subqueries with a query, the speed of getting a result can be improved dramatically. To do this, the database query component will need to rank the severity of each of the constraints in the query. The query will execute optimally when the most severe constraints are applied first.

Of course, neither of these operations ensure that there will be a result from the query, i.e. there may be no answer to the query. Therefore the database query component will pass as much information to the dialogue manager as possible about a query which failed, so that the DM can decide what to do next.

Ultimately, the database querying module provides a means of separating the actual database query (in SQL, for example) from the internal representation in the DM. This should mean, in theory, that the dialogue manager need not be aware of the database used, and that using a different database management system only involves changing the internal mapping used by the database query component.

17.4.5   Response generation

In spoken language systems, responses are normally thought of as being provided using speech. However, in the case of multimodal systems, the response can also be in a graphical or pictorial form. This section is therefore divided into three parts — the generation of text, the synthesis of speech, and the generation of graphics.

17.4.5.1   Text generation

Text generation is the task of putting into words the information that is to be sent to the user. It can be as trivial or as complex a process as is appropriate for any particular SLS.

At the trivial end of the scale, text generation can be avoided altogether by using canned-text sentences. These work well if the application is very simple and there are only a very limited number of things the system will need to say.

The next step up from this is to have carrier or template sentences where relevant information can be slotted. This allows a little more scope for personalizing the messages to the user and has been used successfully in ELIZA [18], the famous artificial intelligence program which attempts to fool people into believing they are communicating with a psychologist. ELIZA recognizes and scores certain patterns in the user's input and uses the highest scoring pattern as a filler in its template output, transposing personal pronouns and possessives such as 'I' for 'you', and 'my' for 'your'. For instance:

User: I am worried about my mother.
ELIZA: Tell me more about your mother.

The template output sentence is 'Tell me more about...' and ELIZA has transposed 'your' for 'my' in the pattern 'my mother' to give the filler 'your mother'.

The most flexible method is to generate the most appropriate responses as and when necessary. Just as grammars can be used for analysis, most can equally well be used for generation. However, the flexibility of a text generator depends on the size of the grammar and lexicon used. If they are very small, the effect is little better than using template sentences.

Text generation is the reverse process of parsing. A parser converts a string of text to some kind of knowledge representation, whereas a text generator converts a knowledge representation to a string of text. For the e-mail assistant SLS, the RELF which has been instantiated after a successful database query is used for generation.

If the initial database query was unsuccessful, then the uninstantiated RELF and the CoopELF are both used for generation. Furthermore, the presuppositions part of the RELF is used to reply in an appropriate way depending on the assumptions made by the user. Thus, if the user was expecting more than one e-mail as in 'Read me my messages from Peter' and only one exists, then the system can reply 'There is only one, the message reads:...'

17.4.5.2   Graphics generation

Multimodal systems are enhanced by graphs and/or pictures incorporated into the text. A graph or picture can show at a glance what would often take many paragraphs to describe in words.

BusCat (see section 17.3.1) builds WWW pages on-the-fly in order to answer a user's query. The WWW pages include pictures of BT products relevant to the user.

A consideration when (automatically) collating material to be shown on a computer screen is size of images and graphs. It is desirable to try to fit the information within the limits of the screen rather than forcing the user to scroll down to view the whole page. The interface to the BT Business Catalogue uses different sized images to attempt to achieve this.

17.4.5.3   Speech synthesis or text-to-speech

BT's SLS systems all use the Laureate speech synthesizer (see Chapters 5, 6 and 7) whenever TTS is required. When a text for synthesis is being generated automatically by the response module, information about the structure of the text will already be known. This opens up the possibility of providing Laureate with details of what kind of utterance it is (declarative, interrogative, or imperative), where the pauses should be, where the strongest emphasis ought to be placed, and so on. This is an area for future research.

17.5   FUTURES

There is a considerable amount of work to be done in taking the demonstration systems described in section 17.3 and turning them into applications which can be used by customers. To a certain extent this is the normal process of downstreaming, addressing issues such as thorough coverage of the application domain, robustness of software in the hands of naive users, and a well-designed user interface. However, there is also further work to be done on the underlying technology to make continuous spoken-language systems which will be really effective. Three such technology areas are highlighted in this final section. These technologies are all aimed at improving the quality of the system by making it more user-friendly and more intelligent.

17.5.1   Speech recognition

SLSs of the kind described in this chapter place great demands on the recognition component, and these demands will increase as the coverage of the system is extended, because the recognizer must still work with sufficient speed and accuracy with a larger vocabulary and higher perplexity of language. In addition to these fundamental requirements, there are some other features which would help improve performance.

- Spontaneous speech recognition — this involves a number of different areas, each of them trying to enable better recognition of speech from naive users in a real application situation, as opposed to experts running a controlled demonstration. Non-speech sounds, such as breath noises, hesitations, coughs, etc, must be accommodated, together with out-of-vocabulary words (because it will never be possible to create a recognizer vocabulary which covers every word to be spoken by real users), and the disfluencies, hesitations and restarts of spoken language.

- Confidence measures — the recognizer should be able to give the dialogue manager an idea of how confident it is that it has recognized an input utterance correctly, and beyond this, its confidence in particular parts of the recognized string. For example, the recognizer might say that it has recognized 'which films are UNK by Clint Eastwood', in which it is highly confident about everything except the unknown word UNK.

- Speaker adaptation — in many cases it is anticipated that SLSs will be used regularly by the same individual, in which case they may be willing to train a speaker-adaptive system to their voice before using the system. If this is not possible, the dialogues with SLS are usually long enough to make speaker adaptation during the dialogue a useful proposition.

- LM adaptation — it is important that the recognizer can efficiently update its language model during the course of the dialogue, as explained in section 17.3.1.

- Prosodic information — in the context of a spoken dialogue, there is considerable prosodic information in the user's utterances (stress, intonation, etc) which could help the recognizer, but is currently ignored.

17.5.2   User-centred language

The ideal would be for users to be able to speak to an SLS in a language which is natural to them, although it is realized that people will be prepared to adapt to the system to a certain extent. To do this the way real users actually say things must be studied, and incorporated as far as possible into the recognizer and grammar. However, user-centred language also requires more sophisticated processing in analysing the input sentence and translating it to a suitable database query. To take a very ambitious example, a user might say: 'How can we cut the cost of getting bulky documents from Edinburgh to London?', which would ideally be translated into a database query about faxes and ISDNs. This kind of translation requires considerable knowledge bases in the system, both domain-specific and general world knowledge, and it also requires a powerful inference engine to make effective use of the knowledge bases. In the shorter term, such queries may be beyond our capabilities, but users will not expect to have to phrase everything in exactly the terms encoded in the database which contains the required information.

17.5.3   More flexible and intelligent dialogues

The dialogues discussed are already considerably more flexible than the traditional prompt and response dialogues. However, there are two areas where dialogues could be enhanced:

- greater use of mixed-initiative dialogues, where the system sometimes takes the initiative — for instance to give the user some background information it has inferred they may be unaware of, or to make a 'sales pitch' for a particular product or service, if, for example, the user has been spending a lot of time asking questions about it,

- more use of user and life-style profiles — these are knowledge bases which the system has built up about the particular user, and the category to which the user belongs, the dialogue manager then being able to use this information both to help it interpret user input, and to decide on suitable initiatives to take, as discussed above.

17.6   CONCLUSIONS

Spoken language systems are likely to be of great importance in the future development of computer technology and its adoption by large numbers of people in the information age. This is because language is one of our primary means of communicating with each other, and, unless humans change fundamentally in a short time, the use of natural language will be an important element in facilitating communication between humans and computers.

This chapter has discussed the components of spoken language systems in general terms, and described three systems which are under development at BT Laboratories. There are a variety of spoken language systems, from pure speech-in/speech-out systems to multimodal systems which aim to combine spoken language with other modalities, such as typed text and mouse clicks, in order to achieve the most user-friendly interface possible.

As for the future of spoken language systems, the current state of the art is that they are on the verge of commercial deployment in some domains. However, it will be some time before we approach the scenarios envisaged in '2001: A Space Odyssey' or 'Star Trek' where one can talk to a computer system in a completely natural way.

REFERENCES

1. Bissell R A and Eales A: 'The set-top box for interactive services', in Whyte W S (Ed): 'Multimedia Telecommunications', Chapman & Hall, pp 116-138 (1997).

2. http://www.cm.cf.ac.uk.movies

3. Gazdar G and Mellish C: 'Natural language processing in Prolog', Wokingham, England, Addison-Wesley, pp 169-174 (1989).

4. Rayner M and Wyard P: 'Robust parsing of n-best speech hypothesis lists using a general grammar-based language model', 4th Eurospeech Conference on Speech Communication and Technology, EUROSPEECH '95, pp 1793-1796 (1995).

5. Special Issue on 'Ill-formed input', American Journal of Computational Linguistics (1983).

6. Hodges W: 'Logic', Hamondsworth, England, Penguin Books (1977).

7. Warren D H D and Pereira F: 'An efficient, easily adaptable system for interpreting natural language queries', American Journal of Computational Linguistics, $\underline{8}$, No 3-4, pp 100-119 (1982).

8. Fillmore C: 'The case for case', in: 'Universals in Linguistic Theory', New York, Holt, Rinehart and Winston, pp 1-90 (1968).

9. Beardon C, Lumsden D and Holmes G: 'Natural language and computational linguistics', Chichester, England, Ellis Horwood, pp 150-156 (1991).

10. Allen J: 'Natural language understanding', The Benjamin/Cummings Publishing Company, p 41 (1987).

11. Chomsky N: 'Syntactic structures', The Hague, Mouton Publishers, p 15 (1957).

12. Cann R: 'Formal semantics', Cambridge, England, Cambridge University Press, p 5 (1993).

13. Alshawi H (Ed): 'The core language engine', London, England, MIT Press (1992).

14. Carbonell J G and Hayes P J: 'Robust parsing using multiple construction-specific strategies', in Bolc L (Ed): 'Natural language parsing systems', Springer-Verlag, Berlin, pp 1-32 (1987).

15. Stallard D and Bobrow R: 'Fragment processing in the DELPHI system', Proc of Speech and Natural Language Workshop, DARPA, pp 305-310 (1992).

16. Earley J: 'An efficient context-free parsing algorithm', Communications of the ACM, 13, No 2, pp 94-102 (1970).

17. Grice H P: 'Logic and conversation', in Cole P and Morgan J L (Eds): 'Syntax and Semantics', 3, Speech Acts, pp 41-58, New York, Academic Press (1975).

18. Weizenbaum J: 'ELIZA — A computer program for the study of natural language communication between man and machine', Communications of the ACM, 9, No 1, pp 34-45 (1966).

18

WHITHER SPEECH? — THE FUTURE OF TELEPHONY

A V Lewis and F A Westall

18.1   INTRODUCTION

18.1.1   A cry from the cradle

Aaaa-auh-aaa.... mmnn-mmm.... mmnmAH-aaah ?

Countless human babies have made the same discovery — that simple sounds, joined together, can make something new. To an adult it is just babbling, yet it marks the start of a mental and physical struggle that is long, arduous and mercifully always forgotten — a struggle to master the subtle, highly structured and pivotally potent ability called speech, an ability that becomes effortless and so integral to our being that we accept its enormous power without a thought. Yet its rich complexity makes us unique among living things, lying at the heart of how we become what we grow to be.

Reflect for a moment how unlikely, even bizarre, the mechanism is. We can, at will, vibrate a tiny piece of our body. We can form, shape and colour these faint vibrations, impressing just a few thousandths of a watt of power on to a thin waft of air. Is it really likely that such insubstantial events could carry joy or sorrow, love or fear, laughter or pain across the vast gulf of consciousness between different minds — and do so in just a few heartbeats? Speech is made in just such a way — and in much less than a minute can penetrate to our deepest emotions or fling our thoughts to the uttermost star. Truly, speech is the cradle of our humanity.

18.1.2   Wither speech ?

Yet some observers claim all human communication will be transformed and telephony made obsolete, within a generation, because of the Internet. Campbell, a Managing Director of Australia Telecom, implies the telephone might be killed [1] by the rapid growth of Internet, data and video traffic. Some say telecommunications will become a low-profit commodity, much like water or electricity supply [2].

The mass media may have just discovered the Internet, but it was conceived 28 years ago [3] and has become a prodigious invention. It can put you in touch with people you need to contact, even though you know nothing about them. It has made a sub-culture out of serendipity (the making of happy, chance discoveries). Negroponte [4] shows how a global computer network can have unique properties, wonderfully augmenting the power and range of human skills. But there are detractors, including some serious and hard-nosed users. Stoll [5] speaks from bitter experience — finding information equated to knowledge, meeting human contact without humanity and being drenched in electronic dross. Unfortunately, the growth and evolution of the Internet is currently so frantic that it addles the brains of a few of those it fascinates. When we all have Internet connections, will we stop talking to each other? Should we forget how to walk when we learn how to swim?

18.1.3   Electronic puberty

Yet these wild pundits do have a point — there is drastic change in the air. The authors believe that electronic puberty is a good description of the change that the telecommunications industry has begun to experience on a global scale. All the ground rules have changed, parts of the body have acquired minds of their own, anguish is flavour of the month, and there is no clear idea of what happens next. But new and exciting abilities have appeared, with a power yet to be explored. Once such ability, with enormous potential, is the speech technology described in this book.

18.1.4   What a piece of work is Man

The greatest challenge in further developing and exploiting the potential of speech technology lies in gaining a deeper understanding of how human speech works. Today's adolescents might well describe human skills in speech generation, reception and comprehension as 'pretty neat — yeah, wicked'. This phrase is, in itself, a good illustration of the mutative variability of speech, since the message received depends on the receiver. Shakespeare put it more eloquently when Hamlet described humankind as: '...How noble in reason! How infinite in faculty! In form and moving how express and admirable! In action how like an angel! In apprehension how like a god!' (Act II, ii, 317).

To a signal processing engineer, speech is an awkwardly complicated thing [6] and difficult to classify. It is sometimes ergodic but seldom stationary, being at times quasi-periodic and at others quasi-random. The one dependable property of speech is that it varies, not only with different talkers but also with the same talker at different times or in different surroundings.

The way our brains are wired for speech seems to involve multiple layers of complexity. Medical evidence, from patients with disease or accidental brain injury, provides clues about how these functions are arranged in the brain. Grammar, vocabulary, cognition (understanding) and linguistic (speaking) skills seem to be quite distinct and independent. Literal and metaphorical understanding also seem to be separate abilities. Any one of these functions can be seriously impaired without much impact on the others. For example, some stroke victims suffer 'empty speech', with perfect grammar and linguistic style but severe loss of, say, nouns — despite understanding the 'lost words' when they hear them. Children suffering from a condition known as 'Williams Syndrome' show severely retarded intelligence, but often use erudite vocabulary in an exceptionally skilful way.

This evidence suggests that emulating human speech faculties with a machine might be impossible, but many applications of speech technology have substantial structure or prior knowledge about the task, which makes useful implementations practicable.

18.1.5   The power of speech

Vogt [7] has analysed the dimensions of human cognition as the key skills for his future breed of 'knowledge workers'. From Fig. 18.1, it is clear that eliminating the speech-related aspects and using only text, data and image-based features would seriously damage or even remove much of the power of human thought. One picture can indeed be worth a thousand words, but the mental pictures that form inside our heads are more potent still — and the most effective expression of those images is through the fast and flexible power of speech.

Speech is more information-rich and revealing than is generally appreciated. Wiseman conducted a test [8] on the relative accuracy of different ways of detecting lying. Observers who just listened to the potential liar speaking did significantly better than those who also watched his/her picture or those who only read his/her words. In this large-scale test, the observers with the most clues (those with sound and picture) produced the least effective detection result[1].

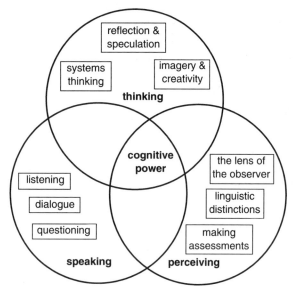

Fig. 18.1   The dimensions of human cognition, from Vogt [7].

The ability to invent metaphor is often considered one of the hallmarks of genius — and linguistic metaphor is, as any good politician knows, the most powerful and memorable of all. The power of speech is something that advocates of new technology for data and image networking, in their enthusiasm for new technology, occasionally seem to forget. From the 'Grunters' of the Pleistocene age to the 'Mass Mediators' of today [9], communication by speech has fuelled the ascent of humankind. It is absurd to predict the death of the telephone, but plausible to predict its transformation into something that will look and feel quite different from the device we know today.

18.2   THE GROWTH OF TELEPHONY

18.2.1   Birth

The notion of transmitting sound to a distant place goes back to antiquity. The world's first practical electric device called a telephone (from the Greek,

---

[1] Detection by a total of 41 471 television, newspaper and radio observers was 51.8, 64.2 and 73.4% correct respectively (50% by chance) in the 'Megalob Experiment' during BBC Television's 'Tomorrow's World', 25 March 1994.

meaning 'far sound') operated digitally. In 1861, Johann Philipp Reis, a Professor in the Garnier Institute of Frankfurt-am-Main, demonstrated an apparatus to his local Physical Society that he called a telephone [10]. It could transmit the pitch of speech sounds, but not their intensity or articulation. It had no true microphone, but used the vibrations of sound to switch an electric current rapidly on and off by a delicately adjusted, diaphragm-driven contact. It was probably David Hughes of Kentucky who proposed the first true microphone, using the variable electrical resistance between iron nails, or carbon sticks, placed loosely in contact. However, Alexander Graham Bell and Thomas Augustus Watson (Figs. 18.2 and 18.3) are universally acknowledged as the true fathers of the 'articulating electric telephone'.

Fig. 18.2   Alexander Graham Bell.

18.2.2   Childhood

Some key events in telephone history are shown below.

- 10 March 1876 — in a Boston attic, Bell shouts: 'Mr. Watson, come here, I want to see you' — the first intelligible sentence ever transmitted by telephone (Fig. 18.4). Two days later, Bell wrote in his notes: 'The effect was loud, but indistinct and muffled', [11].

- 1881 — the British Post Office grants the first licences for commercial telephony in the UK, limiting service to a five mile radius and extracting a 10% royalty [12].

Fig. 18.3   Thomas Augustus Watson.

- 1884 — Clément Ader demonstrates binaural telephony, using two microphones placed some distance apart near the stage of the Grand Opera in Paris. Listeners in the nearby Exposition Hall use two Bell receivers, one to each ear, to experience spatial sound reproduction [10].

- 1891 — an undertaker in Kansas, Almon B Strowger, patents the 2-motion selector for automatic telephone switching [13].

- 1892 — the first automatic telephone exchange opens in La Porte, Indiana — using the Strowger system.

- 1893 — the 'Electrophone Service' sells entertainment by telephone, from London theatres and churches. The service grows modestly until radio broadcasting brings closure in 1926.

Fig. 18.4 The receiver of Bell's original telephone

- 1912 — the first public automatic telephone exchange in the UK opens at Epsom in Surrey, using the Strowger system.

- 1927 — the first transatlantic telephone calls are made, by radio between Rugby in the UK and Rocky Point in the USA.

- 1943 — the first amplified undersea telephone cable links Anglesey in Wales with the Isle of Man.

- 1956 — the first transatlantic telephone cable links Clarenville in Newfoundland with Oban in Scotland.

- 10 July 1962 — Lyndon B Johnson makes the first public telephone call via an active satellite in space and says: 'You're coming in nicely...' [14].

- 14 February 1989 — the first satellite telephone call from an aircraft in flight is made, live on BBC Television's 'Tomorrow's World', using the BT Skyphone™ speech coder.

The earliest days of telephony were very inventive, with experiments in stereo and the commercial distribution of entertainment; but it soon became clear that the most important work was to find efficient solutions for the problems of transmission and switching — tasks which were to occupy generations of engineers for most of the 20th century. As a result, transmission and switching are commonly regarded as the heart and soul of telecommunications.

Transistors replaced thermionic valves in the 1960s, but the manufacturing tolerances of affordable inductors and capacitors still limited the practical complexity of analogue signal processing. In the 1980s, the precision and repeatability offered by digital signal processing (DSP) devices finally removed the limitations of analogue signal processing. This has enabled the economic use of a wide range of complex, adaptive or nonlinear processing algorithms. These ideas had previously been too cumbersome or expensive even to contemplate. This renaissance has been breathtaking [15] in its speed and impact on areas as diverse as communications, weapons, entertainment and transport.

18.2.3   The child today

In the past decade, digitization has enormously improved the economics and flexibility of switching systems. Semiconductor lasers and optical fibres have revolutionized network transmission, and copper is proving resilient in the local loop [16]. Speech coding (see Chapter 3) has helped reduce the cost of international telephony, and is the economic basis of value-added resellers such as Concert™. Global growth in mobile telephony depends on speech coding, through Skyphone™ [17] aeronautical or GSM digital mobile systems (see Chapter 4). Over the same decade, the telephone has been transformed in appearance (Fig. 18.5) and greatly reduced in cost. Modern electroacoustic transducers and integrated circuit technology have been incorporated, without fundamental change in specification or performance — but radical change now awaits.

Fig. 18.5   Telephones from Gower-Bell to Slimtel.

18.3   COMMUNICATION CULTURE

18.3.1   Convergence

The computing, entertainment and communications industries are widely forecast to converge and practical evidence of this trend abounds, but observers disagree about the ultimate result. Personal computers, especially portable ones, have recently absorbed the modem and dialler into their hardware and software. Rapid growth in the use of facsimile and caller identification technology has prompted the addition of these functions. The next stage in this integration may see the computer swallow the whole telephone. One manufacturer [20] already offers a computer with facsimile, voice mail, caller ID, paging and duplex hands-free telephony.

All sorts of new services are becoming practical, such as teleconferencing, teleworking, teleshopping and video-on-demand [21-23]. These services raise new and sometimes conflicting issues. For example, guaranteed privacy [24, 25] is needed to prevent commercial or government abuse of personal data without letting criminals exploit secure communications.

The authors believe that the most rapid and dramatic effects of convergence will be seen first in education [26], because of the breadth and depth of information that will become available, which no single family or school could otherwise afford. Children are by nature 'early adopters', with an eager taste for knowledge — who knows where they will lead with the tools we give them ?

18.3.2   Cultural change

The pundits are polarized about the future cultural impact of telecommunications, predicting a world ruled by either Mephistopheles or Gabriel. The most dire prediction, of the censoring and monitoring of everything with zero personal privacy, makes Orwell's '1984' seem sunny and transforms the Dark Ages into a jolly utopia. The opposing view predicts a triumphant victory of the democratic, libertarian market-place, where the Government disappeared after there was nothing for it to do. Kinney [27] debates both views. Personally, the authors suspect that improving communications will continue to mildly foster democracy and that the politicians have more to fear than the people.

We seem set to enter a 'knowledge worker' world [28], where new technology alters the nature of work itself [7], 'work tribe' alliances blur national allegiances or lead to the death of the nation state [27], and society divides into information-access 'haves' and 'have nots'. The Internet is widely cited as an engine of such cultural change, but with an impersonal or even dehumanized outlook. More natural and versatile telephony, carrying more of the nuances of 'being there', will drive the same cultural changes but with an emphatically human face.

18.3.3   Bit-bare or bountiful?

Lewis et al [29] have previously discussed two imaginary futures — one was a 'bit-bare' world, with all connections mobile and the transmission of uncompressed signals illegal; the other 'bit-bountiful' scene was the opposite, with mobile telephony illegal and any kind of signal compression laughable. Since that time, the bit-bare future has grown closer, with a marked increase in the use of bit-rate compression for mobile telephony and network economy.

However, there is also evidence of a bit-bountiful trend, with the growth of entertainment services on fixed networks. Negroponte [25] speculates that, in future, everything which now comes over the air will come by wire, while all that now arrives by wire will come over the air. The future may turn out to be both bare and bountiful in bits.

18.3.4   Specific fallacy

Some say that telecommunications companies will become transporters of bits, without any need or way to add value to the process. Some observers [2] go further and suggest that: 'They are no different than electrical, water, sewer or any other utility competing for the right to push stuff through a pipe'. The authors find this outlook bizarre. Do you tell your water company which specific droplets of water, of all the many in the reservoir, you would like to use next? When poised at a light switch, do you tell the electricity company where you would like the energy to come from ? When you make a one minute telephone call, are you happy to receive 3.75 million of whatever bits first come to hand?

Telecommunications is profoundly unlike water or electricity supply, because of its specificity. Bits are not 'stuff' in the sense that water, electricity or sewage is. They are not even 'stuff' in the sense of a postal service, since in telecommunications an order of arrival different from that of despatch destroys the message. Bits are not the 'stuff' of the product. Meeting human communications needs, in all their rich variety and with planet-wide, technology-independent interoperability — that is the product.

The supply of telecommunications is like the supply of water or electricity in the same way that painting a picture is like painting a warehouse.

18.3.5   Clairvoyance

'Around the world thought will fly, in the twinkling of an eye,' — Mother Shipton, in the 16th century.

Unlike Mother Shipton, ordinary mortals only ever see the future through the shifting and cloudy veil of today. William Preece's famous remark about messenger boys (see Appendix) shows how poorly the future of telecommunications has been forseen. When some new technology arises it is tempting to think how, in isolation, it could change things. This leads equally easily to wild optimism or blind pessimism. What it is difficult to do [30] is to remember that everything else can change or interact too. The authors suggest that focusing on the human benefits, rather than the technical capacity, of a new idea provides the best foresight — because human needs change much more slowly than technology. Therefore, with almost no sense of guilt, the authors invite the reader to seek truth in unreality and to 'boldly go' to the next section.

18.4   LEARNING FROM FANTASY

18.4.1   Enterprising communications

When Gene Roddenberry [31] and his team of script-writers created the TV series *Star Trek* (Fig 18.6) in 1966, they little knew how frequently or how closely their work would be scrutinized. By a combination of imaginative flair

Fig. 18.6   The crew of the *USS Enterprise* in *Star Trek*.

and an eye for detail, they transcended the limits of a small budget to produce science fiction that still has something to tell us today. It is not the power of the photon torpedoes nor of the warp drive that astonishes the authors, it is the communications system of the *USS Enterprise*. Its performance so completely realizes the potential that today's speech technology has just begun to reveal.

If Mr Spock saw a 1997-vintage telephone in a time-warp, he would no doubt say: 'It's communication, Jim, but not as we know it.' There are no telephones aboard the *Enterprise*. There are no handsets, no keypads and no bells — so what is left? In a sense, the *Enterprise* is one huge telephone. Speech recognition has replaced the keypad and the telephone number — the crew speak to distant people by saying their name.

In the first *Enterprise*, small boxes on the wall use acoustic echo cancellation to provide two-way, wideband hands-free speech. On the surface of a planet, the crew carry a communicator. This is still not a handset — it is simply opened and held in the hand to both listen and speak. This communicator is more than a mobile telephone. The matter transport system needs an extremely precise three-dimension location to 'beam someone up' and this is greatly assisted if they carry a communicator (Fig. 18.7).

In the second *Enterprise*, technology has moved on. Multi-microphone, beam-steering arrays are installed on the ceiling of every room, corridor, nook and cranny. These allow unhindered and natural hands-free communication, from any point in the ship to any other.

Fig. 18.7   An *Enterprise I* communicator.

The speech technology in the *Enterprise II* is provided by hugely distributed, massively parallel DSP computers, with a staggering number-crunching capacity. Each microphone array has over 500 tiny diaphragms, with an organic semiconductor DSP device painted on the back of each one. There are a total of 94 248 such arrays, fitted at 6 m intervals throughout the ship, with a combined processing power of approximately 2.4×10^{15} operations per second. The computational requirements of the speech synthesis, recognition and interpretation systems are, of course, very much lower per channel. But these systems have to track many simultaneous conversations, resulting in a broadly similar processing load.

The badges worn by the crew (Fig. 18.8) are not miniature mobile telephones, but merely serve to authenticate the position of their owners to the communication and security systems. If you send a message to a crew member who is asleep, or has said they do not want to be disturbed, the ship automatically stores the message. Unless, of course, the ship's speech recognition and interpretation systems detect a matter of urgency. In that case, the crew member is woken or alerted with a chime — the last ghostly echo of the telephone.

It is interesting to note that despite the ship's ability to hear every word said on board, no person of malign intent has yet breached the security software which protects personal confidentiality.

Fig. 18.8 The crew location badges in *Enterprise II*.

18.4.2 Comfort and convenience

Staggering speech technology is used abundantly on the *USS Enterprise*, often to add mere convenience, not facility. Some readers might feel this is a gross abuse of engineering. Will speech technology in future be reduced to the status and functionality of paint? In desktop computers, Microsoft has recently realized what Apple has known since birth — that a machine interface which closely matches natural human behaviour brings benefits that can justify the complexity. Bell knew this too; in 1878 he said: 'The great advantage [the telephone] possesses over every other form of electrical apparatus consists in the fact that it requires no skill to operate the instrument.' The designers of the *USS Enterprise* show a deep and subtle understanding of this point. They have chosen to put the interests of the crew first, whenever and wherever possible. Speech technology is indeed used abundantly on the *Enterprise*. It makes the lives of the entire crew more comfortable and efficient, by providing a natural and simple match between human and machine skills.

The crew can talk to each other or to the ship's records, wherever they are, with the same effortless ease as to someone standing alongside. Plotting a new course, or targeting an enemy vessel, needs little obvious effort from the crew — because the *Enterprise* can recognize and interpret the relevant words in the speech of the Captain or other key people on the bridge. The ship's computer talks in a voice that not only sounds entirely human, but shows due deference to the Captain's rank — except that, in times of external threat or internal stress, the computer's vocal inflexions, intonation and even choice of phrase change, if necessary, to chide or even rebuke the Captain, as an equal. The synthetic speech system aboard the *Enterprise* clearly embodies aspects of a synthetic persona.

18.4.3   Optimum interface

However, the ship's designers also show a sharp appreciation of the occasions when safety or security demands a different approach. If you fire a photon torpedo at someone, you had better mean it. The safe operation of the *Enterprise's* weapons and navigation system demands the utmost level of reliability and integrity, as well as an unambiguous audit trail to show who did what, where and when. That is the reason why the weapons are fired by the antique technology of the push-button and why the helm is not voice-operated but controlled by tactile feedback. Of course, those buttons and levers take your finger-prints and capture a tiny DNA sample, should you operate them. Sometimes, buttons are best.

18.4.4   People in charge

It may have been compelled by a low budget, but all the technology in *Star Trek* — even the most awe-inspiringly powerful — remains reassuringly human-sized or even invisible. Furthermore, technology is constantly taken for granted in *Star Trek* and is always secondary to the main theme of interplay between minds — human and non-human. *Star Trek* machines have to adapt to their users' needs. How different from the viewpoint thirty or forty years before, in films such as

Chaplin's '*Modern Times*' or Lang's '*Metropolis*', where people are subservient, powerless slaves to the great machines.

This inversion of emphasis has already started in telephony. Fifty years ago, customers had to contact the operator for long-distance calls and accept a radical reduction in speech quality. Today, customers can dial calls regardless of distance, with similar expectations of quality. In future, the needs of users will define the machine, rather than the needs of the machine dictating the behaviour of its users. As Cochrane and Westall say [32]: 'Stop bending people into the technology, and start bending the technology into people.'

Rothenberg [33] uses the phrase 'deep technology' to mean a philosophy of deployment that emphasizes the inter-relatedness of humanity, nature and technology, leading to a potential reformation of cultural attitudes. He stresses how technology, used correctly, can be transparent, extending human insight and enhancing our appreciation of value, in each other and in the natural world. With an outlook that encapsulates the same attitude to technology as Roddenberry's *Star Trek*, he says: 'Technology is empty if it considers only itself. Deep technology ... does not hide us from the world and does not teach us to feel more powerful than we really are.'

18.5   TELEPHONE EVOLUTION

18.5.1   Alien thought

From this 'deep technology' perspective, how well does today's telephone perform? Familiarity breeds difficulty in just thinking about, let alone answering, this question. So, shed all previous experience. Shed all humanity and step into an alien's skin and mind. What might you see and say about telephony?

> '*Remote communication is their biggest machine, but frustratingly we have not identified the tribe who built it. Its connectivity is impressively rapid and planet-wide, using symbol sequences called telephone numbers. As humans forget these numbers, they often carry a device for linking them to other sequences they do remember, called names. Curiously, their remote communication machine is entirely unresponsive to these names. It is also odd that their remote speech uses a much smaller range of sound than their native mode.*
>
> *When remotely communicating, humans hold an object against one side of their heads. This is shaped like that strange 'bah-nah-naah' fruit, which Om*rty foolishly tasted, although more sombre in colour. This object is common in their communication mythology. The simple task of holding it fully occupies one of their upper extensor protrusions, of which*

they have only two, for the entire communication. Humans willingly accept this functional loss. Om*rty is convinced this behaviour symbolizes an earlier tradition of ritual sacrifice, perhaps in payment for knowledge lying beyond their native senses.'

Handsets are such a familiar and integral part of the telephone that, to some people, the idea of not having one is almost as alien as Om*rty. Handsets are clearly preferable for tasks needing privacy or mobility. But not having to hold a handset can be more convenient in some applications and closer to the experience of face-to-face contact. Yet loudspeaking telephones, available for over forty years, have had only modest market success — because of acoustic limitations.

18.5.2   The first and last metre

When optical fibres reach that final domestic doorstep, some communications engineers might feel their dreams have come true — perfect transmission at last; but the real channel in human communication stretches from one mind to another.

Sound propagation through the air can be a very nasty channel indeed (Figs 18.9 and 18.10), with hundreds of times more fluctuating, multipath propagation than a radio engineer's worst nightmare. This is the technical strength of the handset, placing the electroacoustic transducers so close to the mouth and ear that acoustic transmission is largely free of noise and multipath effects. In future telephony systems, the properties of the acoustic channel will be the dominant technical impairment.

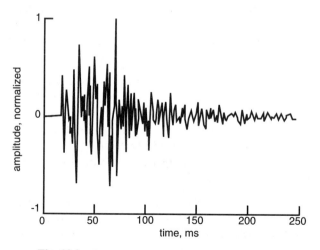

Fig. 18.9 Typical acoustic response between transducers.

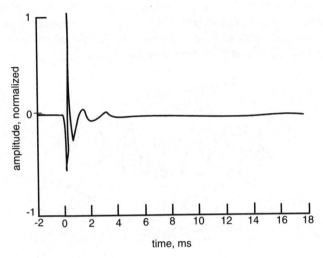

Fig. 18.10   Typical electrical analogue of response in Fig. 18.9.

In loudspeaking or hands-free telephones, for desktop or room conferencing, the transducers are usually better coupled to each other than to the user. This causes two problems. Firstly, a serious risk of loop oscillation and, secondly, multipath contamination of the outgoing speech signal — the 'bathroom effect'. Incoming speech also suffers similarly, but the listener's ear and brain processing essentially removes such contamination.

In 1979, South et al [34] proved the feasibility of solving the first problem with echo-cancelling adaptive filters. This technique [35] provides the same duplex (both-way simultaneous) transmission as a handset and avoids the syllable-clipping degradations of earlier voice-switching methods. The impact of the second problem can be reduced by using a beam-steering microphone array, using an associated set of adaptive filters. This can produce a highly-directional pick-up beam [36], rejecting angularly separate multipath and noise signals, to give a speech quality like that of a handset microphone, but from a distance. Such array microphones could also reduce the impact of environmental noise or nearby conversations on speech recognition systems.

$$f_{max} < \frac{c}{2d} < \frac{(n-1)}{2}f_{min} \qquad \{n \geq 3\} \qquad \qquad \dots (18.1)$$

Equation (18.1) relates the number of microphones n in a linear array of spacing d with the frequency range f_{max} to f_{min} when the speed of sound is c. The directionality of a seven-element, experimental array is shown in Fig. 18.11. In practical teleconferencing systems, with wideband hands-free audio, a physically large array of many microphones is implied by equation (18.1). Such an array could be wall-mounted in a decorative mural, painting or poster. Logarithmic spacing, array weighting or sub-array techniques can be used to avoid an undesirable increase in array directionality with frequency. Efficient algorithms for beam adaption and talker steering in practical environments are currently a subject of international research and development.

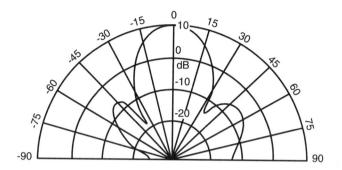

Fig. 18.11 Polar response of an experimental array at 1 kHz.

There are simpler alternatives to holding a handset, such as lightweight cordless headsets, cordless lapel microphones with loudspeakers or cordless earphones with beam-steered microphones. Low-power radio or optical transmission [37] could provide such cordless links.

18.5.3 Reality limited

In some situations, a significant delay in replying to a question can imply the opposite of the verbal meaning. As other impairments in speech, data and video communications are reduced or removed, future customers will come to regard excessive delay as infuriating [38] and a significant barrier to natural interactions. This is likely to reinforce the current economic trend towards the use of ground-based links for long-distance and international calls, freeing satellite bandwidth for mobile services.

The quality of conventional telephone sound — fit for purpose, but hardly high fidelity — might seem a major barrier to greater realism in communications. However, wideband speech coding (with a bandwidth of 50-7000 Hz and a bit rate of 64 kbit/s — ITU-T Recommendation G.722) has been technically feasible for many years. It requires new telephones and exchange equipment, but is compatible with existing switching and transmission systems. It can provide much more natural sounding speech, with a useful improvement in intelligibility. The effects of competition and the enormous inertia of the telephone market have prevented its widespread use. The computer and entertainment industries have used wideband audio-coding techniques for many years.

18.5.4   Variety and adaptability

In the light of these potential improvements, today's telephone is nonetheless a remarkably successful device. Anyone daring to predict its radical evolution must analyse what customers think is good, and not so good, about the thing. Human needs vary and variety enriches life. Different messages, for different purposes, suit different media. We are remarkably adaptable creatures, moulding our behaviour to the circumstances. We unconsciously adopt quite different psychological attitudes in a hand-written letter, a business facsimile, an e-mail, with a voice mail message, in a telephone call or during a videoconference.

Only in the videoconference do these attitudes approach the way we communicate face to face, but not every message needs or suits face-to-face contact. Television has become an enormous success. But it has not killed radio, not least because radio listeners are mentally captive while physically free — to get on with something else. Less can be more. Much of the power of conventional telephony, and the appeal of its mobility, lies in the spontaneity and casualness with which calls can be made and received — wherever we are, whatever we are doing and however sloppily or scantily we are dressed.

The authors are sure the evolution of the telephone will be just that — evolution, rather than revolution. New forms and functions will be added, without necessarily abandoning the power and availability of the existing ones. Customers will be most likely to embrace changes which offer greater convenience, improved ease of use or more naturalness in communication. Cochrane [39] describes an ultimate kind of telephony interface that is '...so kind, so beguiling to humans that they can't resist it...'.

18.5.5   Plural not singular

The experience of telephony today is rather like buying a piece of fruit — in a world where every food market sells the same, single kind of fruit. But in future, the shopkeeper is going to enquire about the precise nature and purpose of the 'fruit experience' which you had in mind. What sort of taste, nutrition, texture, vitamin content, colour and size? When, where and with whom are you thinking of eating the fruit? You reply and the shopkeeper continues: 'Ah, an apple then. Two of the nineteen varieties of apples we sell meet your requirements. Take your pick.'

The authors believe the future of telephony will be plural not singular. The low cost of new speech and video DSP technology, added to the effects of computer convergence, competition, market liberalization and high-speed network technology is likely to fragment the current single-specification market. A burgeoning variety of customer equipment and services would then arise, to meet the previously suppressed variety in the range of human communication needs.

Singular uniformity of specification in telephony has gone unchallenged for so long it seems like bedrock — it was always absurd to consider anything else. This has inhibited improvements in sound quality and limited variety in modes of use. These were tiny sacrifices compared with the mammoth task of achieving efficient global connectivity. Furthermore, uniformity led to one of the most important achievements of telephony today — the universal interoperability of any two telephones anywhere in the world.

A plural set of telephone-with-computer-with-network specifications seems fatal to this vital achievement. Fortunately, the same DSP speech technology that economically enables pluralism can also provide future-proof interoperability, by signal conversion between otherwise incompatible systems. These conversions can add value — for example, a customer using a standard mobile telephone could be a participant in a wideband videoconference. Variety need not spell exclusion, merely 'mix-and-match' freedom. It may seem odd, but the future of telephony is not just the telephone.

18.5.6   Face to face and eye to eye

New ways of using plural mixtures of technology for telephony in the future [40, 41] will need new names, to describe the different but interoperable services that will arise. Such services might include:

- shared-computer-desktop cordless-headset narrowband telephony;

- cross-computer-platform white-boarding with hands-free narrowband telephony;

- wideband hands-free audio and video multi-person-to-person calls;

- high-quality, spatial audio and videoconference calls.

These new options arise when multimedia technologies [42], such as shared computer workspaces, eye-to-eye video and immersive conferencing, are integrated with conventional telephony. A remote location could virtually 'appear' through one wall of a conference room (Fig. 18.12). Additional hardware or new delegates could readily be added to a meeting, even though their need had not been anticipated — something impractical in a remote physical meeting.

Developments in speech synthesis and recognition could replace and augment the role of a human operator in simple and well-defined tasks. An electronic 'secretary' could spot key words and recognize important action points in a business teleconference. This 'secretary' would then summarize the items of interest for each participant and prepare a customized text file, annotated with the relevant speech fragments. Speech technology could also extend the human-computer interface, long dominated by the keyboard, button and screen, by adding an

Fig. 18.12   High-quality immersive teleconference.

intuitive speech interface. For example, a WEB browser might gain a 'voice-driven, voice-responding' fuzzy search engine, particularly sensitive to the expression of uncertainty. A 'voice helper' might bring conventional display technology closer to the power and flexibility of paper books, by aiding navigation and browsing.

Looking further into the future, imagine using eyes-and-hands-free natural speech, in any language, to instruct intelligent agents about the kind of Internet information you seek. Now imagine roaming seamlessly from reality into a shared, virtual workspace — and being able to see, hear and touch synthetic objects or distant people, as easily and intuitively as in the real world.

All these dreams lead the same way, to a kind of remote communication more like natural face-to-face contact. This is the true destiny of telephony — not in the single mode of narrowband handset speech but in a choice of mode, spanning multiple human senses if appropriate, that addresses previously unsatisfied human needs in communication.

18.6   PLOTTING A COURSE

18.6.1   Where are we going ?

If we stand on the threshold of a dream, how do we get there? The world has changed in all sorts of ways. The global future of telephony is increasingly dominated by energetic commercial competition and multinational business structures. Younger industries, with more flexible and customer-centred behaviour than has been traditional in telecommunications, look hungrily at the potential profits [19]. This situation is quite unlike that which applied during most of the 120-year history of the telephone. A multitude of factors, some new and some not, need fresh examination or reappraisal in this new world.

18.6.2   Changing expectations

One of the authors can remember his intense frustration, as a child, on opening a plastic 'Meccano-type' construction set. There were lots of straight, flat bits suitable for building houses, cars and bridges — but no curved pieces at all, which made it really difficult to build a good chicken. Children can take technology and use it in ways that parents do not dream of.

Equally, telecommunications products are now used in places and in ways that the original designers did not consider. High-speed data modems and facsimile machines are commonplace today, but did not exist when the first national networks were developed. Telephones are now used casually in any room of the house, rather than reverentially in the hall. Public payphones are expected to work well in the noisiest of bars and public places. Yet the electroacoustic characteristics of telephones are largely unchanged from those laid down generations ago.

In an increasingly competitive environment [43], subject to rapid changes in technology, customer expectations may rise or become harder to define precisely before launching a new product or service. But this kind of environment is also one where customers' views will spell the difference between commercial success and failure. Prudent designers of future telephony systems, despite cost implications, always consider that 'Meccano chicken' kind of question.

18.6.3   The mind-set menace

During the multi-faceted and multi-party development of Skyphone™ technology [17], the authors learnt the benefits of a holistic outlook — meaning that the whole can be more than the sum of its parts. This idea sounds more like marketing hyperbole than real engineering, when considered from the 'linear systems' viewpoint that is widespread in telecommunications.

When the principle of superposition [44] applies to a system, the whole is always precisely equal to the sum of its parts (Fig. 18.13). The telephone's history has been dominated by essentially linear and time-invariant systems. That is why it was possible to neatly subdivide telephony into the independently developed domains of switching, transmission and customer equipment. Any one part could be amended or rearranged, independently of the rest, with largely predictable results. This 'linear' mind-set has become so widely and deeply ingrained in the industry that it often excludes any other outlook [45].

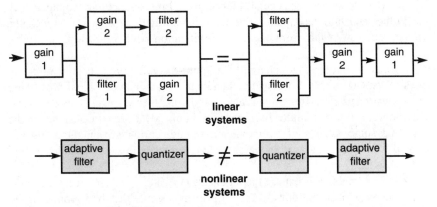

Fig. 18.13   Cascading linear and nonlinear systems.

In business, as in cooking, no one would dispute that the order in which you do things changes the outcome. Such non-commutative behaviour is a hallmark of nonlinearity. As explained in section 18.2.2, new DSP technology has opened a breathtaking panorama of both linear and nonlinear design opportunities. Most of speech technology, such as speech coding, context-dependent speech recognition or sequence-dependent phoneme pronunciation, deals with nonlinear systems. New transmission and switching technology, using asynchronous transfer mode or frame-relay techniques, can also show significant nonlinear effects [46].

The preconceptions of the 'linear' mind-set can be a dangerous handicap when dealing with nonlinear technology. If any one part is amended or rearranged, the results may not be predictable. This nonlinear world is reminiscent of the Discworld® novels of Pratchett [47]. There, life is conventional most of the time except when weird magic breaks out, with unpredictably pleasant or disastrous impact.

There is a significant risk underlying the piecemeal use in telecommunications of systems that are, to varying degree, nonlinear (see Chapters 3 and 4). Those previously independent domains of switching, transmission and terminal design could become uncomfortably interdependent or confusingly less distinct [48, 49]. Delivering predictably good service to customers would then become more complicated.

18.6.4   Hybrid vigour

Someone who studies several disciplines, rather than specializing in one, is sometimes described as a 'Jack of all trades and Master of none', but the disadvantages of over-specialization were known as early as the 17th century. René Descartes, in his 1629 treatise 'Rules for the Direction of the Mind', wrote: 'If, therefore, anyone wishes to search out the truth of things in serious earnest, he ought not to select one special science, for all the sciences are cojoined with each other and interdependent.' The world of horticulture also knows well the vigour that can result from the cross-fertilization or hybridization of living things.

Speech technology is inherently a multi-disciplinary subject, embracing physical acoustics, psychoacoustics, linguistics, mathematics, signal processing in hardware and software, human factors and computer science. Does the growing importance of such multi-disciplinary technology [50] require a change in the way that some aspects of communications engineering are taught in future? Should it even alter the way in which certain academic disciplines are classified and subdivided?

When an engineer is exposed to a new idea or gadget, the classical reaction is a fascinated 'Ooh, what does it do?' However, those with a background in the arts or humanities have a typical response which is cooler and ultimately more powerful — 'Ah, what can it do for me ?' Those two words at the end of that question signal a world of difference, intimately connected with hybrid vigour.

18.6.5   Mindshift

A plural future for telephony has major repercussions on telecommunications planning and design. Instead of solely transporting and routeing bits, networks will need to manipulate them, processing signals to meet customer requirements for interoperability. Pluralism also requires the integration and hybridization of previously disparate engineering disciplines, in speech, video, human factors, networking, terminal design, information visualization and more. A radical mindshift in telecommunications engineering will be needed, changing ways of thinking that worked well in the past, but which may work badly in future.

18.6.6   Ways of thinking

'Empires of the future are empires of the mind' — Winston Churchill.

Challenges to formerly immutable fundamentals should not really surprise us, because all revolutions inspire chaos. Science is concerned with the pursuit of knowledge, but engineering makes a business of accommodating uncertainty. That is why good engineering is an art as well as a science. If electronic puberty is a good description of our situation, then perhaps we should examine how adolescents cope with the real thing. Is it foolish to suggest that a 'hang loose' or 'let the force be with you' outlook might sometimes be valuable?

In letting go and embracing subjectivism, perhaps the world of science-fiction can teach hard-bitten, analytic technologists a few useful tricks [51]. When IBM executives first saw the ENIAC computer, they were blind to its potential. Their thoughts were dominated by the physical embodiment — a huge room crammed full of large, heavy, hot and expensive equipment. They thought of the manifestation and not the metaphor. They could not separate the function of the device from its form. But such alternate ways of thinking are the foundation of science fiction, which makes logical, rational and self-consistent explorations around imaginative leaps in culture or technology. It is probably no accident that two of the most penetratingly accurate pieces of foresight this century came from the best of science fiction writers — H G Wells about the start of World War II and A C Clarke about geostationary communications satellites. The fantasy discussion of *Star Trek* technology in section 18.4 had a serious purpose.

Telephony has an exciting and challenging future, in which we will have to learn how best to exploit the transforming potential of speech technology — but much more than that, we may need to re-examine, or even re-learn, some of our fundamental ways of thinking[1]. That will be part of both the price and the prize of getting through electronic puberty and becoming adult.

18.7   CONCLUSIONS

The childhood of telephony was a time of slow but progressive growth on reliable foundations, with strict parental control by government monopolies. The child was well taught, but adolescence has already arrived — with its rapid, chaotic, miserable and exciting, sometimes frightening, painful and wonderful changes. It is already hard to tell where the former child is going or quite how things will turn out. Strict control is counter-productive [52] and other stronger influences abound — in the effects of fierce competition and peer pressure, accelerating technological change and the convergence of previously distinct markets.

In the face of these developments, the authors hope to have challenged the reader's preconceptions and stimulated fresh thinking about the form and function of the telephone. It is impossible to over-emphasize the scale of the potential changes and almost as difficult to grasp their impact. Imagine a medieval knight trying to understand a machine gun by studying his sword. Logic and prudent judgement are no help here — a leap of imagination is needed.

The original purpose of telephony, 'far speech', has become uniquely linked to the current form of the telephone. That form has been so successful that the fundamental metaphor has become obscured. The authors have tried to restore that metaphor to its prime place. Telephony is about the abolition of distance, between one person and another or between people and machines — but, more than that, its true meaning is satisfying the subtle and richly varied needs of human communication.

---

[1] Vogt [7] says: 'While convergent technologies show significant promise in supporting a radical rethinking of (these) traditional concepts and processes, the effort must begin with our thinking, not with the seductive technologies'.

Predicting the future is usually fraught with failure, but it seems safe to bet that telephones with the appearance of less will accomplish more. Layer upon layer of complex but increasingly affordable speech processing will offer a simple and progressively human-centred interface to a widening range of services.

An information revolution is forecast for the coming century, with communications technology at the forefront of all-embracing change. But this revolution will not stop at better information management. It will reach to the very roots of our humanity, affecting how we individually and collectively learn, think, work and grow. It will change our culture and reshape our concepts of community.

Adopting both the fervour and the bias of the evangelist, the authors feel confident that speech technology will be the most powerful engine of widespread embrace in this revolution. The authors also believe that processing signals will become just as important as transmission and switching are today in telecommunications. Meeting and matching the true needs of customers, and of society as a whole, will become a force stronger than any commercial deal, seductive technology or regulatory restriction. The telephone is on the threshold of multiple transformations — in user-friendliness and convenience, in sound quality and naturalness, and in shape and appearance. These transformations will lead to the largest and most radical shift in the educational, commercial and cultural impact of communications since the telephone was born.

The emotion and magnitude of this prophesy is encapsulated by an episode from cinema history. *The Jazz Singer* was the first commercial talking feature film, using manually-synchronized 78 RPM gramophone records to play a few brief songs. Al Jolson and ad libbed between songs with his stage patter, chancing to say something prophetic about the future of cinema sound reproduction. Warner Brothers dared to leave these snippets of speech in the film they released. Almost seventy years ago, in cinemas around the world, people sat open-mouthed in astonishment or dewy-eyed with emotion as Jolson's image waved its arms and bounced on its toes, shouting [53]: 'Wait a minute: Wait a minute. You ain't heard **nothin**' yet !'

The authors believe that this remark will prove to be even more securely prophetic about the future of telephony. It is indeed the essential message of this chapter.

APPENDIX

Early vision of telecommunications

William H Preece was the Post Office Engineer-in-Chief from 1892 to 1899. From a Welsh Non-Conformist background, he was much influenced by Faraday, Thomson and Airy. He gave testimony about the telephone to a House of

Commons committee in 1879, saying [12] '...Here we have a superabundance of messengers, errand boys and things of that kind.... Few have worked at the telephone much more than I have. I have one in my office, but more for show. If I want to send a message — I use a sounder or employ a boy to take it.'

This quote maligns his vision. He was probably giving his political masters the kind of message he knew they wanted to hear, emphasizing short distance operation to deflect criticism. It seems the Treasury wanted the telephone to fail, to defend revenue from the State-owned telegraph service. He had previously been enthusiastic about Bell's telephone. He subsequently co-wrote 'The Telephone' — the first authoritative book on telephone engineering published in the UK. Political interference in telecommunications is not new.

REFERENCES

1. 'Telecom's phone revenues will fall to zero says Campbell', Editorial,   Exchange (1 July 1994).

2. Malamud C: 'Viewpoint', IEEE Spectrum, p 32 (January 1995).

3. Cerf V G: 'The Internet's 26th anniversary.. we think!', Telecommunications, pp 182-183 (September 1995).

4. Negroponte N P: 'Get a life?', Wired, 3, No 9, p 206 (September 1995).

5. Stoll C: 'Silicon snake oil — second thoughts on the information highway', Macmillan (1995).

6. Syrdal A, Bennett R and Greenspan S: 'Applied speech technology', Ch 1, pp 1-45, CRC Press (1995).

7. Vogt E E: 'The nature of work in 2010', Telecommunications, pp 21-34 (September 1995).

8. Wiseman R: 'The Megalab truth test', Nature, 373, p 391 (2 February 1995).

9. Wheddon C, OBE: 'The seven ages of communication', BT Lecture (March 1995).

10. Fagen M D (Ed): 'A history of engineering and science in the Bell system — the early years (1875-1925)', Ch 3, Bell Telephone Laboratories, Inc (1975).

11. Bruce R V: 'Alexander Graham Bell and the conquest of solitude', p 181, Gollancz (1973).

12. Young P: 'Power of speech', Allen and Unwin (1983).

13. Young P: 'Person to person — the international impact of the telephone', pp 11-12, Granta Editions (1991).

14. Solomon L: 'Telstar', Constable Young Books, London (1963).

15. Westall F A and Ip S F A (Eds): 'Digital signal processing in telecommunications', Chapman & Hall (1993).

16. Foster K T, Young G and Cook J W: 'Broadband multimedia delivery over copper', BT Technol J, 13, No 4 (October 1995).

17. Lewis A V et al: 'Aeronautical facsimile — over the oceans by satellite', BT Technol J, 12, No 1, pp 83—97 (January 1994).

18. Willis P and Dufour I G: 'Towards knowledge-based networks', in Dufour I G (Ed): 'Network Intelligence', Chapman & Hall, pp 133-143 (1997).

19. Sterling B: 'Dropping anchor in cyberspace', Telecommunications, p 115 (September 1995).

20. Globalyst™ TPC, AT&T Global Information Solutions (1995).

21. Marshall I W and Bagley M: 'The information services supermarket — an information network prototype', in Dufour I G (Ed): 'Network Intelligence', Chapman & Hall, pp 212-229 (1997).

22. Lynch T and Skelton S: 'Teleworking: a necessary change', Br Telecommunications Eng J, 14, No 2, pp 122-130 (July 1995).

23. McClelland S: 'Telework's global reach', Telecommunications, pp 184-188 (September 1995).

24. Eder P E: 'Privacy on parade', The Futurist, 28, No 4, pp 38-42 (July-August 1994).

25. Negroponte N P: 'Being digital', Hodder and Stoughton (1995).

26. Halal W E and Liebowitz J: 'Telelearning: the multimedia revolution in education', The Futurist, 28, No 6, pp 21-26 (November-December 1994).

27. Kinney J: 'Anarcho-emergentist Republicans', Wired, 3, No 9, pp 90-95 (September 1995).

28. Toffler A: 'Powershift: knowledge, wealth and violence at the edge of the 21st Century', Bantam Books (November 1990).

29. Lewis A V et al: 'DSP in network modelling and measurement', in Westall F A and Ip S F A (Eds): 'Digital signal processing in telecommunications', Chapman & Hall, pp 434-473 (1993).

30. Rosenberg N: 'Why technology forecasts often fail', The Futurist, 29, No 4, pp 16-21 (July-August 1995).

31. Roddenberry G: 'Star Trek — the man trap', Paramount Television (September 1966).

32. Cochrane P and Westall F A: 'It would be good to talk!', 2nd Lang Eng Convention, London (October 1995).

33. Rothenberg D: 'Deep technology', Wired, 3, No 10, p 121 (October 1995).

34. South C R, Hoppitt C E and Lewis A V: 'Adaptive filters to improve loudspeaking telephone', Electronics Lett, 15, No 21, pp 637-634 (October 1979).

35. Lewis A V: 'Adaptive filtering — applications in telephony', in Westall F A and Ip S F A (Eds): 'Digital signal processing in telecommunications', Chapman & Hall, pp 111-138 (1993).

36. Flanagan J L: 'Beamwidth and usable bandwidth of delay-steered microphone arrays', AT&T Technical Journal, 64, No 4, pp 983-995 (April 1985).

37. Smyth P P et al: 'Optical wireless local area networks — enabling technologies', in Smith D W (Ed): 'Optical network technology', Chapman & Hall, pp 98-118 (1995).

38. Cochrane P: 'A three click, one second world', Electronics & Comms Eng J, 7, No 4, pp 138-139 (August 1995).

39. Cochrane P: 'Experience required', Wired (UK), 1, No 2, pp 72-76 (May 1995).

40. Bowles B A: 'Collaborative working and integrated communications services in the UK manufacturing sector', BT Technol J, 12, No 3, pp 12-38 (July 1994).

41. Jabez A: 'Real work in the virtual office', The Times (4 August 1995).

42. Whyte W S: 'The many dimensions of multimedia communications', in Whyte W S (Ed): 'Multimedia telecommunication', Chapman & Hall, pp 1-27 (1997).

43. Quittner J: 'Talk gets very cheap', Time, p 59 (27 March 1995).

44. Dertouzos M L: 'Systems, networks and computation: basic concepts', McGraw-Hill, pp 118-121 (1972).

45. Cochrane P: 'Copper mind-sets', British Telecommunications Eng J, 13, Part 1, pp 10-19 (April 1994).

46. Alley D M, Kim I Y and Atkinson A: 'Audio services for an asynchronous transfer mode network', BT Technol J, 13, No 3, pp 80-91 (July 1995).

47. Pratchett T: 'The colour of magic', Colin Smythe (1983).

48. van Landegem T, de Prycker M and van den Brande F: '2005 — a vision of the future network', Telecommunications, pp 119-130 (September 1995).

49. Pontailler C and Vilain B: 'Telecom networks: the walls break down', supplement to Telecommunications, pp S10-S17 (September 1995).

50. Geppert L: 'Educating the renaissance engineer', IEEE Spectrum, 32, No 9, pp 39-43 (September 1995).

51. Wagar W W: 'A funny thing happened on my way to the future or, the hazards of prophesy', The Futurist, 28, No 3, pp 21-25 (May-June 1994).

52. Vallance I: 'Customers must be the industry's driving force', IEEE Communications Mag, 31, No 12 (December 1993).

53. Jolson A: 'The Jazz Singer', dialogue before the song 'Toot-toot-tootsie', Warner Bros Vitaphone (6 October 1927).

Index